electronic components and measurements

PRENTICE-HALL ELECTRICAL ENGINEERING SERIES

William L. Everitt, *Editor*

electronic components and measurements

Bruce D. Wedlock

Department of Electrical Engineering
Massachusetts Institute of Technology

James K. Roberge

Department of Electrical Engineering
Massachusetts Institute of Technology

Prentice-Hall, Inc., Englewood Cliffs, N.J.

PRENTICE-HALL INTERNATIONAL, INC., London
PRENTICE-HALL OF AUSTRALIA, PTY. LTD., Sydney
PRENTICE-HALL OF CANADA, LTD., Toronto
PRENTICE-HALL OF INDIA PRIVATE LTD., New Delhi
PRENTICE-HALL OF JAPAN, INC., Tokyo

Current printing (last digit):
10 9 8 7 6 5 4 3 2 1

Library of Congress Catalog Card Number 69-15046
Printed in the United States of America

13-250464-2

to our children

KAREN AND WALTER
ANNE AND JIMMY

preface

The purpose of this text is to provide the student with familiarity with basic electronic components and sufficient understanding of modern measurement equipment and techniques that he may intelligently undertake experimental investigations. To this end, the experimental exercises are designed to focus attention on the individual components or the measurement equipment, and to emphasize particularly their departure from ideal behavior. By demonstrating that an instrument does not always report a circuit's performance accurately, the student learns which factors must be considered in choosing an appropriate piece of equipment for the particular measurement at hand. As a result of this approach, the exercises are not intended to provide examples of basic electrical circuit performance or to complement classroom instruction in elementary circuit theory. In fact, little theoretical background is required to accomplish most of the experimental exercises in this text. The material has been successfully taught to sophomores who were concurrently taking a first course in circuit theory and to freshmen who have had "hobby" experience with electronics.

This material was originally developed for use as a first course in laboratory instruction covering the broad spectrum of electronic measurements as preparation for a subsequent series of project laboratories. However, it may also be used selectively with a series of laboratory subjects, the balance of the material provided by the individual instructor as he sees fit. Thus, the individual instructor is freed from the burden of developing material to present the basic measurement instruments and techniques, without encroaching on his right to determine how individual topics are best illustrated in the laboratory. Finally, much of the material on the more advanced topics should provide a valuable reference and review for graduate engineers when they face a new measurements problem. To these ends, as many practical considerations have been included as possible without cluttering the descriptions with infinite detail.

The organization of the text departs from the tradition of placing all the exercises at the end of each chapter. Rather, most exercises immediately follow the descriptive material to which they pertain. Moreover, they usually pose a question to be answered through an experimental investigation. However, they also usually require some specific result which involves a simple theoretical calculation, normally carried out right after the data are obtained. For those whose students require additional written drill in connection with the subject material, a typical set of experimental data can be substituted for the actual data so that the exercise may be performed as an outside homework assignment. With regard to examinations, the authors have found that an oral performance examination of about 15 minutes' duration is an excellent teaching tool. Not only is the student's mastery of the material quickly determined, but the opportunity to immediately correct his errors and to strengthen his areas of weakness is much more productive than a few cryptic notes in the margin of his notebook.

It is impossible to acknowledge the help of all who have contributed to this text. Numerous faculty members and countless graduate teaching assistants have made many suggestions for improving the presentations. However, special mention must be given to Professor Richard D. Thornton, who first suggested that the introductory laboratory course should be based on the concept of "driver education" applied to electrical measurement equipment. Also, credit must be given to Professor Peter Elias, who, as head of the Electrical Engineering Department at M.I.T., provided the opportunity and encouragement for this subject development. Special thanks are due to the National Science Foundation and the Ford Foundation for grants which supported in part the production of this laboratory subject.

Bruce D. Wedlock
James K. Roberge

Cambridge, Massachusetts

contents

1

safety in the laboratory

1.1 introduction

Equally as important as performing accurate electronic measurements and constructing neat circuit breadboards is the execution of this work in a safe manner. Therefore, we begin the study of experimental work with a discussion of the ever-present hazard of electric shock and a review of some of the important safety considerations that must be observed at all times in the laboratory and shop.

1.2 the fatal current*

Offhand it would seem that a shock of 10,000 volts would be more deadly than one of 100 volts. This is not true! The real measure of a shock's intensity is the amount of current forced through the body. Although any amount of current over 10 mA is capable of producing painful-to-severe shock, currents between 100 and 200 mA are lethal. In this current range, ventricular fibrillation of the heart occurs. Above 200 mA, the resulting muscular contractions are so severe that the heart is forcibly clamped during the shock, and ventricular fibrillation is prevented. Thus, although severe burns, unconsciousness, and stoppage of breathing occur, the shock is not usually

* This section is based on an article entitled "The Fatal Current," originally published by Fluid Controls Corporation of Ridgewood, N.J.

fatal if the victim is given immediate resuscitation or artificial respiration. Figure 1.1 summarizes the physiological effects of various current levels.

Although it requires a voltage to force a current through the body, the magnitude of the current depends on both the voltage and the body resistance between the points of contact. The actual body resistance varies depending on the points of contact and the condition of the skin. The total resistance may be as low as 1 kΩ for wet skin and as high as 500 kΩ for dry skin. Death by electrocution has been recorded from a voltage as low as 42 volts DC, implying a resistance as low as 400 ohms. Thus, the only conclusion with respect to voltage is that *50 volts can be just as deadly as 500 or 5000 volts.*

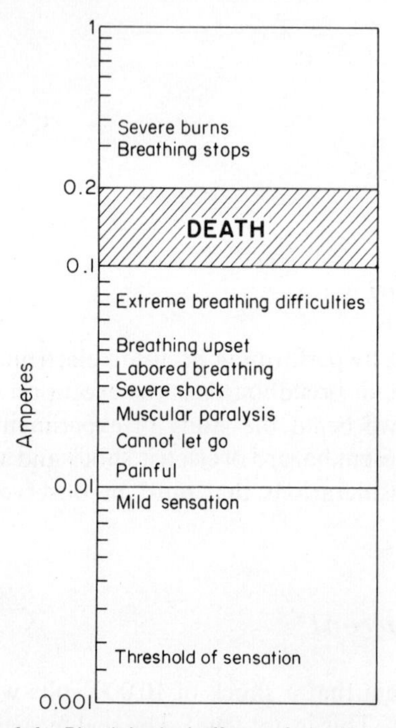

Figure 1.1. Physiological effects of electric currents.

1.2.1 action in case of shock

If an electric shock occurs, cut the power and/or remove the victim as quickly as possible without endangering yourself. If the power switch is not readily accessible, use a length of dry wood, rope, clothing, belt, or other insulating material to pull or pry the victim loose. The resistance of the victim's contact decreases with time so that the fatal 100- to 200-mA current may be reached if action is delayed.

If the victim is unconscious and has stopped breathing, start artificial respiration *at once. Do not stop resuscitation until a medical authority has pronounced the victim beyond help.* It may take as long as eight hours to revive the patient. There may be no pulse and a condition similar to *rigor mortis* may be present; these are manifestations of the shock and are not necessarily an indication that the victim has succumbed.

1.3 shop tools

Breadboard construction usually involves a variety of hand and power tools. Hand tools in good condition and properly used are relatively safe. However, dull cutting edges which require excessive force, or improper use of tools (such as prying with a file) can result in serious personal injury.

One hand tool that requires special mention is the soldering iron. Careless use can result in painful burns and damage to property through fire. Hot irons should be rested only in an approved holder designed to shield both the operator and the bench top from contact. When an iron is not being used it should be turned off. The short warm-up time is a small price to pay for the prevention of potential destruction by a fire caused by an unattended hot iron.

Power tools such as hand-held drills, drill presses, lathes, and milling machines, should not be operated without first obtaining instruction in their safe use. Some general guides are to avoid loose clothing such as ties which can become entangled in the work; to clamp work securely, especially in a drill press; and to be sure to wear safety glasses or eye shields.

Metal objects, especially rings on fingers, should be kept out of regions of strong AC fields. Induction heating can cause very painful burns.

1.4 chemical hazards

Frequently, the use of cleaning solvents such as trichloroethylene is required following construction of circuitry. In semiconductor laboratories many powerful acids and etchants are present. Be sure to provide adequate ventilation for fumes, such as a chemical hood, and to wear suitable personal protective clothing, including an apron, gloves, and eye or face protector.

Improper disposal of chemical wastes can create a safety hazard for others. Acids and other liquids should be poured into sinks with plumbing designed for corrosive materials and the pouring accompanied by a large flow of water for dilution. Solid chemicals and empty chemical containers should be placed in trash recepticals designated for that purpose and not in an unmarked wastebasket.

1.5 some safety rules

1. Be sure there are at least three people in the laboratory or shop: one to aid the victim and one to obtain additional help.
2. Be sure that instruments and power tools connected to the 120-volt AC power line are grounded via a three-wire power cable. (See Section 2.5.2.)
3. Kill the power and ground high-voltage points before touching wiring.
4. Never work on live equipment when tired.
5. Move slowly with feet firm to maintain balance. Never lunge for a falling object.
6. Avoid standing on metal surfaces or wet concrete. Keep your shoes dry.
7. Never handle electrical equipment with wet skin.
8. Do not wear loose clothing around machines. Wear protective clothing when using chemicals. Always wear safety glasses or goggles when using power tools and handling chemicals.
9. Hot soldering irons should be rested in an approved holder. Never leave a hot iron unattended.
10. Dispose of chemical waste in an approved trash container.

2

basic laboratory practice

2.1 introduction

To the student entering an electronics laboratory for the first time, the wide range of components, instruments, and vocabulary which confront him can produce a bewildering experience. In the laboratory assignments which follow, these items will be studied in detail. However, before beginning any experimental work, there is some basic information on experimental procedure and practice that should be explained to the new student. The object of this discussion is to point out some of this information necessary for accurate and effective experimental work.

2.2 the laboratory notebook

The laboratory notebook is a diary or log of all work pertaining to the experiment. There is no one best way to keep a laboratory book; rather, there are many good ways. The following comments are considered to be generally applicable to laboratory notebooks since they are based on practical experience.

A guiding principle to keep in mind throughout laboratory work is that an outsider with a similar technical background should be able to duplicate your entire experiment, data, and conclusions by following your laboratory book. A year, a month, or even a few days after the experiment, you yourself may be a stranger to it. Do not trust memory to fill in details. Put everything directly in the notebook. Scratch paper has no place in the laboratory. Why run the risk of losing loose sheets that might turn out to be valuable? The left-hand pages of the notebook are ideal for doodling and rough calculations.

Organization is important if someone else is to follow your work. Descriptive headings should separate the various parts of the experiment. Record data in chronological order. The laboratory notebook is not to be written as a report after the completion of the experiment but as a blow-by-blow account direct from the laboratory bench. A neat, organized, and complete record of an experiment is as important as the actual experiment.

2.2.1 heading

Your name and the date should be at the head of the first page of each day's experimental work. This may seem trivial but is a good habit to acquire since patent laws require it.

2.2.2 object

A brief but complete statement of what you intend to find out or verify in the experiment should head the entry in the notebook. While mainly for your reader's benefit, you will also find that it is a great help in clarifying the problem in your mind.

2.2.3 reference reading

Take notes on reference material read in preparation for an experiment directly in the laboratory book. This is a matter of convenience in keeping things together. You may need to refer to these notes during the experiment. Also include the circuits to be investigated and a rough outline of the proposed procedure to be followed in the experiment.

2.2.4 wiring diagram

The wiring diagram should be drawn and labeled so that the actual experimental circuitry could be quickly duplicated in the future. Be especially careful to record all circuit changes made during the experiment.

2.2.5 apparatus list

The apparatus list need include only those items that have a direct effect on the accuracy of the data. Usually, the meters from which readings were taken are the important items. The serial number of each item listed must be included so that the equipment can be located for rechecks if discrepancies develop in the results. Adopt some sort of keying system that clearly shows which instrument measured each column of data and exactly how the instrument was connected in the circuit.

2.2.6 procedure

In general, lengthy explanations of procedure are unnecessary. Running commentaries of a sentence or merely a phrase alongside the corresponding data are in order. Keep in mind the fact that your experiment *must be reproducible from the procedure you give.*

2.2.7 data

Data tables must be clearly labeled so that it is obvious what circuit yielded which data. Each column of data should be headed by the proper units (i.e., volts, millivolts, amperes, etc.).

2.2.8 graphs

Graphs are used to present data in a form that shows the results obtained as one or more of the parameters is varied. The purpose of a graph is, therefore, to present large amounts of data in a concise visual form. Data to be presented in graphical form *must be plotted in the laboratory.* This permits checking of questionable data points while the experiment is still set up. Graphs must be plotted on graph paper. The gross rulings in a laboratory notebook are not sufficiently accurate for careful graphing of data. Keep a supply of various types of graph paper with your notebook.

Give all graphs a short descriptive title. Avoid titles like "Current vs. Voltage" or "Transistor Curves"; instead use a title like "Collector Characteristics of a 2N696 Transistor." It is often convenient to avoid the problem of coded curves and keys by labeling each curve. This simplifies the job of reading the graph.

2.2.9 calculations

At this point be specific. Do not head a section "Calculations"; calculation of what? Use, for example, a heading such as "Calculation of Saturation Current From Data in Table 3." If repetitive calculations are called for, a sample is sufficient. The rest may be done on the left-hand pages.

2.2.10 results

One of the primary objectives of laboratory work is to clarify theory and indicate when and how well it applies in practical situations. Plot theoretical and experimental results on the same graph whenever possible, clearly showing which is which. If only a few points are to be compared and a graph does not seem to be called for, adjacent tables of results make it easy to see the correlation between the theory and the experiment.

2.2.11 conclusions

Particular attention should be paid to the notebook section on conclusions. This is where you interpret the results as an engineer rather than comment on them as a technician. Avoid saying things such as:

1. "The theoretical and experimental results check well." Present the two quantitative results side by side so that the reader may see how well they check.

2. "The graph of A versus B is linear over the range from $B = 0.1$ to $B = 100$." (Any idiot can see this for himself if you show him the graph.)

Do include things such as:

1. "The *reason* for the large discrepancy between the calculated upper cutoff frequency and that observed by experiment is the large amount of stray and interwiring capacitance present. The value of this stray capacitance, estimated from the size of the discrepancy, is 4.2 pF."

2. "If further work can be done along these lines, I *propose* to measure the stray capacitance directly by using a capacitance comparison bridge."

Notice that reasons for results, proposals for future investigations, and general comments are in order in the conclusions. Try to be brief but complete. This takes practice.

2.3 abbreviations, multipliers, and prefixes

The SI system of units (*Système Internationale d'Unités*) defined at the 11th General Conference on Weights and Measures (1960) has established a set of standard symbols for electrical quantities. These are summarized in Table 2.1.

Table 2.1 / Standard Abbreviations for Electrical Units

Quantity	Unit	Abbreviation
Time	second	s
Frequency	hertz	Hz
Power	watt	W
Charge	coulomb	C
Voltage	volt	V
Current	ampere	A
Resistance	ohm	Ω
Conductance	siemens	S
Capacitance	farad	F
Inductance	henry	H

In order to easily describe the power-of-ten multiplier associated with electrical quantities and component values, a standard system of prefixes and symbols has also been established. These are listed in Table 2.2.

Table 2.2 / Prefixes and Abbreviations for Decimal Multipliers

Multiplier	Prefix	Abbreviation
10^{12}	tera	T
10^{9}	giga	G
10^{6}	mega	M
10^{3}	kilo	k
10^{2}	hecto	h
10	deka	da
10^{-1}	deci	d
10^{-2}	centi	c
10^{-3}	milli	m
10^{-6}	micro	μ
10^{-9}	nano	n
10^{-12}	pico	p
10^{-15}	femto	f
10^{-18}	atto	a

Their use is illustrated in the following examples:

$$1 \text{ million ohms} = 1 \text{ megohm} = 1\ \text{M}\Omega$$

$$\tfrac{1}{1000} \text{ of an ampere} = 1 \text{ milliampere} = 1 \text{ mA}$$

$$1000 \text{ volts} = 1 \text{ kilovolt} = 1 \text{ kV}$$

2.3.1 *the decibel*

When measurements are made on individual electronic components, the required data are usually the absolute values of the parameters associated with the component. However, when components are connected to form linear systems, such as amplifiers and filters, the usual measure of the system's performance is the *ratio* of the output signal to the input signal rather than the absolute magnitude of the signal level. This ratio is termed the *gain* of the system if the output signal is increased over the input, or the *loss* if the output is diminished by passing through the system. If the input and output signals are measured in terms of electrical power, then the gain or loss is frequently expressed in terms of the logarithm of the ratio, called the *decibel*:

$$\left.\begin{matrix}\text{Gain } (+)\\ \text{Loss } (-)\end{matrix}\right\} = 10 \log_{10}\left(\frac{P_o}{P_i}\right) \text{ decibels} \tag{2.1}$$

If the output power exceeds the input power, then the logarithm is positive, whereas if the output power is less than the input power, the logarithm is negative. Thus gain is associated with positive decibels, or "dB," and loss measured by negative decibels. A factor of 10 in the power ratio equals 10 dB.

The use of the logarithm in (2.1) provides two advantages. First, when several systems are connected or cascaded as shown in Figure 2.1, the gain

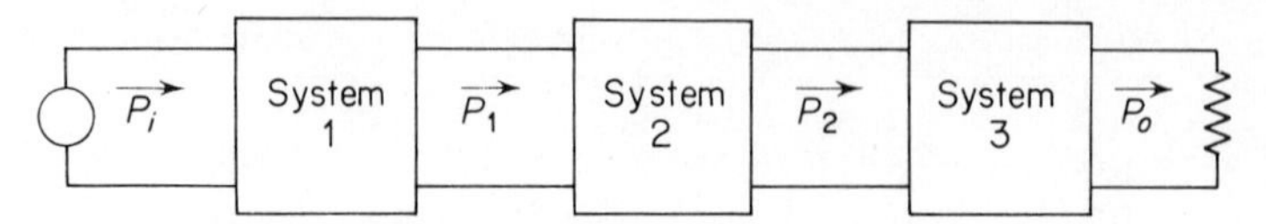

Figure 2.1. Cascaded systems.

of the overall system is the product of the gains of each individual system.

$$\frac{P_o}{P_i} = \frac{P_1}{P_i} \times \frac{P_2}{P_1} \times \frac{P_o}{P_2} \tag{2.2}$$

However, if the individual system gains are expressed in decibels, the overall gain is obtained by simple addition of the decibels of gain of each system rather than by multiplication:

$$10 \log_{10}\left(\frac{P_o}{P_i}\right) = 10 \log_{10}\left(\frac{P_1}{P_i}\right) + 10 \log_{10}\left(\frac{P_2}{P_1}\right) + 10 \log_{10}\left(\frac{P_o}{P_2}\right) \tag{2.3}$$

The second advantage is that system gains may range from very small numbers to very large numbers. By using a logarithmic scale, this wide range is compressed to a more convenient numerical range, generally not exceeding ± 100 dB for an entire system.

Equation (2.1) is the basic definition of the decibel. Since electrical signals are more frequently measured in terms of voltage or current rather than power, there are some equivalent expressions for the system gain. Figure 2.2 illustrates the relationships between voltage, current, and power at the system input and output. In terms of voltage,* the gain becomes

$$G = 10 \log_{10}\left(\frac{V_o^2}{R_L} \times \frac{R_i}{V_i^2}\right) = 20 \log_{10}\left(\frac{V_o}{V_i}\right) + 10 \log_{10}\left(\frac{R_i}{R_L}\right) \tag{2.4}$$

If the input and load impedances are equal, then $\log_{10}(R_i/R_L) = 0$ and (2.4) reduces to

$$G = 20 \log_{10}\left(\frac{V_o}{V_i}\right) \tag{2.5}$$

* The symbol convention used throughout this text is as follows:
Uppercase variables; amplitudes of DC and sinusoidal quantities.
Lowercase variables: arbitrary functions of time. *Example*: $v_i(t) = V_i \sin \omega_1 t$.

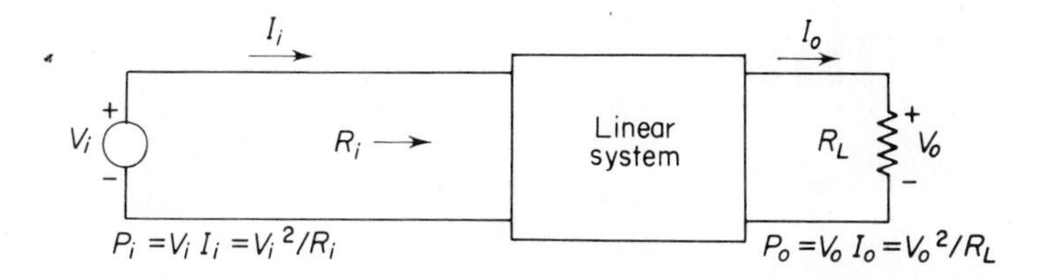

Figure 2.2. Input and output power of a linear system.

which is the frequently used expression for voltage gain in decibels. Thus a factor of 10 in voltage gain equals 20 dB. Note, however, that (2.5) is only correct if the impedances R_i and R_L are equal. Because of the convenience in using the decibel, it has become common, though incorrect, to discuss a voltage ratio in terms of decibels as expressed in (2.5) and to ignore the values of the resistances across which V_o and V_i are developed. Under these conditions, the information about power gain, the basic definition, is lost. Thus, when discussing voltage gain in terms of decibels, one should assume that (2.5) is the definition rather than (2.1) and specifically inquire whether information concerning power gain is required.

The use of the decibel has also spread to other applications, where the reference level is not always the input signal level. For example, the output of an amplifier as a function of frequency with constant input level is often expressed in decibels where the reference level is the output power or voltage at some specified frequency, say, 1 kHz.

In telephone transmission applications, the power level of 1 milliwatt has been accepted as a convenient reference. Thus a signal level of +10 dBm denotes a signal whose power is 10 mW. Since the impedance level used in telephone transmission systems is usually 600 ohms, this impedance is often assumed when "dBm" are specified. Many signal generators and voltmeters are calibrated in decibels directly as well as in absolute voltage. The tacit assumption is usually that the signal level in decibels is referenced to 1 mW in a 600-ohm load, or 0 dB $= \sqrt{0.6} = 0.775$ volt.

2.4 component identification

Before constructing any circuits, it is necessary to identify the required circuit components, as to both type and value. In the following chapters the various components will be described in detail, and the information below is intended only for basic identification purposes.

2.4.1 resistors

The most common type of resistor is the carbon-composition resistor. It is readily identified as a brown cylinder with axial wire leads. The body

A B C D

$R = AB \times 10^C, D$

A, B, C digits

Black	0	Green	5
Brown	1	Blue	6
Red	2	Violet	7
Orange	3	Gray	8
Yellow	4	White	9
Gold (*C* only)	−1		
Silver (*C* only)	−2		

Example

A Red *C* Orange
B Violet *D* Gold

$R = 27 \times 10^3$, 5 %
= 27,000 ohms, 5 %

Tolerance (*D*) code

No band	20 %
Silver	10 %
Gold	5 %

Figure 2.3. Resistor color code.

of the resistor carries three or four color bands to identify the resistance value in ohms and the tolerance of the resistance value. Figure 2.3 illustrates the formula for reading the color code of a resistor.

The other important resistor parameter is the power rating, which measures the amount of voltage or current the component can handle without being destroyed. Carbon-composition resistors are normally supplied in $\frac{1}{4}$-, $\frac{1}{2}$-, 1-, and 2-watt sizes, which are identified by the dimensions of the cylindrical body. Figure 2.4 illustrates these dimensions.

Before choosing a resistor power rating, estimate the voltage V or current I that it will be expected to handle. Calculate the power level

$$P = I^2R = \frac{V^2}{R} \tag{2.6}$$

and choose the next highest standard power rating. For a conservative design you may wish to choose a power rating perhaps 5 to 10 times larger than the actual dissipation in the resistor.

Resistors with power ratings over 2 watts are normally wirewound on ceramic cores. These resistors usually have their resistance value and power rating printed directly on their body.

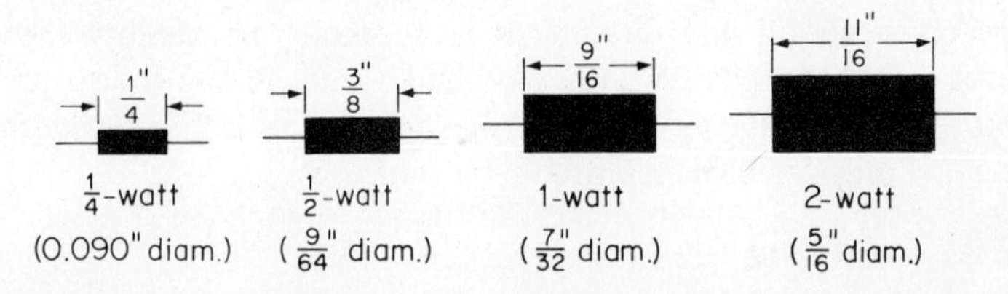

Figure 2.4. Full-size body dimensions of carbon-composition resistors according to power rating.

2.4.2 potentiometers

Frequently it is desirable to have a resistor whose value can be continuously adjusted. This function is provided by the potentiometer or "pot", consisting of a carbon or wirewound resistor with a movable tap which can be positioned by rotating a shaft. As the tap or slider changes position along the body of the resistor, the value of the resistance between the ends of the resistor and the tap changes. Figure 2.5 illustrates a potentiometer with its circuit symbol.

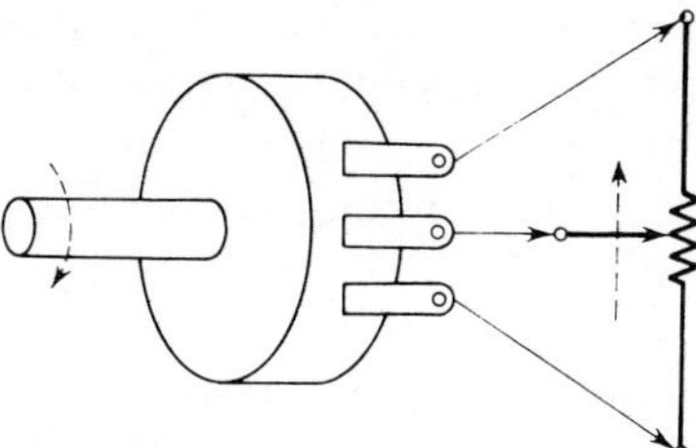

Figure 2.5. Potentiometer and circuit symbol showing relative motion of shaft and slider.

The stated resistance value and power rating of a potentiometer is for the entire resistance between the two ends. This information is usually stamped on the metal case. Occasionally, a three-digit code is used to give the resistance. The digits correspond to the letters *A*, *B*, and *C* in Figure 2.3 and are interpreted in the same manner. Thus a potentiometer carrying the digits 104 would have a resistance of 100 kΩ between the end terminals.

One precaution that must be observed in using potentiometers is to be careful not to inadvertently exceed the power rating by motion of the slider. In some circuits, with the slider in one position the dissipation level may be well within limits, but moving it to another position will cause excessive power to be dissipated over a portion of the resistance. This will result in a burned spot, which will cause poor contact of the slider or even an open circuit. If a potentiometer is suspected of being burned, and the odor is distinctive, it is best to discard it.

2.4.3 capacitors

Capacitors come in a wide variety of shapes. Electrolytic and paper dielectric types are usually cylindrical with axial leads. These normally carry the capacitance in microfarads (μF or, occasionally and incorrectly, mF) and the voltage rating written directly on the case.

Mica dielectric capacitors are round, rectangular, or irregular in shape and usually carry the capacitance written in picofarads (pF or, incorrectly, mmF). The round and rectangular types are sometimes color-coded to indicate their capacitance, tolerance, and temperature coefficient or "characteristic." The presently used color code is illustrated in Figure 2.6. One may encounter mica capacitors which carry an obsolete three-dot or six-dot code. If there is doubt about a capacitor's value, the best approach is to measure the capacitance directly. Occasionally, capacitance values in picofarads are shown by a three-digit code which gives the values of *A*, *B* and *D* in Figure 2.6. In these cases, the tolerance is shown by the letter code listed in the table of Figure 2.6. Thus, a mica capacitor carrying the code 233K would be 22,000 pF, 10%.

Ceramic capacitors are usually in the form of either a disc or a cylinder. The disc types may carry the capacitance written on the body, with whole numbers in picofarads and decimal fractions in microfarads (33 pF; 0.01 μF), or they may carry a three-digit or color code. Tubular ceramic capacitors are normally color-coded. These color codes are also summarized in Figure 2.6.

The voltage rating and polarity are most critical with the electrolytic capacitor. It must be *strictly* observed. If the reverse polarity or excessive voltage is applied, substantial leakage current will flow and heating will destroy the capacitor. Paper capacitors usually have voltage ratings of 100 volts or more. A black band indicates the outer foil terminal and this is usually placed at the lower voltage. Ceramic and mica capacitors are usually rated at 100 volts or more, and there is no preferred voltage polarity.

2.4.4 *inductors*

Inductors are the least used circuit element. Usually they carry their values listed directly on the case, which may take many different shapes. The only possible confusion will come with very small inductors (microhenry range) which are packaged in cylinders similar to a resistor or ceramic capacitor. A quick check with an ohmmeter (the inductor will have a very low DC resistance while the capacitor will be an open circuit to DC) will resolve the question.

2.4.5 *semiconductor diodes*

Semiconductor diodes come in a variety of shapes. The smallest sizes are usually cylindrical with axial leads and resemble a $\frac{1}{4}$-watt resistor, including a color-code identification. However, the body color for such a diode is usually black, compared with the resistor's usually brown body. Some small diodes are enclosed in clear glass cylinders and hence are readily identified. Other diode packages include colored plastic capsules and a variety of metal cases.

Identification of semiconductor diodes may be by color code or by numbers printed on the diode case. The color-numeric code is the same as for resistors. For color-coded diodes, the 1N prefix is assumed and the

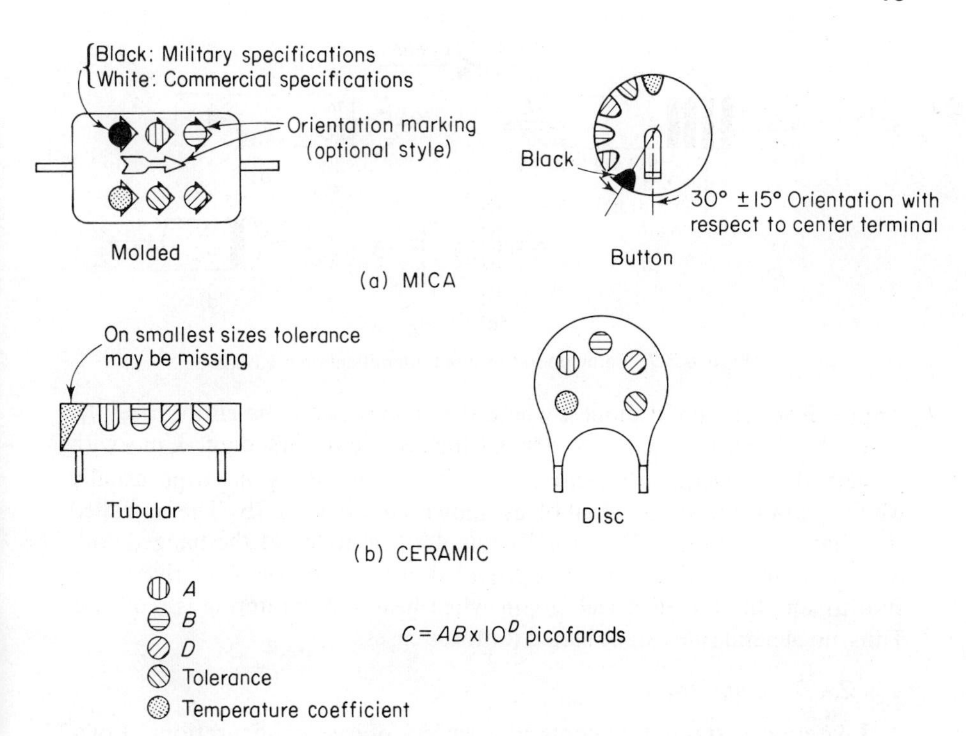

Color	A, B	D	Tolerance %	Temp.Coeff. ppm/°C	CERAMIC D	CERAMIC Tolerance %(>10 pF)	CERAMIC Tolerance pF(<10 pF)	CERAMIC Temp. Coeff. ppm/°C
Black	0	0	±20 M		0	±20 M	±2.0 G	0
Brown	1	1	±1 F	±500	1	±1 F	±0.1 B	- 33
Red	2	2	±2 G	±200	2	±2 G		- 75
Orange	3	3		±100	3	±3 H		-150
Yellow	4	4		-20 to +100	4			-220
Green	5		±5 J	0 to +70		±5 J	±0.5 D	-330
Blue	6							-470
Violet	7							-750
Gray	8				-2		±0.25 C	-1500 to +150
White	9				-1	±10 K	±1.0 F	-750 to +100
Gold		-1	± 0.5 E					
Silver		-2	± 10 K					

Figure 2.6. Capacitor color codes.

remaining digits indicated by color bands as shown in Figure 2.7a. For example, a 1N645 would be marked as follows: first digit blue, second digit yellow, third digit green. There would be no fourth-digit color band in this case.

The polarity of a diode is indicated in several ways, as shown in Figure 2.7. For color-coded diodes the "cathode" end is marked with the color

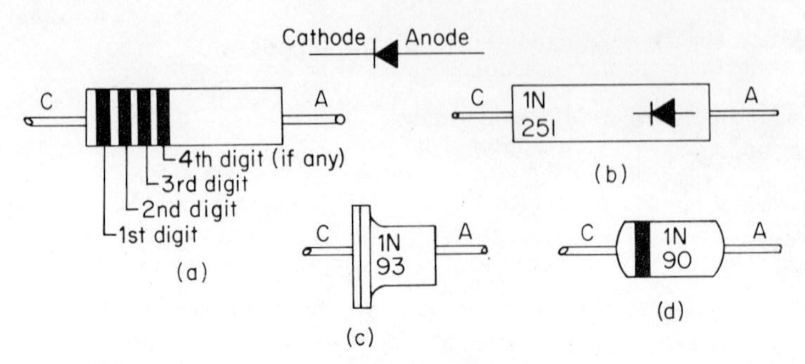

Figure 2.7. Semiconductor diode identification methods.

stripes. For very small diodes, where the stripes cover the entire case, the "cathode" is marked by a wide stripe, which is also the first digit. Units with numerical identification indicate the "cathode" end by a wide stripe, usually white, and/or the diode symbol as shown in Figure 2.7b. The so-called "top hat" case, Figure 2.7c, usually has the "cathode" at the flanged end. Stud-mounted rectifiers are often provided in both polarities relative to the stud to simplify the electrical circuit when heat sink mounting is required. Thus, no general rule can be stated for these cases.

2.4.6 transistors

Like diodes, transistors come in a variety of case configurations. Low-power (under 1 watt) units are generally mounted in metal cans or encapsulated in plastic. High-power devices are contained in cases with large contact areas or studs to facilitate mounting to heat sinks. Some of the more common shapes and lead arrangements are illustrated in Figure 2.8.

Transistors are specified by a prefix 2N followed by a number. Devices such as unijunction transistors, field-effect transistors, and silicon-controlled rectifiers also use the 2N prefix and the same case types. Therefore, when in doubt about a device, consult a reference manual.

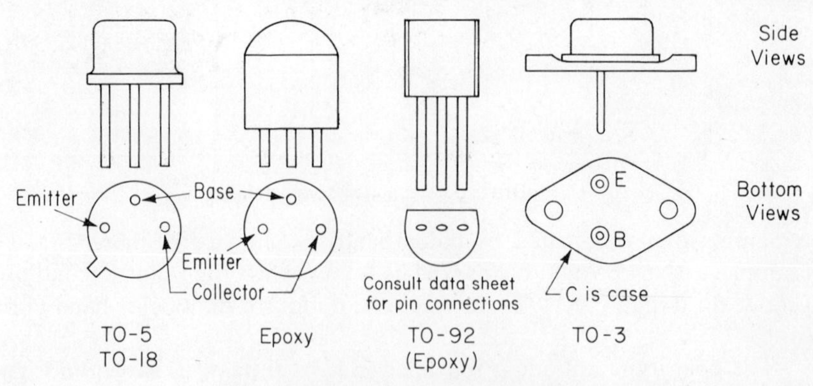

Figure 2.8. Common transistor cases.

2.5 breadboard construction practice

2.5.1 layout and lead lengths

In laying out a circuit for testing, neatness is a primary consideration. To the farthest extent possible, the component locations should correspond to their location in the schematic diagram. This facilitates locating test points. If a solderless method of interconnections is used, be sure all connections are secure. A loose connection can waste many hours of laboratory time.

Some method of mechanical mounting is desirable to keep components securely connected and in fixed location. In simple networks there is a temptation to save time by hanging a network together on the bench without a proper mechanical support. Generally, resulting insecure connections waste more time than would have been required for proper breadboarding.

In choosing leads for connections within the circuit and for connections to auxiliary equipment such as signal generators, meters, and scopes, the rule of thumb is *The shorter the better*. Besides making a neater arrangement, short leads reduce the parasitic capacitance, which is troublesome in high-gain amplifiers, and minimize noise pickup in low-level circuits.

2.5.2 grounding

The voltage reference potential of zero volts is commonly called *ground* potential. This terminology stems from the fact that one side of the 120-volt AC power line is fixed at the potential of the earth's surface by means of low-resistance connections to rods and plates buried under the earth's surface. However, it has become common to refer to any zero-voltage reference as "ground" even though it is not directly connected to earth potential. For example, the metal chassis on which an electronic circuit is built often serves as the ground of the circuit.

The symbol for a ground point is ⏚. This means that the particular circuit point is connected to some reference potential point—the chassis or perhaps a heavy bus wire. If there is no direct connection to the earth potential, this is often indicated by using ⊥ as the symbol for the reference potential.

In the laboratory, several connections to the earth's potential, or true ground, are usually provided. As mentioned above, one side of the AC power line provides this function. However, since this ground wire carries AC current, the series resistance between the lab bench and the actual ground results in an AC voltage at the bench terminal. Therefore, a third connection on the AC power receptacle is provided to furnish a non current-carrying path to true ground. This arrangement is shown in Figure 2.9. Power panels at the laboratory bench will usually have several terminals labeled GROUND or GND which are interconnected and ultimately tied to true ground.

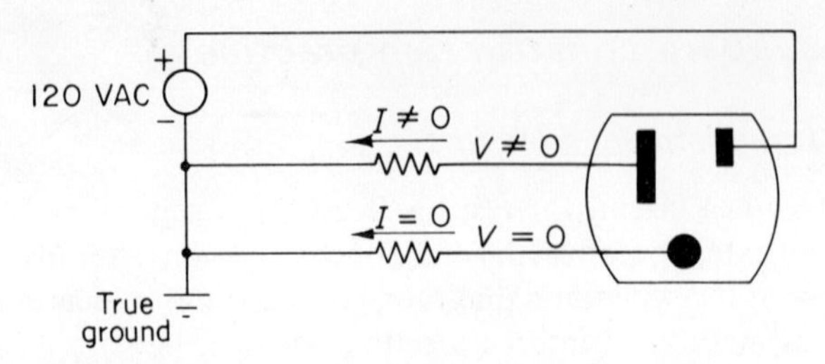

Figure 2.9. Ground terminals on an AC receptacle.

Most instruments and power supplies which connect to the AC line are provided with a three-terminal plug. The "third" or "ground" terminal (the round, non current-carrying contact) is normally connected to the instrument chassis and case. Thus, all instruments which have a three-terminal AC plug have their individual chassis interconnected through the grounding system in the power line. These connections are often the source of improper operation of a circuit which otherwise appears to be wired correctly, since by not paying careful attention to the ground system, short circuits can accidentally result. An example of this is shown in Figure 2.10.

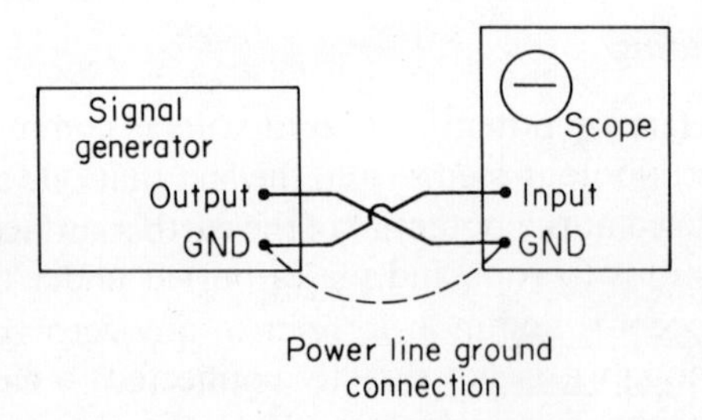

Figure 2.10. Inattention to ground will short circuit the generator output and scope input.

2.5.3 isolation from ground

Frequently in measurements work it is necessary to have an instrument or power supply whose terminals are *floating*; that is, neither terminal is connected to ground. Many instruments and supplies have a three-terminal arrangement, in which two terminals provide the appropriate output isolated from ground, while the third terminal is connected to the chassis and the third-terminal ground of the power cord. In this way, either output terminal can be grounded, or both left floating as the particular situation warrants. An example of this arrangement is shown in Figure 2.11. Notice in Figure 2.11 that the circuit connections are always made to the floating output terminals. If a connection is made at the + terminal and the ground terminal, *with no connection to the* − *terminal*, no power will be applied to

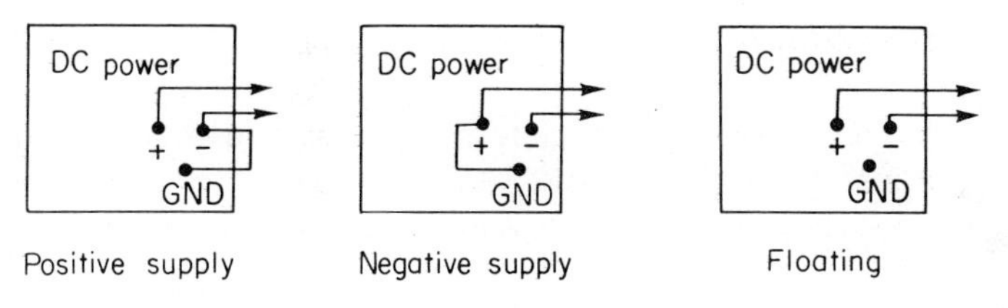

Figure 2.11. Proper connections to an isolated output.

the circuit. The ground terminal is provided for convenience in grounding one of the floating outputs and does not carry power itself.

Many instruments, however, do not have an isolated output. In order to "float" these units one of two methods can be employed. The most straightforward method is to use an isolation transformer, which is placed between the instrument and the power source as shown in Figure 2.12. If the instrument is equipped with a three-wire power cord, care must be taken to insure that the third-wire ground is not connected. This is most easily accomplished by means of a three- to two-terminal adapter on the AC power plug.

In some instruments, the incoming 120-volt AC power connections are made directly to the primary of a power transformer through perhaps a power switch and fuse. If neither side of this incoming power line is connected to the chassis or circuit ground through any type of component, then the power transformer will play the role of the isolation transformer in Figure 2.12. All that need be done is to insure that the third-wire ground on the power plug is not connected, usually by means of the adapter mentioned above.

One final word of caution: Whenever the third-wire ground is disconnected and an instrument "floated," the instrument case will assume the potential to which the output ground terminal is connected. *Thus, an electrical shock hazard exists.* Use extreme care when connecting instruments in this manner. The safest method is to obtain an instrument with a floating output provision. However, this is not always possible and then the methods described above must be employed.

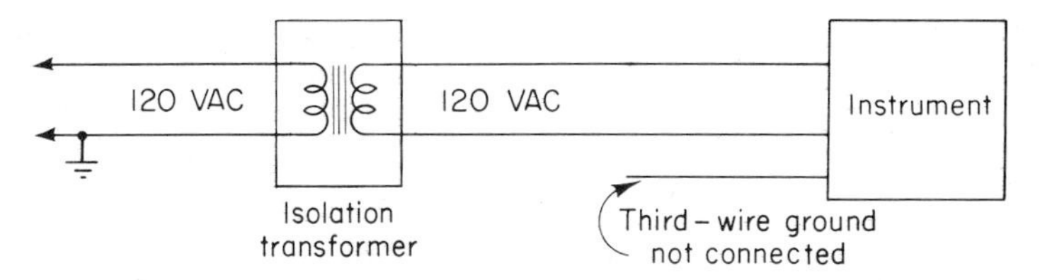

Figure 2.12. Use of an isolation transformer to float an instrument.

2.6 noise pickup

Ultimately, the smallest observable signal which can be experimentally detected is limited by the presence of noise. Therefore, good experimental practice includes taking steps to minimize noise pickup.

2.6.1 capacitive noise pickup

Capacitive noise pickup results from unshielded wires and circuits being placed in regions of unwanted electric fields. In the laboratory, these fields are an unavoidable reality. Therefore, shielded wires should be used for low-level signals.

Exercise 2.1*

To demonstrate the effect of shielding on capacitive noise pickup, connect a 1 × probe to an oscilloscope and set the vertical sensitivity at the maximum gain. With the sweep set at 5 m/cm, observe the relative signal amplitudes (1) when you grasp the shielded cable between the scope and the probe, and (2) when you grasp the plastic sleeve near the probe tip (do not actually touch the probe tip).

2.6.2 inductive noise pickup

Whenever a changing magnetic field links a conductor, voltage is induced in the conductor. In the laboratory, changing magnetic fields are also present, both from the 60-Hz power lines and from high frequency sources. The third-wire ground system can be a source of magnetically induced noise by the creation of "ground loops."

Exercise 2.2*

Connect an oscilloscope and a signal generator with third-wire ground plugs to a three-wire power source. Turn on the scope, but leave the signal generator off. Remove the ground lead from the scope probe and connect the probe to the ground terminal on the signal generator. Look for waveforms at both 60 Hz and in the 10- to 100-kHz frequency range. The ground loop, consisting of the third-wire ground from the generator through the power line to the scope, will have induced signals of several millivolts. When these signals are present at the input of a high-gain amplifier, serious noise levels will appear at the output.

Reconnect the ground lead to the scope probe and connect it also to the ground terminal of the signal generator. Observe the change in the waveforms observed above due to shorting the ground loop by means of the ground in the scope probe.

* Students completely unfamiliar with the oscilloscope should defer these Exercises until completion of Chapter 4.

With both the probe and its ground lead connected to the ground terminal of the signal generator, examine the effect of breaking the ground loop by using a three- to two-terminal adapter on the scope or the signal generator power connection.

Generally speaking, a ground point should be provided near the input terminals of a low-level circuit, and all grounding should be made to this one point only. This avoidance of ground loops plus shielding of low-level signal wires goes a long way toward minimizing noise pickup in the laboratory.

3

elements of data presentation and analysis

3.1 introduction

In scientific and technical work a graphic presentation is an efficient and convenient method of expressing and analyzing data. Graphs are most commonly used in interpolating data, determining a law or relationship between two quantities, discussing error, and visualizing analytic functions. Although there are many types of graphs (bar graphs, pie charts, etc.), the experimentalist is primarily concerned with line graphs. We will begin by describing the preliminary procedure of plotting line graphs, and then examine various types of graph paper.

3.2 preliminary procedure

All graphs must be neat and well labeled. Every graph must be titled, signed, and dated; both axes must be labeled and scaled; all lines must be smooth and uniform (see Figure 3.1). In general, if a draftsman can make a complete graph for publication from the original, without additional information, then the original graph is acceptable.

3.2.1 equipment

Graphs are to be drawn with a sharpened pencil. If a student prefers to use ink, he may do so. However, erasures are nearly impossible with ink. A straightedge and one or two French curves are also required. They will

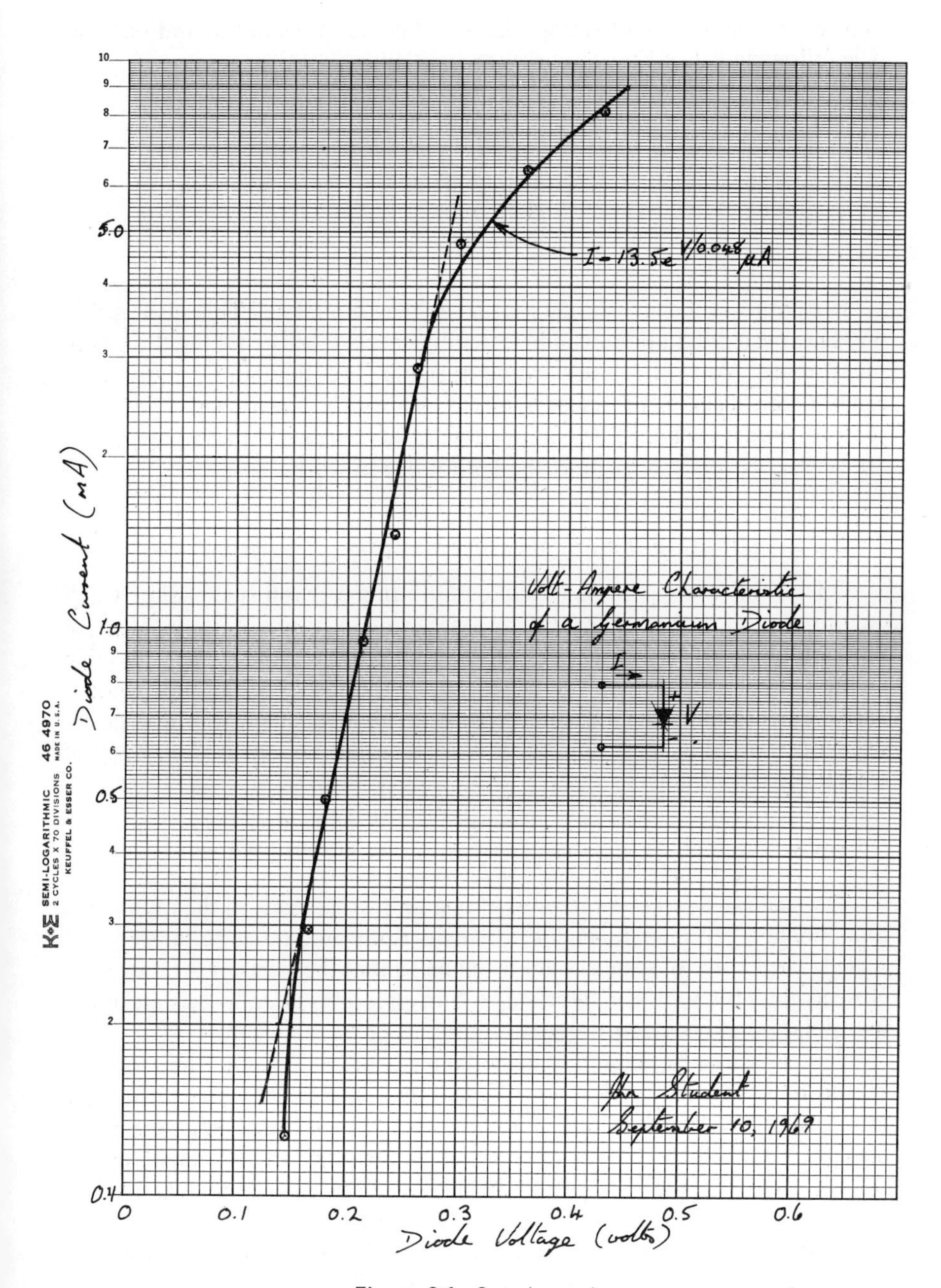

Figure 3.1. Sample graph.

be used to insure that all straight lines and curves are smooth and uniform. The following list of French curves are suitable.

Useful French Curves*

KE 57-1007-5	KE 57-1007-26	KE 57-1600-19
KE 57-1007-13	KE 57-1600-1	KE 57-1600-21
	KE 57-1600-3	

It is important that data be plotted as they are taken. This insures that any unexpected data point can be carefully checked before taking the equipment apart. Also, in regions where variables are changing rapidly, data can be taken at sufficiently close points to insure an accurate curve. For these reasons, students should be equipped with the necessary graph papers and drafting tools when they come to a laboratory to work.

3.2.2 *choosing the ordinate and abscissa*

One of the first decisions to make in plotting a graph is choosing which data to plot on the ordinate (vertical axis) and which to plot on the abscissa (horizontal axis). The usual convention is to plot the independent variable on the ordinate and the dependent variable on the abscissa. The independent variable is normally the measured response to a controlled variation of the dependent variable, a parameter such as the input voltage or frequency of excitation.

3.2.3 *choosing the scales*

The scale of an axis is the ratio of one unit of the data to the number of units (inches, centimeters, etc.) of the axis. Both axes must be labeled and scaled. The choice of scale should be made so that the locus of the data is as close as possible to a 45 degree slope. This preserves equal precision in the interpolation of both coordinates.

The chosen scale should also be convenient to use. One unit of axis should be equal to 1, 2, 4, 5, or 10, etc. units of data. The student should avoid using scales such as one unit of axis equal to 3, 7, 8.5, or 11, etc. units of data. Care should be taken to avoid excessively large or small scales. In the former case the precision of the line would be exaggerated, and in the latter case the precision would be lowered. If the student finds that the data to be plotted are significant over a few scattered ranges but not in between, he should "break" the scale as shown in Figure 3.2.

* Manufactured by Keuffel and Esser Co.

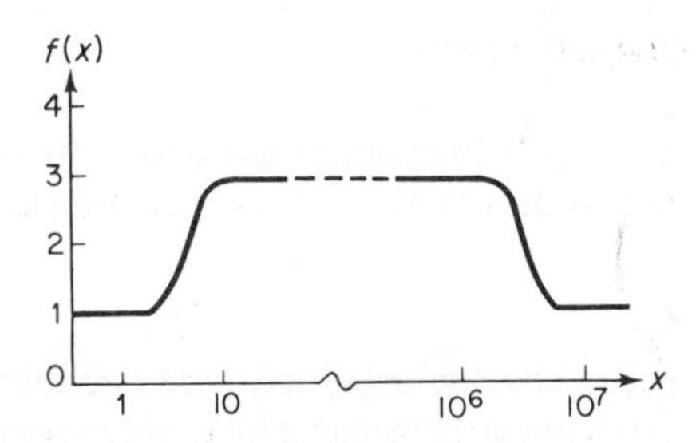

Figure 3.2. Broken scale.

3.2.4 plotting the data and fitting the line

Data points are usually drawn as small circles on the graph. The diameter of the circle can be varied to indicate the accuracy of the data. However, in many graphs, an I bar is used. The top and bottom of the bar can indicate the accuracy of the data point, or it can indicate some other feature of the data such as one standard deviation. This type of data point is particularly useful if one axis is known much more accurately than the other axis. Other symbols such as squares, triangles, diamonds, etc. are used as data points when more than one line is plotted on the graph, each line having its own symbol.

After the data are plotted, the next step is to draw a smooth curve which best fits the data. If the points appear to lie in a straight line, the best fit is usually a line which has data points lying as closely as possible on either side of the line. The data points above the line should deviate from it by about the same amount as the points below. One exact means of locating the "best-fit" line is to use the criterion of least squares (see References).

If the data points suggest that a curve is the best-fit line, then a French curve should be used to draw the curve. The curve is drawn so that again the data points are distributed on either side of the line.

Exercise 3.1

On Cartesian coordinates, graph $1/x$ vs. y and $1/y$ vs. x where the corresponding values of x and y are given in the table below.

x	0.5	0.33	0.33	0.25	0.06
y	2.0	3.0	4.0	6.0	15.0

In each graph draw a "best-fit" straight line.

For $y = 10$ what value of x do you read from each graph? Explain why the value of x at $y = 10$ is different in the two graphs.

3.3 types of graph paper

In addition to the familiar Cartesian graph paper, there are several other types of particular use to the engineer. These are discussed below.

3.3.1 semilog

Semilog graph paper has one linear axis and one logarithmic axis. This type of graph paper is particularly valuable for graphing exponential functions or functions which have one widely ranging parameter. The logarithmic scale is ideal for parameters which have a wide range since it "expands" the function in the low range and "compresses" the function in the high range.

Semilog paper has the property that the analytic function $y = Ae^{x/\alpha}$, where A and α are constants, yields a straight line if y is plotted on the log axis and x on the linear axis. Furthermore, the value of A is obtained from the value of y at $x = 0$, and α may be obtained from the "slope" of the line. Since the complete exponent x/α must be dimensionless, the constant α must have the same dimensions as the variable x. Thus, this constant is often called by its dimension, i.e., *time* constant, absorption *length*, *energy* gap, or Debye *temperature*.

The "slope" of a line on semilog paper differs from the usual $\Delta y/\Delta x$ definition for Cartesian coordinates. If one keeps in mind the reasoning in determining the constant α, no difficulty should arise. First, α is the amount x must change to change y by a factor $e = 2.718$. Thus the appropriate Δy in this case is the linear distance on the y-axis that corresponds to "e" on the log scale. This is most easily found by noting the distance from 1.0 to 2.718 (or 0.1 to 0.2718, etc.). The corresponding Δx is then α. The sign of α is positive or negative according to the sign of the slope of the line in the usual sense. If a decade on the log axis is chosen for Δy, then Δx will be 2.3α, since $10 = e^{2.3}$. These points are summarized in Figure 3.3.

Semilog graph paper may be purchased with the number of cycles in the logarithmic axis ranging between 1 and 5. One cycle is equal to one decade. However, if semilog paper is not readily available, linear paper can be roughly scaled logarithmically.

Table 3.1 / Conversion of Linear Paper to Semilog Graph Paper

Number of Units of Linear Axis	Log Scale
3	2
6	4
9	8
10	10

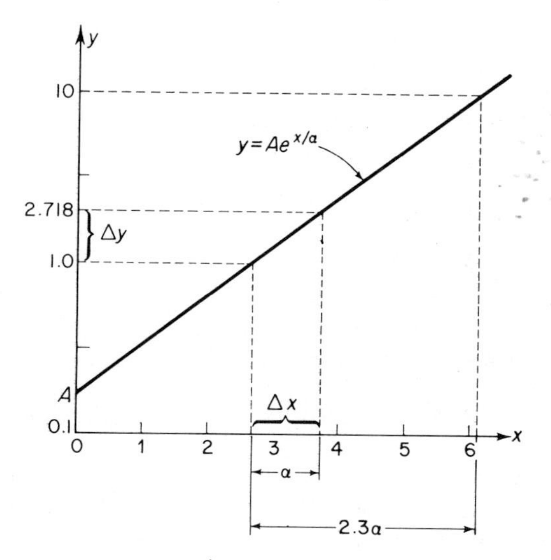

Figure 3.3. Parameter determination on semilog coordinates.

Exercise 3.2

On 3-cycle semilog paper, graph the following data.

x	y
1	0.274
2	0.074
3	0.020
4	0.0055

Analytically, the function may be written as $y = Ae^{-x/\alpha}$. Graphically find A and α.

3.3.2 *log-log*

Log-log graph paper has both axes scaled logarithmically. This type of graph paper is useful when both parameters vary over a wide range, or when one is searching for a power law relationship between two experimental quantities.

On log-log coordinates the functional form $y = Bx^{\beta}$ will plot as a straight line if B and β are constants. The value of B is found from the value of y at $x = 1$. In a manner similar to the semilog case, the value of β is determined from the slope of the line. In this case, β is given by the ratio of the total linear length along the y-axis for a decade change in x, divided by the linear

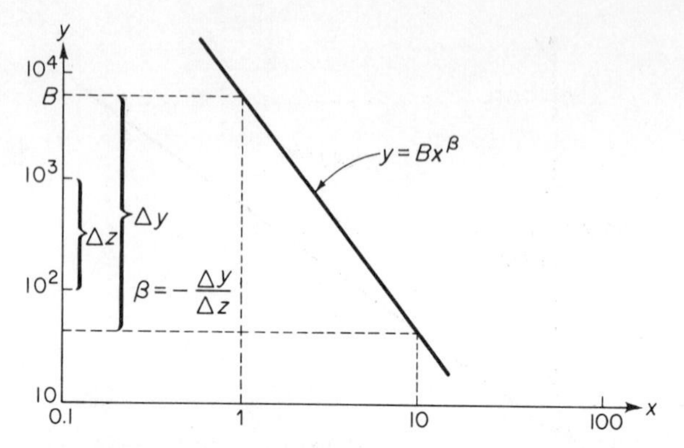

Figure 3.4. Parameter determination on log-log coordinates.

length of one decade change in y. Again, the sign of β is determined from the sign of the slope in the usual way. This calculation is illustrated in Figure 3.4.

If a desired size of log-log paper is not obtainable, one can rescale a logarithmic axis by raising one decade to the new scale power. For example, a change from 3-cycles to 1-cycle is shown in Figure 3.5. Figure 3.6 shows another method of changing scale which is valid for both linear and logarithmic scales is to use the technique of similar triangles.

Exercise 3.3

Graph the following data and convert them to analytic form on 2 × 2 log-log paper.

x	3	4	7	12.5	20
y	26	40	92.5	225	450

3.3.3 other types of graph paper

There are many other types of graph paper. Some examples are polar graph paper which uses polar coordinates; triangle graph paper, primarily used by metallurgists in plotting equilibrium phase diagrams; and probability graph paper, used in statistical analysis.

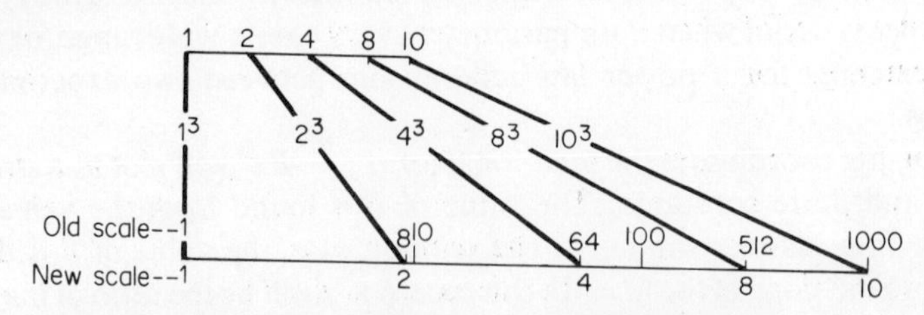

Figure 3.5. Scale change on logarithmic axis.

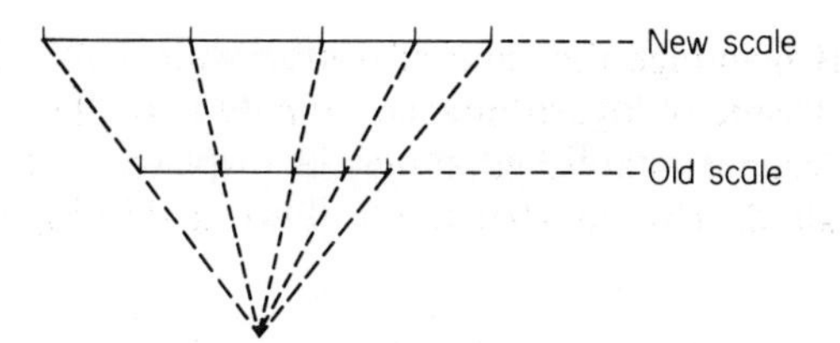

Figure 3.6. Scale change by similar triangles.

Exercise 3.4

Using polar graph paper plot a typical graph showing two equal loudness contours of a loudspeaker in a room.

Contour 1		Contour 2	
Distance (feet)	Angle (degrees)	Distance (feet)	Angle (degrees)
2	110	0	90
4	120	1	120
6	140	2.5	140
7	160	3.5	160
7	180	4	180
6	200	3.5	190
4.5	220	3	210
3.5	240	2	230
2	260	1	240
0	90		

3.3.4 decibel scales

In Section 2.3.1 the decibel was introduced as the measure of the ratio of power or voltage. In this section we will discuss how one may conveniently plot data in terms of decibels.

If data are being measured with a meter that has a decibel scale, then the easiest method is to read the decibel indication directly, subtract the meter reading in decibels that corresponds to 0 dB for the particular measurement, and plot the result on a linear scale. Note that while 0 dB on most voltmeters corresponds to a 1-mW dissipation in a 600-ohm load, the shift to any arbitrary reference level is accomplished by subtracting a constant from the decibel indication of the meter. This constant is simply the meter indication in decibels under the specified conditions for 0 dB for the case at hand.

If the indicating meter does not have a decibel scale, then one may plot the voltage reading on a logarithmic coordinate and then attach a decibel scale. Since one decade on the logarithmic coordinate corresponds to 10 dB in power or 20 dB in voltage, the required decibel scale is obtained by dividing the span of one decade of logarithmic coordinate into 10 or 20 equal parts. Once this linear scale is established, it may be displaced so that 0 dB occurs wherever it is desired. This construction is illustrated in Figure 3.7.

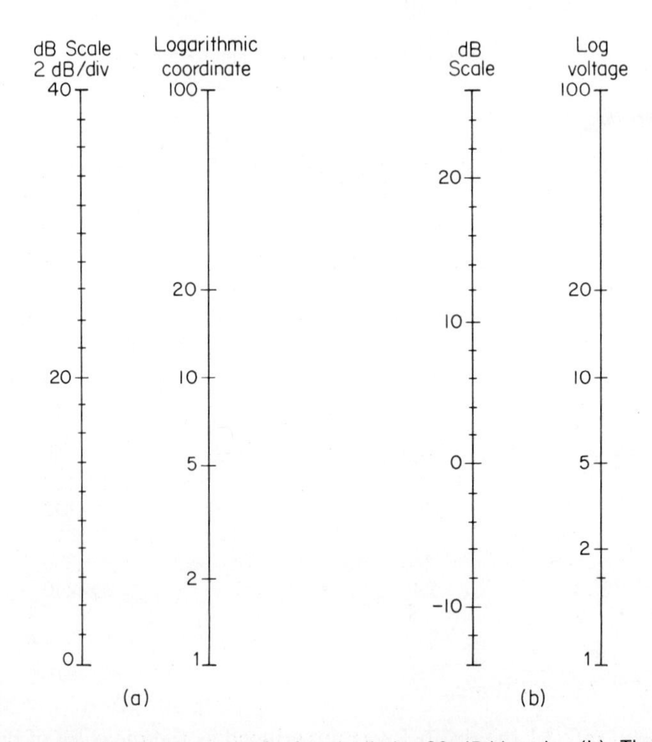

Figure 3.7. (a) A linear scale for decibels, 20-dB/decade. (b) The decibel scale displaced such that 0 dB = 5 volts.

3.4 *tabular data presentation*

In addition to graphs, data are frequently recorded in tables. However, the type of data recorded in tables differs from that recorded in graphs in that tabular data usually consist of pieces of information not functionally related to one another. On the other hand, graphed data are usually intimately related by some continuous function. For example, the data presented in the following graph, Figure 3.8, should be presented in tabular form instead of as a graph.

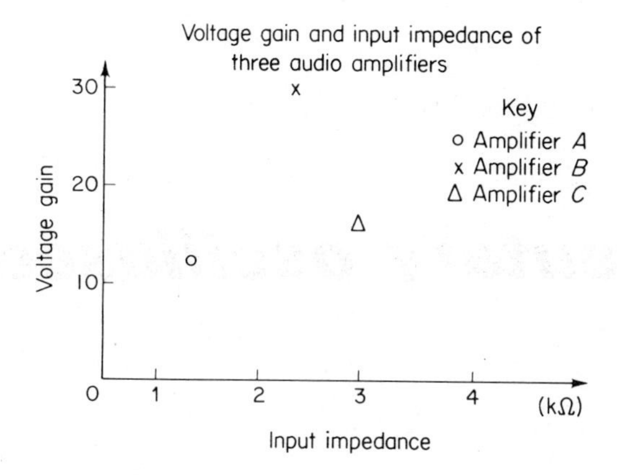

Figure 3.8. Example of data better presented in tabular form.

The required table is as follows:

Table 3.2 / Voltage Gain and Input Impedance of Three Audio Amplifiers

Unit	Gain	Input Impedance (kΩ)
A	12	1.3
B	30	2.2
C	16	3.0

3.5 references

1. H. M. Goodwin, *Elements of the Precision of Measurements and Graphical Methods*, McGraw-Hill Book Company, New York, 1913.

2. W. H. Burrows, *Graphical Techniques for Engineering Computations*, Chapters III and VI, Chemical Publishing Co., New York, 1965.

4

elementary oscilloscopes

4.1 introduction

The oscilloscope is a uniquely useful measuring instrument in that it can display electrical signals in visual form. This chapter will familiarize the student with the operation of a typical elementary oscilloscope, the Tektronix Type 503, and will teach some basic scope measurement techniques. The discussion will begin with a block diagram of the sections of the oscilloscope and an explanation of how these sections interrelate. The operation of each section will then be explained in greater detail.

4.2 oscilloscope block diagram

A typical elementary oscilloscope, the Tektronix Type 503, is illustrated in Figure 4.1. Figure 4.2 is a block diagram of this instrument, including the approximate control locations on the Type 503 pertinent to each block.

The cathode-ray tube (CRT) is the signal display device. Its basic function is to convert an electrical signal into a visual image. A beam of electrons is emitted from a heated cathode and is accelerated and focused so that it strikes a fluorescent screen registering as a small spot. If signal voltages are applied to deflection plates between which the beam passes, the resulting electrostatic forces will cause the beam to move and produce a visible trace on the screen. In this manner, a visual display of an electrical signal is achieved.

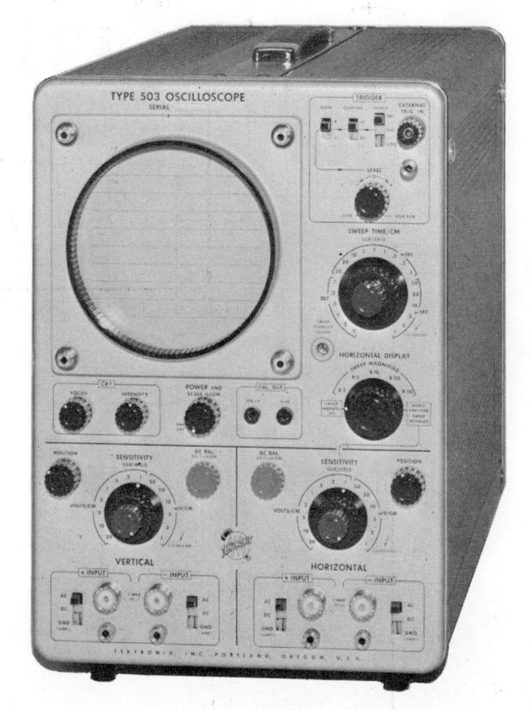

Figure 4.1. The Tektronix Type 503 oscilloscope.

In most scopes the deflection plates are located at right angles so that a Cartesian display is achieved. In this manner, the vertical position of the spot is independent of the horizontal position of the spot. Some CRT displays, such as those used in radar, employ a polar deflection system, but these are not normally found in general laboratory oscilloscopes.

In the CRT section of the scope are located the controls which adjust the intensity and dimensions of the spot on the screen. At the beginning of operation these should be adjusted to produce a readily visible small round spot. *Caution: Permanent damage to the screen may occur if the intensity control is set so bright that a "halo" forms around the spot.*

In most applications, the voltages under observation are not sufficient to produce a measurable deflection of the spot. For this reason amplifiers are inserted between the input signal and the CRT deflection plates. Usually the vertical amplifier is calibrated to produce a certain deflection for a given applied voltage, which is selected by a switch attenuator on the front panel. In some scopes, such as the Tektronix Type 503, the horizontal amplifier is also calibrated; in others, it is not.

Frequently it is desirable to make the horizontal deflection proportional to the parameter time, permitting displays of vertical signal amplitude versus time. To achieve this time axis, a voltage which increases linearly with time must be applied to the horizontal input. This causes the spot to sweep horizontally across the face of the CRT at a constant velocity, its

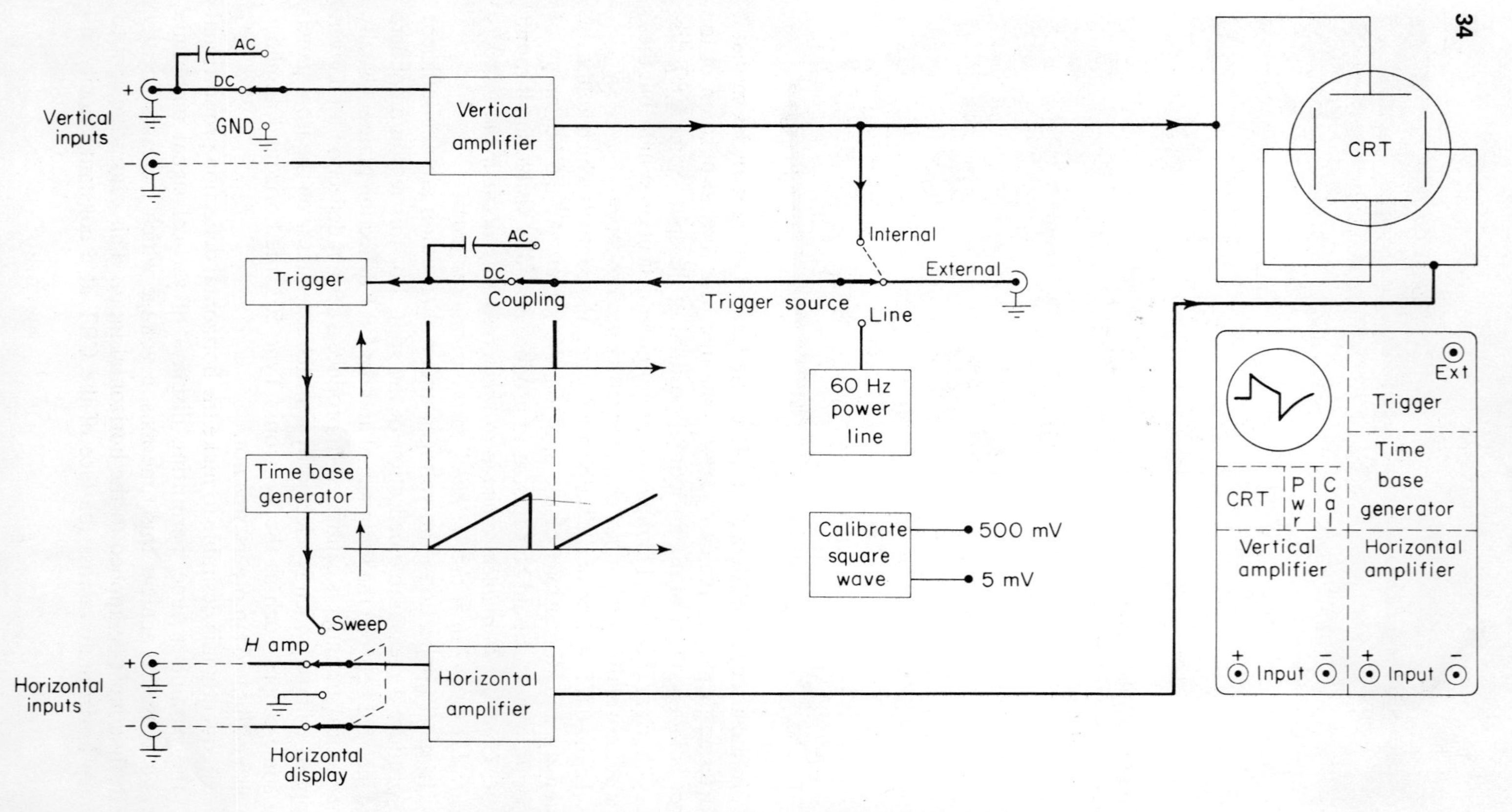

Figure 4.2. Elementary oscilloscope block diagram.

position in direct proportion to time. Thus, by means of a horizontal display selector switch, either an external signal or an adjustable time axis may be chosen for the horizontal spot displacement.

In order to select an appropriate starting point in time relative to the signal under observation a trigger circuit is employed. This circuit in effect establishes the time origin, or $t = 0$ point, which permits coherent superposition of a repetitive input signal on the CRT screen.

4.3 amplifiers

The amount of amplification or gain in the oscilloscope is adjusted by the sensitivity controls. In all scopes, the vertical channel is provided with a calibrated sensitivity control, which is accurate to the order of $\pm 3\%$, and adjustable in discrete steps. Provision is normally made for continuous variation of the gain by a variable sensitivity control, usually located as a knob concentric with the calibrated control. *For the amplifier to be calibrated, the variable sensitivity control must be in a set position, as indicated on the scope panel.* In most scopes the horizontal gain control is uncalibrated and does not provide the range of amplification available in the vertical channel. The Tektronix Type 503, however, has identical horizontal and vertical amplifiers, both calibrated, which are very useful in making X-Y plots of signals as discussed in Chapter 6.

In order to provide for positioning the trace on the screen, adjustable DC voltages are applied to the amplifiers, in addition to the signal voltages. These DC voltages are controlled by the position controls, which are part of each channel's amplifier circuits.

Because the scope amplifiers are direct coupled, small changes in the DC operating points of the input stages due to temperature variations, component aging, etc. result in a displacement of the electron beam which is indistinguishable from a DC voltage applied to the input. In order to cancel out this unwanted effect, each amplifier is provided with a balance control, which is set so that the measured DC voltage does not depend on the sensitivity setting of the amplifier.

Exercise 4.1

Turn the intensity control to the minimum position (full counterclockwise) and turn on the oscilloscope by means of the power switch. Allow a few moments for the scope to warm up.

Set the sensitivity switches to 0.2 volt/cm, calibrated. Set the input switches to the grounded or OFF position. If the inputs do not have switch provision for grounding, connect a short cable from the amplifier inputs to a ground terminal on the scope.

Advance the intensity control until a spot is observed on the screen, or until the control has been advanced about half-way. If a spot is not observed, adjust the position controls to see if the spot is off screen. If still no spot is observed, increase the intensity control a small amount. Continue with the position and intensity controls until a spot is centered on the screen. *Be sure the spot is not too bright. There should be no "halo" visible.*

Adjust the CRT controls, intensity, focus, and astigmatism (if provided) to produce as small a spot as possible.

Increase the vertical sensitivity about 10 times. If the spot moves from the central location, recenter it by means of the balance control. Reset the vertical sensitivity to 0.2 volt/cm and recenter the spot with the position controls. Repeat this procedure until the spot remains centered for all values of vertical sensitivity from 0.2 volt/cm to maximum. Normally it is necessary for the scope to warm up about 15 minutes before a stable balance can be achieved. During experiments in which small DC voltages are being measured, the scope should be periodically checked for balance.

If the horizontal amplifier is calibrated and provided with a balance control, repeat the balancing procedure on that amplifier.

4.3.1 amplifier inputs

The amplifier inputs to an oscilloscope are designed to receive a signal voltage relative to ground. If only one terminal is provided for a channel, it is termed *single-ended* and is capable of measuring only signals relative to ground. Many vertical amplifiers, however, are provided with two inputs, arranged so that the signals relative to ground at each input may be added or subtracted electronically. This type of input is termed *differential* since it permits measurement of the voltage difference between two nodes, neither of which is grounded.

A switching arrangement is provided to select the desired amplifier coupling, AC or DC. This arrangement is illustrated on the +INPUT in Figure 4.2. In the DC position the total voltage at the scope input is amplified and applied to the deflection plates of the CRT. In the AC position, a capacitor will block any DC voltage and AC frequencies below the range of 20 Hz from reaching the amplifier. This feature is particularly useful when small AC signals are superimposed on a large DC signal. If the AC coupling were not provided, the DC voltage would drive the spot off the screen when sufficient sensitivity to observe the AC signal was used. Normally, it is preferable to use the DC position if possible as this provides the best amplifier low frequency response. A third input switch position, GROUND or OFF, is also provided. This position *disconnects* the input terminal while grounding the amplifier input internally. On scopes so equipped, the input switch should be set to ground on any unused input channels. Study this arrangement in Figure 4.2. This switch position *does not ground* the input terminal on the oscilloscope.

4.3.2 input probes

Connections to the amplifier inputs should be made by means of special probes designed to shield against stray signal pickup and to provide maximum frequency response of the amplifier. They are also provided with interchangeable tips which facilitate connections to the circuit wiring. Scope probes are often designed to attenuate the input signal by a factor of 10. These are marked 10 ×, and this factor must be used in calculating the overall sensitivity of the measurement system. The use and adjustment of a 10 × probe will be discussed in detail in Section 11.4.1. For the remainder of this assignment, the 1 × probe, which does not attenuate the signal, should be used.

4.3.3 amplifier limitations

The oscilloscope amplifiers are not ideal, and their performance limitations must be considered when interpreting the observed waveforms. As mentioned above, in the AC coupling position the low frequency response begins to degrade in the neighborhood of 20 Hz. In a scope such as the Tektronix Type 503 the upper frequency response limit is 450 kHz. More advanced oscilloscopes are available with frequency response to 150 MHz, and to the GHz range by sampling techniques.

The input of the oscilloscope will also "load" the circuit under test. For the 1 × probe connected to the Tektronix Type 503, the effect at the probe tip is the equivalent of 1 MΩ in parallel with 67 pF from the probe tip to ground. In many instances this will be negligible; in others it will not, and the presence and effect of this "loading" must be kept in mind. Figure 4.3 illustrates the loading situation.

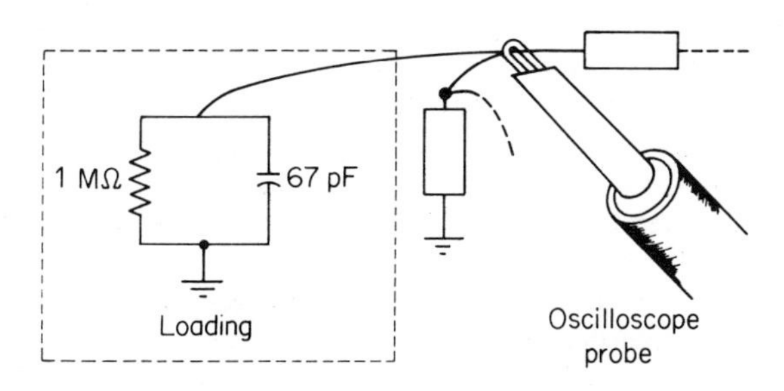

Figure 4.3. Circuit loading by oscilloscope input.

Exercise 4.2

Plot $x = A \sin \omega t$, $y = A \cos \omega t$ on a Cartesian coordinate system (Figure 4.4). Note that the locus of the plot is a circule of radius A.

$$x^2 + y^2 = A^2 \sin^2 \omega t + A^2 \cos^2 \omega t = A^2 \tag{4.1}$$

Figure 4.5 demonstrates this in a graphical way.

Observe that these sine waves, out of phase by $\pi/2$ radians, give rise to a circle. What visual display would you expect for two sine waves in phase? In your notebook, using a construction like that in Figure 4.5, plot y vs. x for two sine waves that are in phase.

What trace would you expect for two sine waves of equal amplitude but one with a frequency twice the other? Find the locus graphically in your notebook.

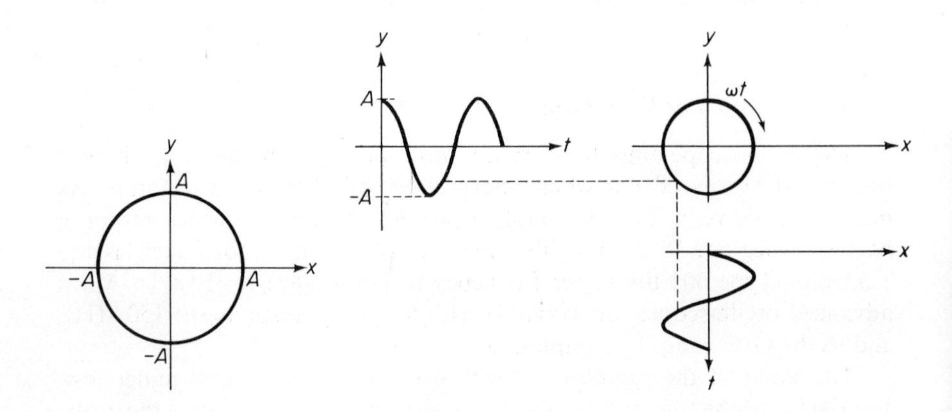

Figure 4.4.

Figure 4.5. Graphical construction of Figure 4.4.

Exercise 4.3

Connect a 6.3-volt, 60-Hz bench supply to the input of the vertical amplifier. Connect a signal generator set to 60 Hz to the horizontal input. Set the input coupling to DC. Adjust the sensitivity controls on the scope and the amplitude and the frequency controls of the signal generator until a straight line, circle, or ellipse is observed on the scope. If the pattern rotates, it is an indication that the oscillator output frequency is not the same as that of the bench supply. The length of time it takes the pattern to rotate 360 degrees is a measure of the difference between the two frequencies. Tune the signal generator to 120 Hz. Do you observe the anticipated pattern? Increase the frequency of the oscillator until another stable pattern occurs. Can you predict what frequency this will be? These patterns are called Lissajous figures in honor of the French mathematician.

In general,

$$\frac{f_{\text{vertical}}}{f_{\text{horizontal}}} = \frac{\text{number of intersections of horizontal axis}}{\text{number of intersections of vertical axis}} \tag{4.2}$$

Verify this relation experimentally.

4.3.4 phase measurement

In general, the phase difference between two signals of the same frequency may be determined from a Lissajous pattern as shown in Figure 4.6. This can be demonstrated as follows:

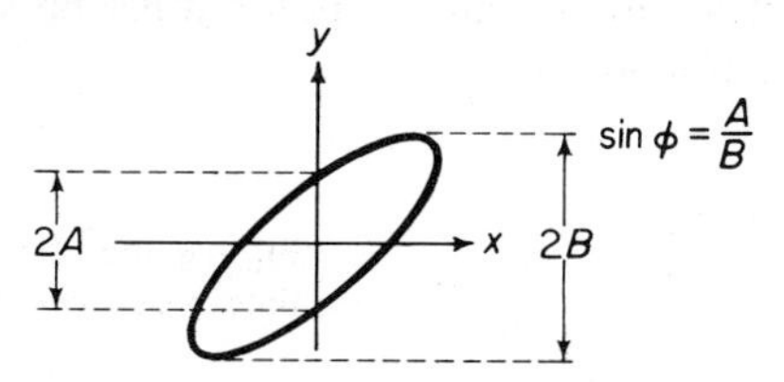

Figure 4.6. Phase measurement.

Let

$$x = X \sin \omega t \tag{4.3}$$

$$y = B \sin (\omega t + \phi) \tag{4.4}$$

and, from Figure 4.6, $y = A$ when $x = 0$, and hence $\omega t = 0, 2\pi, \ldots$. But at $\omega t = 0$,

$$y = B \sin \phi = A \tag{4.5}$$

$$\therefore \quad \frac{A}{B} = \sin \phi \tag{4.6}$$

Exercise 4.4

Construct the circuit shown in Figure 4.7. Measure the phase angle between V_2 and V_1 for various frequencies from 100 Hz to 10 kHz. Plot the data on semilog paper. What is the phase angle at 1590 Hz?

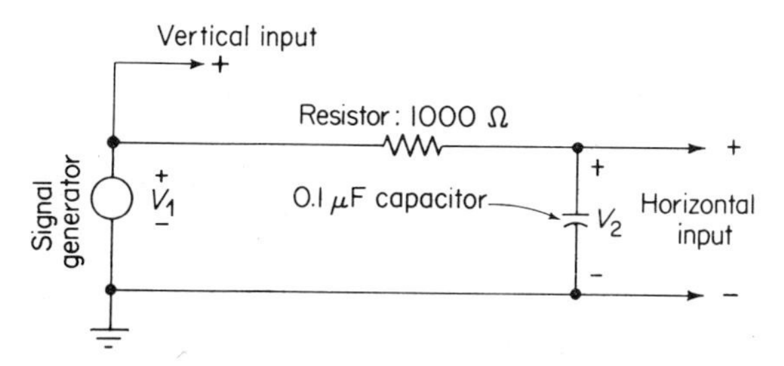

Figure 4.7. Circuit for phase measurement.

4.4 time base

As indicated on the block diagram, one has a choice of applying an external signal or a time-base waveform from the internal time-base generator to the horizontal amplifier. This selection is accomplished by means of the horizontal display switch as shown in Figure 4.2. The time-base generator provides a sawtooth wave having a slope proportional to the time/cm setting on the time-base control as shown in Figure 4.8. The time-base control normally has discrete calibrated positions plus a concentric control to provide for intermediate sweep speeds. As with the input sensitivity controls, the control for intermediate speeds must be set properly for the sweep time to be calibrated.

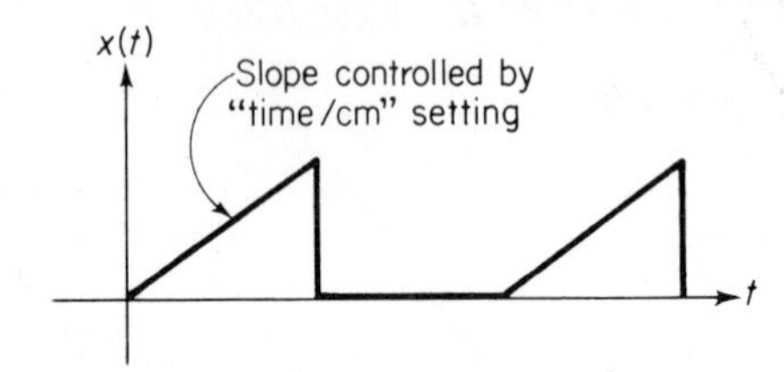

Figure 4.8. Sawtooth waveform from time-base generator.

Exercise 4.5

What trace would you anticipate if a sine wave were applied to the vertical amplifier and a sawtooth to the horizontal amplifier (by way of the time-base generator)? Demonstrate the result by drawing the pattern $y(t)$ vs. $x(t)$ shown in Figure 4.9 in your notebook.

Now, on the same $y(t)$ vs. x plot, repeat the construction with the sawtooth waveform displaced as shown in Figure 4.10. From this construction it is clear that if the repetitive sweeps of the sine wave $y(t)$ are to superimpose, some method must be employed to start the sawtooth wave at the same point on the sine wave. This function is provided by the trigger circuits.

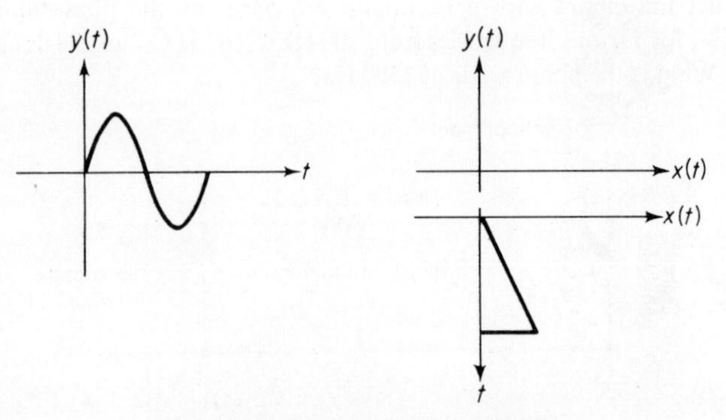

Figure 4.9. Time-base signal display.

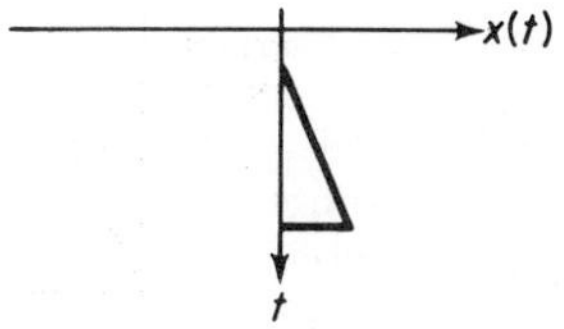

Figure 4.10. Displaced time base.

4.4.1 trigger

In order to obtain a stable display, it is necessary to start the horizontal sweep consistently at the same time relative to recurring cycles of the input waveform. The sawtooth sweep therefore must be started or triggered by the input waveform itself or by some external waveform which bears a fixed time relationship to the input waveform. In most scopes it is possible to select the source of the trigger signal to be either the input signal itself, an external synchronization signal, or the AC line voltage. The function of this source selection switch is shown in Figure 4.2.

For most applications, the sweep can be triggered internally by the vertical input waveform after amplification by the vertical amplifier. The only requirement is that after amplification the input signal provide at least 0.5 cm of deflection on the CRT screen at the sensitivity level for which the vertical sensitivity control is set.

Sometimes it is advantageous to trigger the sweep with some external signal. This is especially true when the input waveform is of small magnitude or when one samples waveforms from several different points within a circuit. The advantages of using the external trigger are twofold. First, the gain setting on the vertical amplifier no longer affects the triggering level. Second, it provides a means of starting the sweep independent of the vertical input signal.

When one is observing a waveform which bears a fixed time relationship to the line frequency (60 Hz), it is convenient to use line as the source of the trigger signal.

The *coupling*, *slope*, and *level* controls provide for selection of the particular point on the trigger source waveform that will start the sawtooth. When the conditions set by these controls are met, a short trigger pulse is sent to the time-base generator and the sawtooth starts. This is illustrated in Figure 4.11. It is important to note that the trigger pulses must not occur more frequently than the time for the completion of a sawtooth; otherwise the time-base generator will not be ready to start a new sweep when the next trigger pulse arrives. Failure to observe this condition will result in double images.

4.4.2 coupling

For most recurrent waveforms, AC coupling of the triggering signal will provide satisfactory triggering of the sweep. DC coupling of the triggering

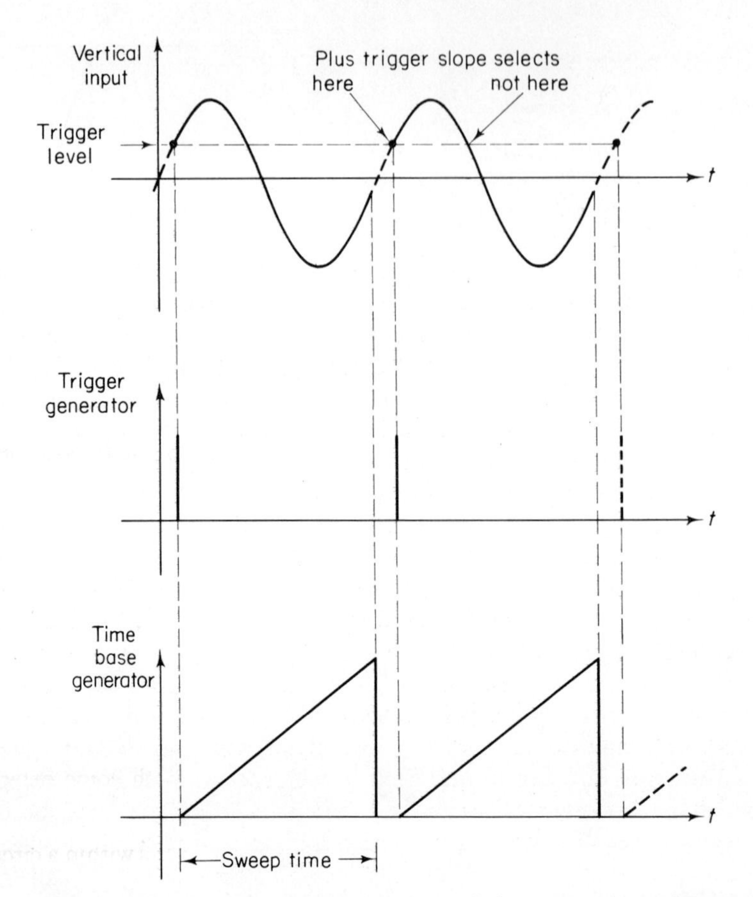

Figure 4.11. Illustration of the functions of the sweep time, trigger slope, and trigger level controls.

signal is particularly useful in triggering on very low frequency waveforms. With DC coupling, the sweep is triggered by an absolute voltage level. With AC coupling the sweep is triggered when the signal reaches a given amplitude from its average DC level.

4.4.3 slope

When the slope switch is in the plus position, the sweep is triggered on a positive-going slope of the triggering signal. When the slope switch is in the minus position, the sweep is triggered on a negative-going slope of the triggering signal. This is also illustrated in Figure 4.11.

4.4.4 trigger level, automatic mode, free-running mode

The level control determines the instantaneous voltage level (AC or DC, depending upon the setting of the coupling switch) of the triggering signal at which the trigger pulse is produced. With the slope switch in the plus position, adjustment of the level control makes it possible to trigger the sweep consistently at virtually any point on the positive-going slope of the triggering signal. Likewise, with the slope switch in the minus position, adjustment of the level control makes it possible to trigger the sweep consistently at virtually any point on the negative-going slope of the triggering signal. The trigger level may have some small frequency dependence such that the actual voltage level at which a trigger pulse is produced will vary somewhat with frequency of the source waveform for fixed level-control setting.

Exercise 4.6

Apply a 500-Hz sine wave to the vertical input. Using both internal and external sources, trigger on various points on the waveform using the slope and level controls. Observe the dependence of the trigger point on the gain of the vertical amplifier when using the internal source and the frequency of the input sine wave.

In connection with the level control there are usually *free-run* and *automatic* positions. For most applications, the automatic position is suitable. In this mode, the triggering signal is AC coupled and the trigger level is automatically set such that any external triggering signal of 1 volt or more, or internal triggering signal which would produce 1 cm or more of deflection on the CRT screen, will trigger the sweep. In the absence of such a triggering signal, the sweep will continue to be triggered automatically at about a 50-Hz rate to produce a base line on the CRT screen, indicating that the instrument is adjusted to display any signal which might be connected to the vertical channel.

Setting the level control to the free-run position produces a free-running sweep, independent of any synchronizing signal. The frequency of the free-running sweep is dependent upon the setting of the sweep time/cm control. This free-running trace is useful as a base line for making DC measurements when the input signal is DC coupled and for observing high-frequency signals by adjustment of the variable sweep time control.

Exercise 4.7

Turn the horizontal display switch to sweep with 1 × magnification, trigger source to internal, and the trigger level to automatic. Connect the 60-Hz signal from the bench supply to the vertical input and adjust the vertical sensitivity and sweep time controls for a trace. The frequency of a signal may be measured by the relationship

$$f = \frac{1}{T} \tag{4.7}$$

where T is the period. Vary the sweep time control. Observe that horizontal magnification (expansion) begins at the beginning of the trace. Turn the horizontal display switch to 2×, 5×, etc. Observe that the magnification begins, not at the beginning of the trace, but at the center of the screen. Using the horizontal display switch in conjunction with the horizontal position control, it is possible to magnify any portion of the displayed waveform.

4.5 additional measurements

4.5.1 differential rejection ratio

The *differential rejection ratio* is normally defined as the ratio of amplitude of a signal applied simultaneously to the plus and minus input terminals to the amplitude of the residual signal indicated on the scope. The differential rejection ratio is a measure of how perfectly the minus input subtracts from the plus input.

Exercise 4.8 (differential input scopes only)

Connect the 500-mV calibration square wave located on the scope to the vertical plus input. Measure the peak-to-peak voltage of this calibrated waveform. Next connect the same waveform simultaneously to the vertical plus and minus inputs. Increase the amplifier gain and measure the magnitude of the residual signal. Calculate the differential rejection ratio.

Note that the differential rejection in a good scope is at least 100:1 provided that the peak-to-peak voltage to the input does not exceed differential amplifier breakdown voltage. (See operating instructions for voltage limitations.)

4.5.2 additional phase measurement

The phase shift between two signals may also be measured by using the external trigger feature of the oscilloscope. This is accomplished by measuring the time difference t_d between an identifiable point on the two waveforms and then calculating the phase angle

$$\phi = \frac{t_d}{T} \times 360 \text{ degrees} \tag{4.8}$$

where T is the period of the waveform.

To measure the time difference t_d, the reference signal is used to trigger the oscilloscope externally. The reference signal is then applied to the vertical amplifier and the location of an identifiable point on the waveform, such as a zero crossing, is noted on the scope face. Next, the reference signal is removed from the vertical amplifier and replaced by the second signal.

Without changing any trigger control settings, the time difference t_d between the corresponding identifiable points on the second signal and the reference signal is measured by multiplying their spacing on the scope face in centimeters by the sweep time per centimeter setting.

Exercise 4.9

Using the circuit of Figure 4.7, remeasure the phase angle between V_1 and V_2 at 1590 Hz using the time difference method described above.

4.5.3 additional exercises

Exercise 4.10

Construct the circuit shown in Figure 4.12. Using a 1 × probe, compare the waveforms V_1 and V_2 at 20 Hz, 200 Hz, and 2 kHz using both AC and DC input coupling. Explain what features are due to the AC coupling capacitor and what features are due to the probe loading, described in Section 4.3.3.

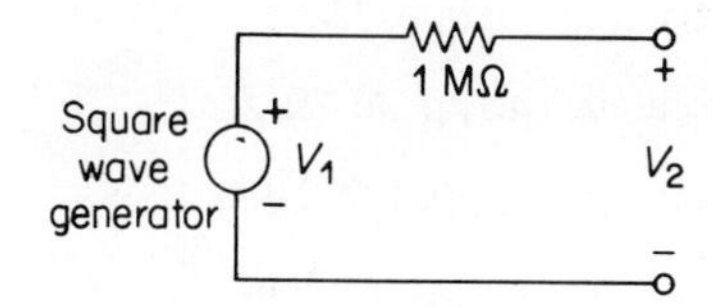

Figure 4.12. Circuit to illustrate probe loading.

Exercise 4.11

Use the horizontal magnification feature of the oscilloscope to examine the rise time of a 1-kHz square wave.

4.6 references

1. H. Carter, *An Introduction to the Cathode-ray Oscilloscope*, 2nd ed., Phillips Technical Library, 1960, pp. 5–31, 92–97.

2. J. H. Ruiter, *Modern Oscilloscopes and Their Uses*, 2nd ed., Holt, Rinehart and Winston, Inc., New York, 1955, pp. 15–60.

3. J. Czech, *The Cathode-ray Oscilloscope*, Interscience Publishers, Inc., New York, 1967, pp. 6–25.

4. J. F. Rider and S. D. Uslan, *Encyclopedia on Cathode-ray Oscilloscopes*, 2nd ed., J. F. Rider Publisher, Inc., New York, 1959, pp. 1–196.

5. *Instruction Manual Type 503 Cathode-ray Oscilloscope*, Tektronix, Inc.

For additional information on the amplifiers see:

Tektronix 503 Instruction Manual, *op. cit.*

J. Ruiter, *op. cit.*, pp. 83–95.

H. Carter, *op. cit.*, pp. 54–63.

J. Czech, *op. cit.*, pp. 81–134.

For additional information on phase and frequency measurements see:

Tektronix 503 Instruction Manual, *op. cit.*

J. Ruiter, *op. cit.*, pp. 130–148.

H. Carter, *op. cit.*, pp. 173–182.

J. Rider, *op. cit.*, pp. 427–475.

J. Czech, *op. cit.*, pp. 141–155, 175–184, 197–210.

For additional information on time bases see:

Tektronix 503 Instruction Manual, *op. cit.*

H. Carter, *op. cit.*, pp. 32–53.

J. Czech, *op. cit.*, pp. 36–80.

J. Rider, *op. cit.*, pp. 197–210.

J. Ruiter, *op. cit.*, pp. 96–108.

5

basic DC and AC meters

5.1 introduction

The aim of this chapter is to introduce the basic DC and AC meters commonly used for making voltage and current measurements in the laboratory. In addition, some of the limitations of these meters will be considered.

Almost all meters used for measuring electrical signals employ current detection as their basic indicating mode. An ideal *ammeter* is a device which measures the current flowing in an electrical circuit. This ideal measurement can be achieved only if the ideal ammeter presents a short circuit to the loop in which it is connected, since only then will the circuit voltages and currents remain unchanged when the ammeter is connected. A dual argument shows that the ideal *voltmeter*, which measures the potential difference (voltage) between a pair of nodes of an electrical circuit, presents an open circuit to the node pair; for only in that case will the circuit voltages and currents remain unchanged when the meter is connected. As with all devices, voltmeters and ammeters can only approach the ideal characteristics in practice.

5.1.1 electrical meter movements

A meter *movement* is the electromechanical device which provides a mechanical motion of an indicator in response to an applied electric signal. This motion is produced by the interaction of electric or magnetic fields produced at least in part by the voltage or current to be measured. Four types of meter movements are in common use for electrical instrumentation. Each is shown in Figure 5.1 and is briefly described below.

The *electrostatic* movement is the only type which employs the forces due to electric fields and charges, and is the only movement that measures voltage directly rather than by the effect of a voltage-produced current. Basically, the movement is simply a variable capacitor, with the attraction between fixed and movable plates balanced by a restoring force. The response is square-law and hence it is a true rms meter. The expected accuracy is 0.5–1%. The movement responds to DC as well as AC, and the frequency response is limited only by the AC current-carrying capacity of the movement or the capacitive loading effect on the circuit under test. Electrostatic voltmeters are normally used for voltages in the 10-volt to

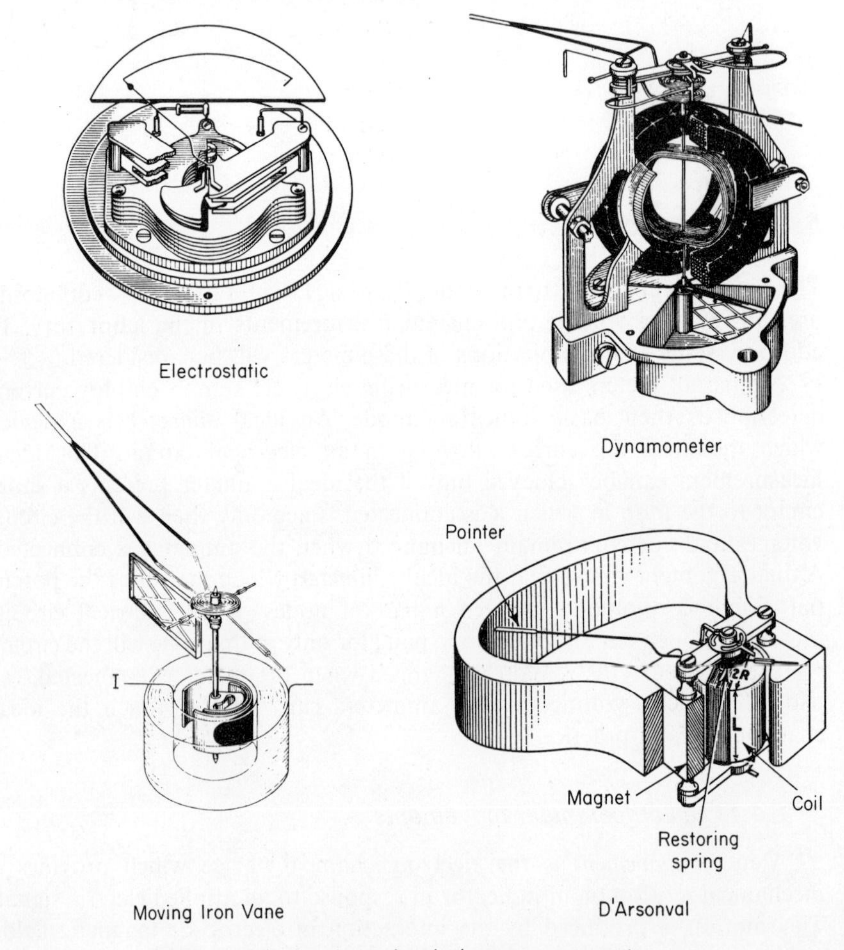

Figure 5.1. Common electrical meter movements.

100-kV range and require infinitesimal power at DC since they are basically a capacitor. For AC voltages, the loading depends on the frequency and the capacitance, which ranges from 10 pF in a 100-kV movement to about 200 pF in a 100-volt movement.

The *dynamometer* movement consists of two coils; one is fixed, the other movable. The measured current flows in both coils, and the interaction of the resulting magnetic fields produces a torque. Since the current flows in both coils, the torque is proportional to the square of the current, and hence the dynamometer is also a true rms movement. The absence of any magnetic materials, with their attendant nonlinearities, makes the dynamometer inherently the most accurate movement; 0.1% accuracy is available. Because of the inductance of the coils, the frequency response of the dynamometer is limited to about 200 Hz for high accuracy. A second limitation is the high power requirement; 1 to 3 watts is the typical loading of a dynamometer movement. Commonly available dynamometers range from 1 to 50 amperes, and with a series resistor, from 1 to 300 volts.

The *moving iron vane* movement is closely related to the dynamometer. In this movement the moving coil is replaced by an iron sheet. The torque in a moving iron movement is produced by the reaction of the magnetic field due to the current coil and the magnetic field due to the current induced in the iron. At low currents, the torque is proportional to the square of the measured current. However, as the iron vane becomes saturated, the response becomes linear in current. Thus, the scales on moving vane movements are compressed at the low end. The frequency response of these movements is limited to about 125 Hz. The best attainable accuracy is 0.5%. As ammeters they range from 10 mA to 50 amp, and voltmeter versions are available from 1 to 750 volts. As with the dynamometer, the power required is on the order of 1 watt. Since they are less expensive, moving iron vane movements are normally employed in place of dynamometers when the highest accuracy is not required.

The most common meter movement is the *permanent magnet–moving coil* type, sometimes known as a d'Arsonval movement. The pointer deflection is accomplished by suspending a coil carrying the current to be measured in a steady magnetic field produced by the permanent magnet. The torque produced by the interaction of the current-produced field and that of the permanent magnet is opposed by a restoring spring. Hence, as the current in the meter coil changes, so does the steady-state angular position of the coil. The pointer, attached to the coil, in turn indicates the amount of current in the coil by motion across a calibrated meter scale. This general arrangement is shown in Figure 5.1. If the permanent magnet field is uniform and the spring linear, then the pointer deflection is also linear in coil current. Since an average torque can be produced only by a DC current, these movements do not respond to AC. However, below about 10 Hz, most d'Arsonval movements will track an AC current to some extent. Generally available sensitivities are in the range of 15 μA to 1 amp, with accuracies to 0.1%. With a series resistance, DC voltages from 1 to 1000 volts may be indicated. D'Arsonval movements are extremely sensitive, requiring only 1 to 100 μW of power.

5.2 DC ammeters

Most DC ammeters employ a d'Arsonval movement. The meter coil has an associated resistance which may be represented in a circuit model by a resistor in series with an ideal ammeter as shown in Figure 5.2. For the actual ammeter to approach the ideal ammeter, the coil resistance

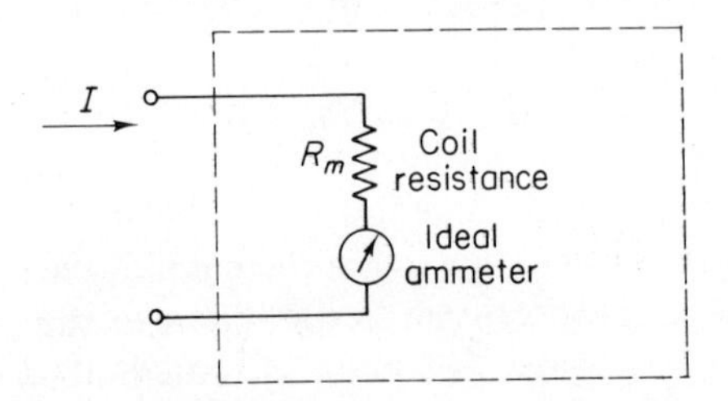

Figure 5.2. Equivalent circuit for D'Arsonval meter.

should be much less than the resistance in the loop whose current is being measured.

The commonly used designation of ammeter sensitivity is the current required for full-scale deflection. The more sensitive the ammeter, the smaller the full-scale current value.

A basic d'Arsonval movement can be used to indicate currents higher than the full-scale current of the movement by inserting a shunt resistor R in parallel with the meter movement as shown in Figure 5.3a. By using several shunts and a selector switch, a d'Arsonval movement can be made to indicate a wide range of current. This arrangement, shown in Figure 5.3b,

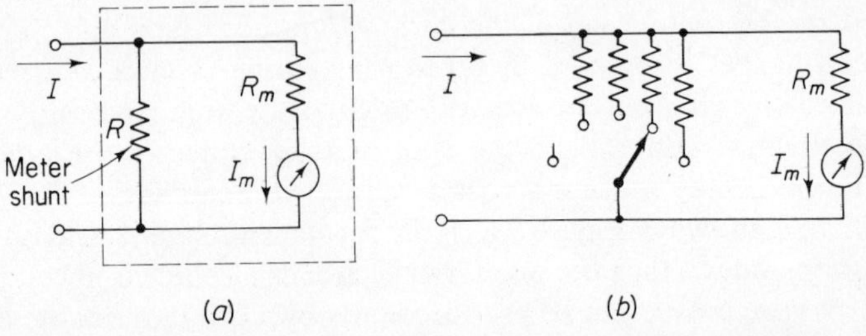

Figure 5.3. (a) Ammeter with shunt. (b) Multirange ammeter.

is employed in the common multitester, where a selector switch is used to choose a full-scale current deflection.

Exercise 5.1

What is the relation between the terminal current I and the meter movement current I_m for the circuit in Figure 5.3a?

When the shunt resistor R is disconnected from the circuit of Figure 5.3a, the basic meter movement has a full-scale deflection of 100 μA and a resistance of 900 Ω. When R is switched into the circuit, the ammeter has a full-scale deflection when I equals 1 mA. What is the resistance value of R?

5.3 DC voltmeters

A D'Arsonval movement can be used to indicate voltage by inserting a resistor in series with the movement as shown in Figure 5.4. Since ideally the voltmeter presents an open circuit to its environment, these resistances should be much larger than the impedances of the circuit being measured, and they are usually much larger than R_m. As with the ammeter, a multirange voltmeter can be constructed by using several values of series resistance and a selector switch.

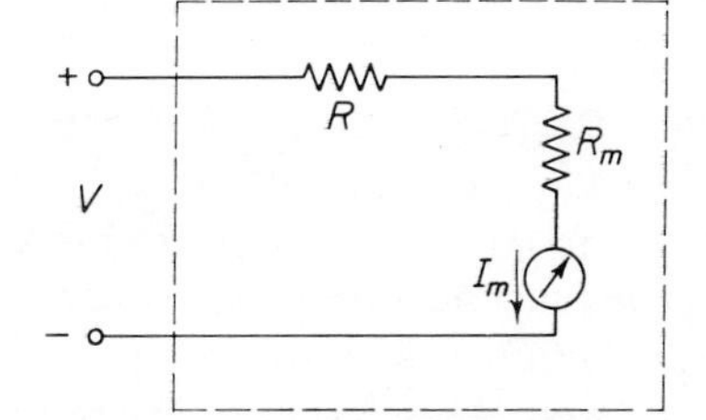

Figure 5.4. D'Arsonval meter as a voltmeter.

A common voltmeter sensitivity rating is the voltage required at the meter terminals for full-scale deflection. Another commonly used voltmeter sensitivity measure is the ohms/volt rating. The ohms/volt rating is obtained by dividing the series resistance (in ohms) of the voltmeter on a given scale by the *full scale* voltage indication of that scale. This rating essentially gives a measure of the nonidealness of a particular voltmeter due to series resistance, and is always equal to the reciprocal of the full-scale d'Arsonval meter current. Thus, a voltmeter rated at 20,000 ohms/volt uses a 50-μA meter movement.

Exercise 5.2

What is the algebraic relation between terminal voltage V and meter current I_m of Figure 5.4?

A multitester has a 30,000 ohms/volt rating. What series resistance does it present when the selector switch is on 1-volt DC scale? On 2.5-volt DC scale?

In the voltmeter of Figure 5.4, the given ohms/volt rating is 10,000 ohms/volt. What is the value of I_m for full-scale deflection?

Exercise 5.3

Using a small power supply or a battery, a multitester, and composition resistors, set up the circuits in Figure 5.5. Compare the actual voltmeter readings with the results expected if the voltmeters were ideal.

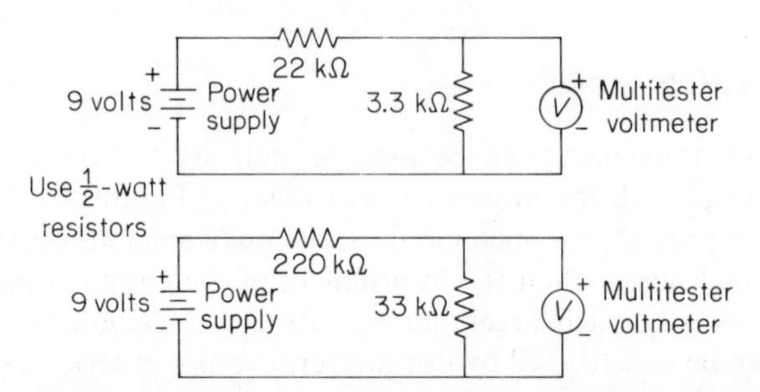

Figure 5.5. Use of the multitester voltmeter.

5.4 calibration and meter errors

The *accuracy* of a meter is its ability to indicate true voltage or current, and depends not only on external effects, such as circuit loading, observational errors, and temperature, but also on the basic accuracy of the meter movement itself. This basic accuracy is normally specified as a percentage of full-scale deflection; and hence, at small deflections, the meter reading can be in doubt by a large absolute value. For example, if a meter accurate to 5% full scale were used on the 100-volt scale to measure 20 volts, the actual measurement would be in doubt by ±5 volts or ±25%! It is therefore clear that to minimize error, the meter scale should be selected to provide the maximum possible deflection.

This problem of poor accuracy can be partly overcome by having the meter calibrated. A simple method of calibrating a meter against a highly accurate meter is shown in Figures 5.6 and 5.7. In Figure 5.6 the potentiometer should have a resistance of about one-tenth or less of the combined resistance of the voltmeters. This insures that the current flow in the potentiometer will be large compared with the current flow in the meters. The power-supply voltage V should be adjusted to about 10% above the maximum value required for the calibration measurement. If a variable supply

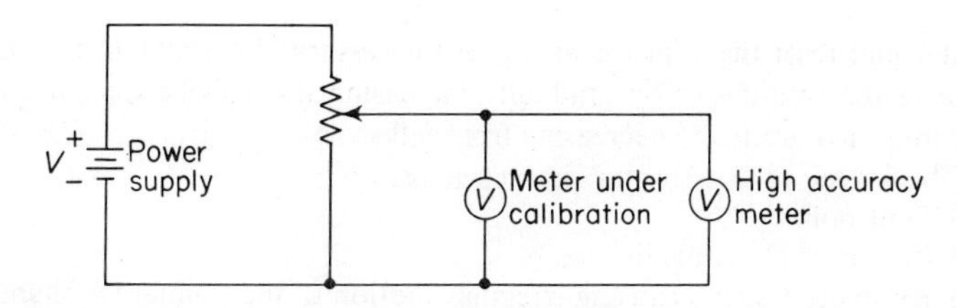

Figure 5.6. Voltmeter calibration.

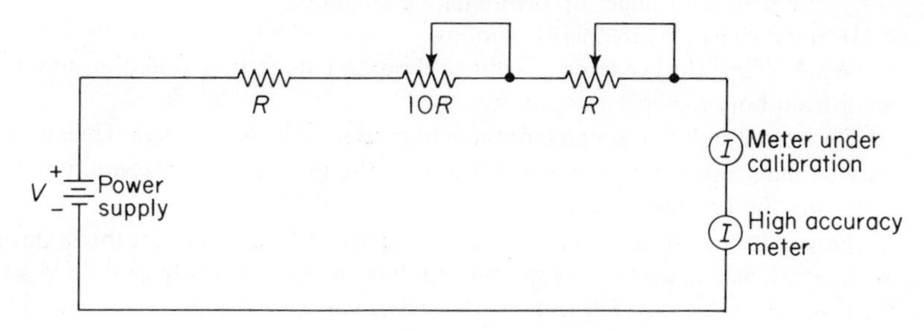

Figure 5.7. Ammeter calibration.

is available which can cover the required range, the potentiometer can be omitted and the meters connected directly to the supply.

In Figure 5.7, the values of V and R should be chosen such that R is large (10 times) compared with the combined meter resistances, and the current V/R is about 10% greater than the required full-scale deflection.

Other calibration methods are described by Frank in *Electrical Measurement Analysis*, Chapter 5 (see Reference 1). A sample of a calibration curve is shown in Figure 5.8.

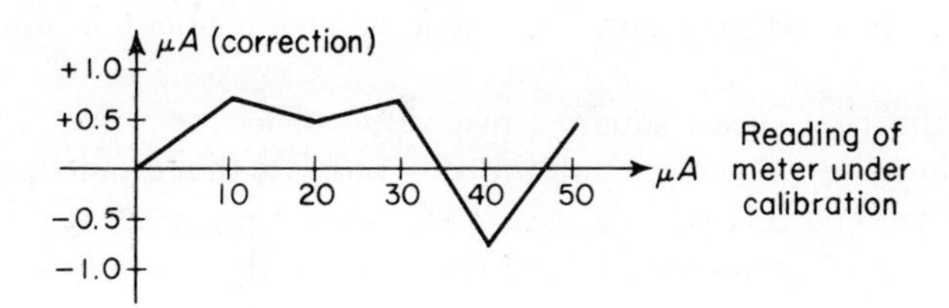

Figure 5.8. Sample calibration curve.

Exercise 5.4

The emphasis of this exercise is on the realization and estimation of error as well as the calibration of an instrument.

Calibrate either a voltage scale or a current scale of a multitester using a precision meter of the appropriate range. Use extreme care with the precision meters. *Do not exceed their full-scale voltage or current.* First, check the zero position of the pointer and reset the adjustment screw if necessary. Then obtain the calibration curve of the multitester by gradually increasing the voltage (or current) to full scale and then gradually decreasing from full scale.

Check your multitester for signs such as:

(a) Bent pointer.
(b) Pointer rubbing against scale.
(c) *Excessive* friction causing irregular motion of the pointer. A slight amount of friction is always present and may be minimized by tapping *gently* on the case with your finger tip before taking a reading.
(d) Static charge on the plastic window.

What affects the accuracy of your calibration curve? (Consider the amount of error from both meters.)

The repeatability of your measurements is referred to as *precision*. Thus, a meter may be inaccurate and yet precise. Estimate the precision of your multitester by repeating the previous calibration curve.

Now switch your multitester to the next higher scale and calibrate this scale only at major divisions up to the full-scale reading of the previously calibrated scale. *Caution: Do not exceed full scale on the precision meter.* Compare this calibration curve with the original curve and explain the differences.

5.5 AC measurements

Although several types of AC meters which respond directly to the alternating waveform are described in Section 5.1.1, they are usually limited in frequency response to a few hundred Hz or less, or are of low sensitivity. The majority of AC voltmeters, however, convert some feature of the AC waveform into a DC current so that it may be indicated on a d'Arsonval meter. The meter scale may be calibrated in terms of an AC voltage, but the meter is actually responding to a DC current. This general arrangement is illustrated in Figure 5.9.

Since the root mean square (rms) value of a sine wave is frequently required, most AC meters are calibrated to indicate that quantity. However, except in a few special cases, the actual AC meter does not respond to the rms value of the applied waveform, and the meter scale will give incorrect rms readings for other than sine waves. In order to interpret AC meter readings for other waveforms, one must know what waveform property the converter circuit utilizes and then compute a correction factor.

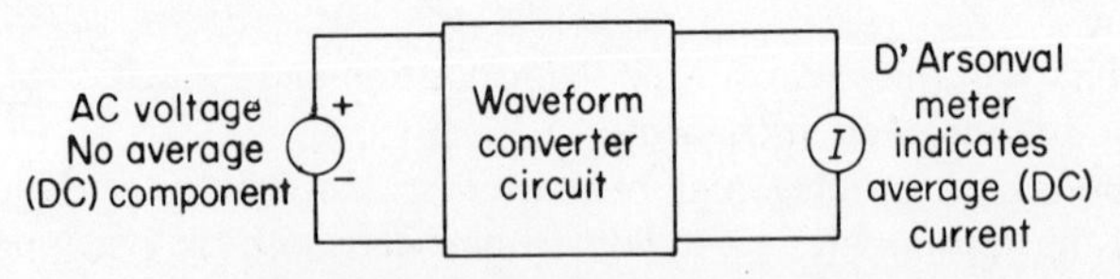

Figure 5.9. General AC voltmeter.

5.5.1 *waveform converter circuits*

There are several waveform properties which may be used to indicate the amplitude of an AC waveform. Some of the most common ones are described graphically in Figure 5.10. It should be clear from the drawings that different waveform converter circuits will deliver different amounts of DC current to the d'Arsonval movement, depending upon how the circuit operates. Thus, if the meter responding to the half-wave rectified average is calibrated in terms of the rms value of a sine wave, it will *not* indicate the rms value of a square wave, since the relationship between the rms value and the half-wave rectified value of a sine wave and square wave are different. If the actual response characteristic of the AC meter is known, the correction factor can be calculated.

For example, Figure 5.10 shows that the rms value of the sine wave is $A/\sqrt{2}$ and the half-wave rectified average value is A/π. Thus, a meter that responds to the half-wave rectified average of a waveform but is calibrated in terms of the rms of a sine wave actually indicates $\pi/\sqrt{2} = 2.22$ times the true half-wave rectified average of the input waveform. If a square wave of amplitude A were measured on such a meter, the indication would be 2.22 times the actual half-wave rectified average value ($A/2$) or $1.11A$.

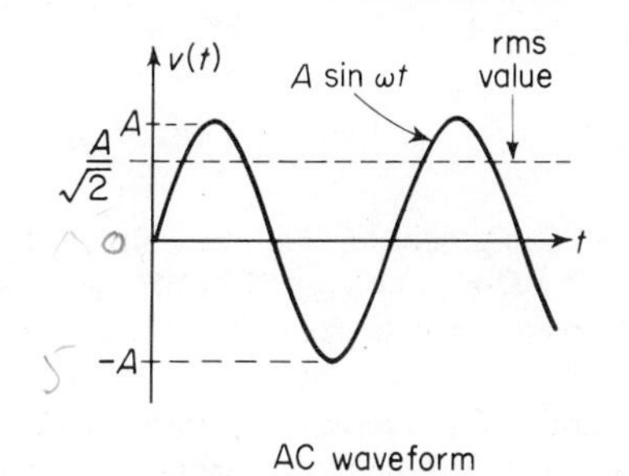

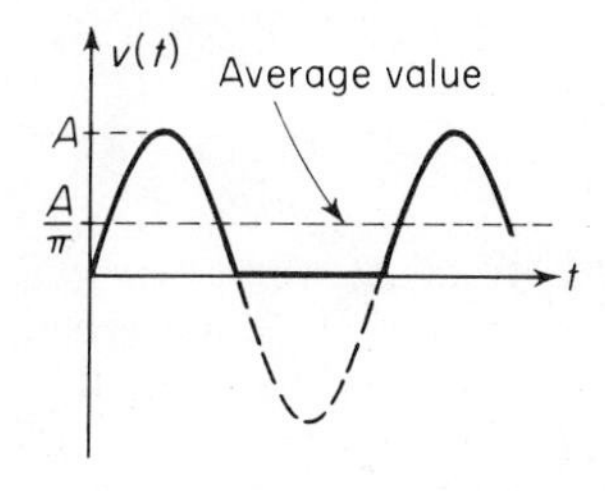

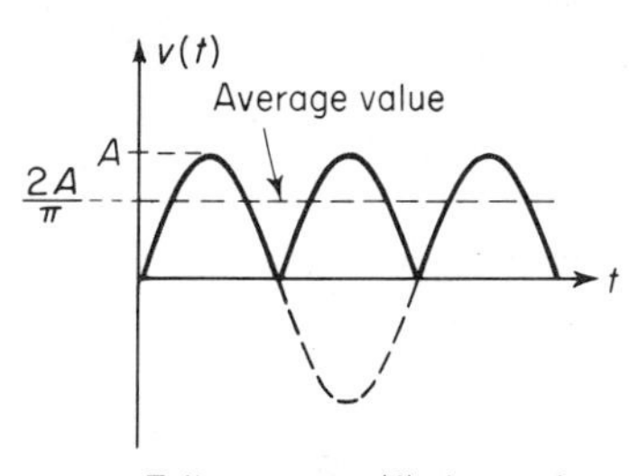

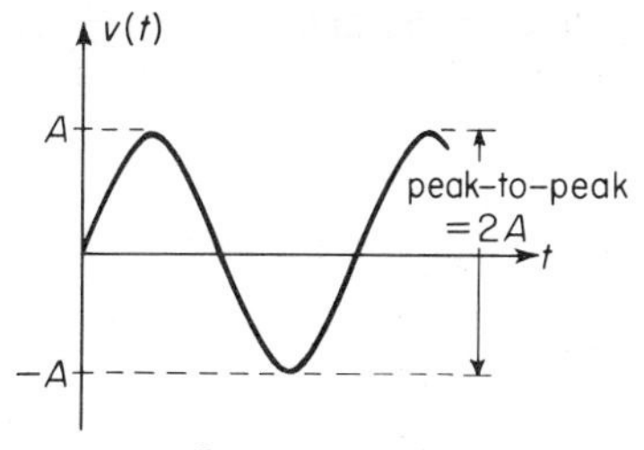

Figure 5.10. Examples of waveform characteristics used in AC meter converter circuits.

An additional factor which must be considered is whether or not the original AC waveform has a DC component. Some waveform converters use a capacitor in series with the input to block any DC that might be present. Others do not, and this fact must be considered when calculating the correction factor as described above.

Some examples of AC voltmeters and the type of converter circuits employed are listed in Table 5.1.

Table 5.1 / Examples of AC Voltmeters

Type	Waveform Converter
Hewlett-Packard Model 400D	full-wave rectified average blocks DC component of input
RCA Senior VoltOhmyst	peak-to-peak blocks DC component of input
Simpson Model 250 Multimeter	full-wave rectified average includes DC component of input
Triplett Model 630 Multimeter	half-wave rectified average includes DC component at "V-Ω-A" input, blocks DC component at "output" input

Exercise 5.5

Connect several AC voltmeters to a sine-square generator and an oscilloscope as shown in Figure 5.11. Set the generator for a 100-Hz sine wave of 10 volts peak-to-peak as indicated on the scope. Record the readings of the voltmeters and compare them to the expected value.

Insert a 1.5-volt battery at the point marked X and repeat your measurements. Reverse the polarity of the battery and again record the meter readings. Explain your results in terms of the information given in Table 5.1.

Remove the battery and set the generator for a 100-Hz square wave of 10 volts peak-to-peak as indicated on the scope. Explain the variation in meter readings by finding the appropriate scale conversion factors. Again insert the 1.5-volt battery (either polarity) and explain any changes in your readings.

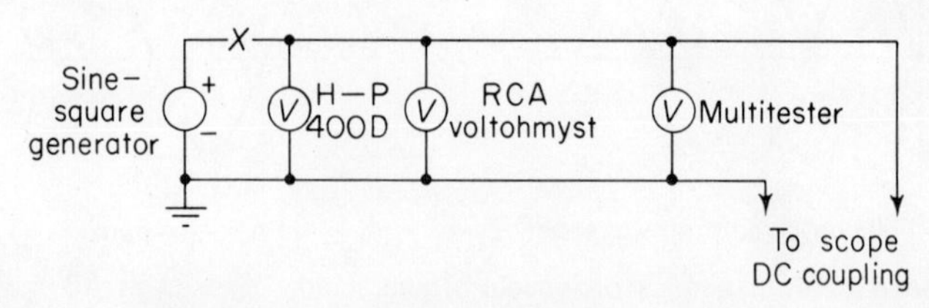

Figure 5.11. AC meter comparison circuit.

5.5.2 frequency limitations of voltmeters

Because the circuits used in AC voltmeters are made of components which exhibit frequency limitations in performance, it is not surprising that meters have a frequency response that is not constant. Therefore, in making AC measurements one must be aware of the frequency limitations of the measurement devices used.

Exercise 5.6

Connect the circuit shown in Figure 5.11. Measure and plot on semilog paper the frequency response of the meters and the scope. Plot the magnitude of each meter reading (in AC volts specifying which definition) versus frequency (20 Hz to 500 kHz) on the logarithmic scale. In general the source magnitude will vary as a function of frequency, but the HP 400D VTVM measures accurately to within 2% over the above frequency range. Therefore, use that VTVM as the standard and adjust the source output voltage to keep the reading of the HP 400D constant.

5.6 references

1. E. Frank, *Electrical Measurement Analysis*, McGraw-Hill Book Company, New York, 1959.

2. M. B. Stout, *Basic Electrical Measurements*, Prentice-Hall, Inc., Englewood Cliffs, N.J., 1950.

3. *Measurements and Data*, Vol. 1, No. 4, "Electrical Movements," pp. 65–96, July–August, 1967.

6

graphical display of two- and three-terminal characteristics

6.1 introduction

Description of two- and three-terminal devices in terms of their volt-ampere characteristics is usually the first step toward the useful application of the device as a circuit element. In the case of linear circuits, this description is often accomplished through a set of parameters (z's, y's, h's, etc.) which relate the various terminal voltages and currents of the circuit. In the case of nonlinear devices, a single set of numbers cannot adequately describe the electrical performance, and a graphical display is usually employed as the most efficient means to present the related data. Common examples of these cases are diode and transistor characteristics. This chapter discusses the production of such graphical displays on an oscilloscope and a transistor curve tracer, and with an X-Y recorder.

Because displays such as these are obtained "instantaneously" as compared with point-by-point DC measurements, they are frequently referred to as "swept" displays. That is, by "sweeping" one variable through its entire range, a complete display is obtained. In this chapter we will be concerned with only swept V-I characteristic displays. Chapter 15 will discuss swept-frequency displays.

6.2 graphical oscilloscope displays of two-terminal characteristics

The basic circuits used to provide graphical oscilloscope displays of the V-I characteristic at a terminal-pair are shown in Figure 6.1. In Figure 6.1a the

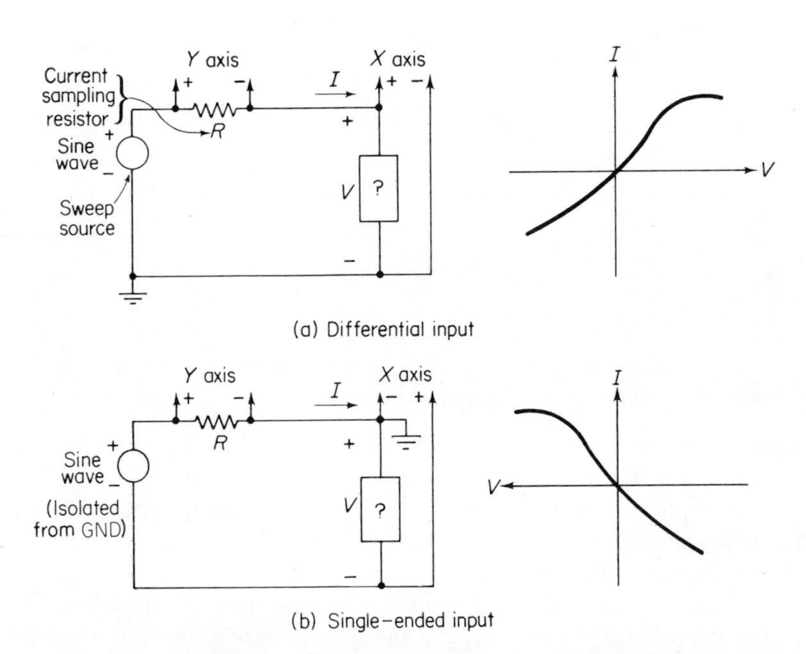

Figure 6.1. Basic circuits for oscilloscope display of V-I characteristics.

oscilloscope has a differential vertical input and the actual display appears with the axes in the normal directions. If a scope with a single-ended input is used, then the connection shown in Figure 6.1b is employed, because the – terminal of the X and Y inputs must be common. Note that this has the effect of reversing the direction of the voltage axis of the display and also requires the use of a signal source that is *isolated* from ground. (See Section 2.5.3.)

The signal source or *sweep source* normally has a sinusoidal output since this provides a single-frequency excitation. In most applications where a "static" or DC characteristic is desired, a 60-Hz signal is employed. Besides being readily available, it is fast enough to provide a flicker-free display and usually slow enough that high-frequency effects in the unknown do not enter the measurement. This latter point can be checked by varying the frequency and noting whether there is a change in this displayed characteristic. If the display is independent of frequency, then it is safe to assume that a static characteristic has been obtained. A second test of static display is that the voltage and current always be in phase. In terms of the oscilloscope display, a phase difference between the voltage and current will produce a V-I characteristic that is a closed curve rather than a single line. This effect is illustrated in Figure 6.2.

It is important to distinguish between the phase angle introduced by the unknown under test and the possible phase difference between the X and Y

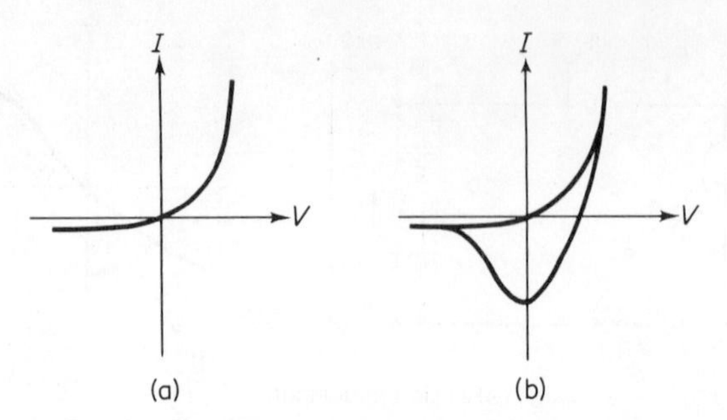

Figure 6.2. Effect of phase angle on *V-I* characteristic. (a) Desired static characteristic. (b) Characteristic with phase angle between *V* and *I*.

amplifiers of the oscilloscope. If the oscilloscope has identical *X* and *Y* amplifiers, then normally the phase difference is negligible. However, if the *X* and *Y* amplifiers are not identical, as is often the case, the phase difference should be checked by displaying the characteristic of a composition resistor. If the resistor's *V-I* characteristic is not a straight line, but appears as an ellipse, then there is a phase difference in the oscilloscope amplifiers. Note that it is the phase *difference* in the *X* and *Y* amplifiers which causes difficulty, not the absolute phase shift through an individual amplifier. If both amplifiers have the same phase shift, then the net phase shift between *X* and *Y* is zero. In using a scope with identical *X* and *Y* amplifiers, such as the Tektronix Type 503, it is essential that the input coupling be the same for both amplifiers. If one is set at AC and the other at DC, a phase shift between *X* and *Y* will result, as seen from the description of the input coupling switches in Section 4.3.1. In most situations, DC input coupling will be used for *V-I* characteristic displays.

The current-sampling resistor, *R* in Figure 6.1, provides two functions. First, it produces a voltage proportional to the input current to drive the vertical amplifier of the oscilloscope. Generally a value which is a power of 10 is chosen. If a value of $R = 1\ \text{k}\Omega$ is used, then the *Y* deflection factor in volts per division is automatically converted to milliamperes per division.

The second function of the resistor is to limit the power dissipation in the unknown, an important safety consideration when studying an unfamiliar device. If the maximum voltage of the sweep source is V_m, then the current into the unknown is limited to V_m/R. In the *V-I* plane, the choice of V_m and *R* provides a boundary for the characteristic curves as shown in Figure 6.3. As *R* is decreased, the maximum possible current increases, and the slope of the boundary increases, the line always passing through $V = \pm V_m$.

It is also possible to use a current probe, described in Section 11.4.2, to sample the input current of the device under test. However, they provide less sensitivity than can be normally obtained with a sampling resistor. Also, at low frequencies these probes may have undesirable phase shift. If measure-

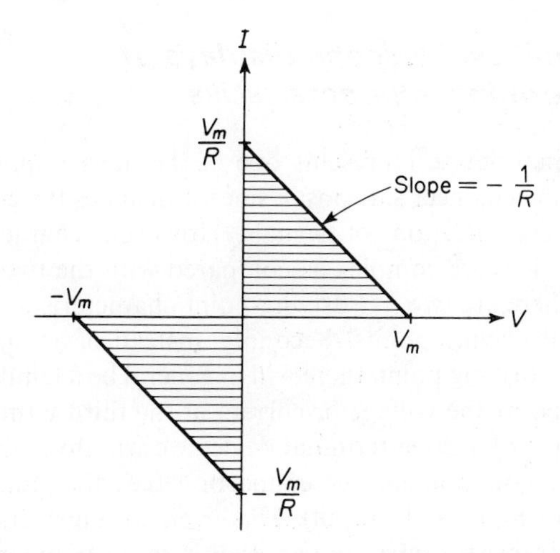

Figure 6.3. Region of *V-I* display limited by V_m and *R*.

ments are to be made at high frequencies, the current probe may be a better choice than the sampling resistor, since the latter will introduce stray capacitance and inductance.

Exercise 6.1

Construct the circuit of Figure 6.1 using a current-sampling resistor $R = 1$ kΩ. Display the *V-I* characteristic of a second 1-kΩ resistor on the oscilloscope. Vary the frequency of the sweep source from 20 Hz to 2 kHz with various combinations of the input coupling switches to observe the effects of phase difference in the amplifiers.

6.2.1 diode characteristics

A germanium diode begins to conduct current at a forward voltage of about 0.2 volt, whereas a silicon diode does not begin to conduct until the forward voltage increases to about 0.6 volt. This is due to a property of the semiconductor material and hence is a useful rule of thumb in deciding whether a particular diode is a silicon or germanium device. Also, the current flow under reverse bias in a silicon diode is usually much less than in a germanium diode of comparable cross-sectional area.

Exercise 6.2

Using DC input coupling and a frequency low enough to provide a "static" display, compare the *V-I* characteristics of a germanium diode and a silicon diode. Observe the approximate voltage at which the diode current is 5 mA in the forward direction. Estimate the diode current in each at a reverse voltage of −3 volts (a change in the current-sampling resistor may be helpful).

6.3 graphical oscilloscope displays of three-terminal characteristics

The three-terminal device represents one of the most important classes of components at the engineer's disposal, since it includes the control elements which permit amplification of signals. However, characterizing three-terminal devices is more complex as compared with the two-terminal component. First, there are two *V-I* driving-point characteristics since there are two independent driving points. Secondly, instead of a single *V-I* characteristic curve at a driving point, there will in general be a family of parametric curves, controlled by the voltage or current at the third terminal.

As an example of a three-terminal device we will investigate the bipolar transistor in the common-emitter connection (i.e., the emitter terminal is common to both input and output). However, we must first learn how to distinguish one transistor type from another in terms of polarity (*npn* or *pnp*) and material (silicon or germanium).

6.3.1 determination of transistor type

In order to determine the type of an unknown transistor, the characteristic between the collector and base terminals may be measured with the emitter open-circuited. Under this condition, the resulting characteristic curve will be that of a simple diode. If the equivalent diode is pointing from base to collector, the unit is *npn.* If the equivalent diode is pointing from collector to base, the unit is *pnp.* Also, the forward voltage at which current begins to flow will tell if the transistor is made from silicon or germanium as described in Section 6.2.1. This measurement is summarized in Figure 6.4.

Exercise 6.3

Using the circuit of Figure 6.1 with $R = 1\ \text{k}\Omega$, check several transistors to determine their type and material of construction. For terminal diagrams of transistors see Section 2.4.6.

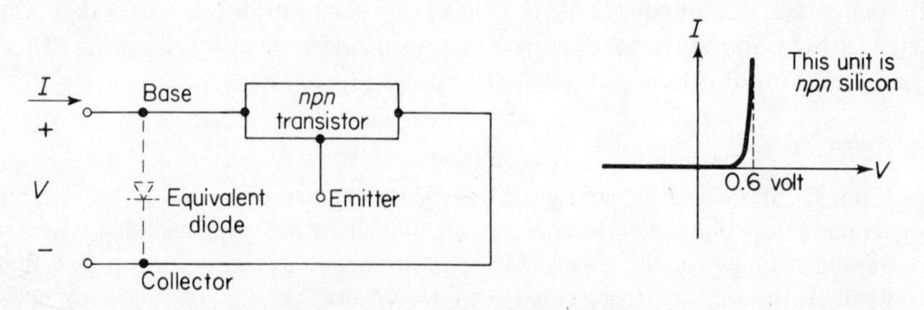

Figure 6.4. Determination of transistor type.

6.3.2 *common-emitter transistor characteristics*

The terminal variables for a transistor along with the usual circuit symbols are defined in Figure 6.5. The positive convention for voltage and current remains as shown in Figure 6.5, even though the actual currents and voltages will reverse direction for the different polarity types in the normal operating region.

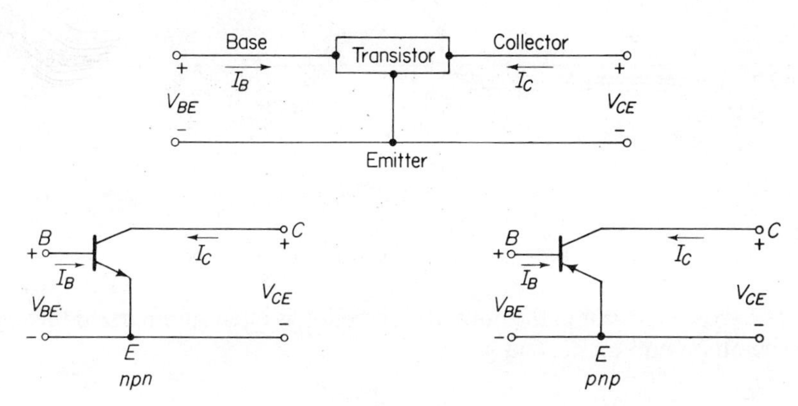

Figure 6.5. Terminal variables and circuit symbols for a bipolar transistor.

Since there are, in Figure 6.5, four terminal variables, six characteristics curves are possible, and for each of these six there are two choices for parametric variation, making a total of twelve possible characteristic curves. To construct all twelve would result in an overspecification, since in general two sets of parametric curves are sufficient to describe a three-terminal device. The choice of which two of the possible twelve are to be employed will depend both on the properties of the device and the circuit application at hand. However, for most transistor applications the curves of collector current I_C vs. collector-to-emitter voltage V_{CE} with I_B as the parameter are the most useful. A typical set of I_C vs. V_{CE} characteristics (frequently called the *output curves*) for an *npn* transistor is shown in Figure 6.6.

Notice that the "spread" of the curves in the first quadrant is larger than in the third quadrant. The first quadrant is the *normal* region of operation since it provides the maximum amplification or gain. Conversely, the third quadrant is known as the inverted region. The complementary *pnp* device has the normal operation region in the third quadrant and the inverted region in the first quadrant, the parametric variable being negative values of base current. Manufacturer's data sheets invariably show only the normal region of the characteristics.

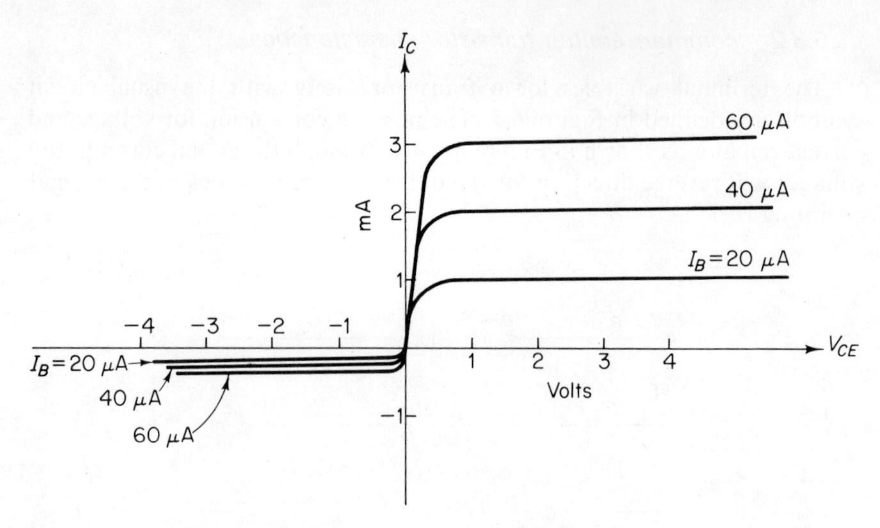

Figure 6.6. Typical output curves for an *npn* transistor.

The gross behavior in the normal gain region is often summarized through the common-emitter current gain

$$\beta_F = \left.\frac{I_C}{I_B}\right|_{V_{CE}} \tag{6.1}$$

This parameter is generally dependent on both I_C and V_{CE}, but over a wide range of operation this dependence may be only a small percentage change. A method of determining β_F from a set of characteristic curves is shown in Figure 6.7.

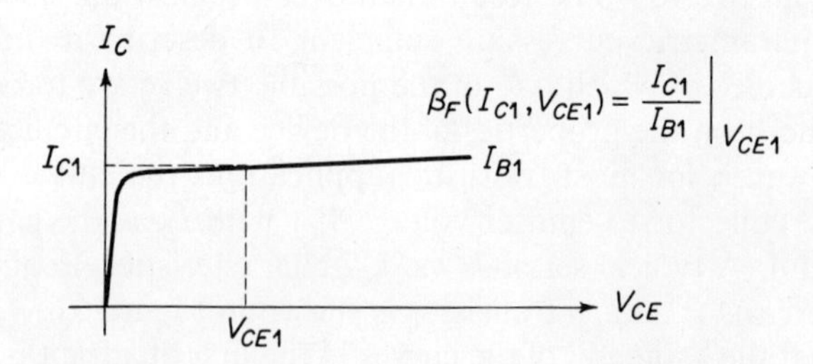

Figure 6.7. Determination of common-emitter current gain.

Exercise 6.4

The circuit of Figure 6.8 may be used to graphically display the I_C-V_{CE} characteristics of a transistor. The full-wave bridge rectifier circuit has been added to the sweep source in order to provide a symmetrical load to the signal generator. Without this circuit, the assymetrical *V-I* characteristic of the transistor causes

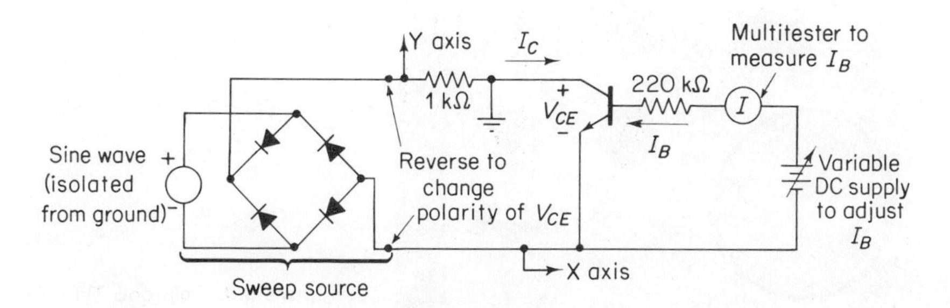

Figure 6.8. Circuit for display of transistor output characteristics.

difficulty in signal generators with capacitively coupled outputs. (See Section 14.2.1.) A second feature of the bridge rectifier circuit is to apply only one polarity of sweep voltage to the device under test. This is often desirable since the maximum voltage levels for devices are normally different for opposite polarities. In order to reverse the polarity of voltage applied to the device, it is necessary to reverse the output leads from the bridge rectifier.

The variable DC supply is used to fix the value of base current during the sweep. Note that only one curve, determined by the value of I_B, will be displayed. The value of I_B must be varied to obtain the parametric family. In order to reverse the polarity of I_B it is necessary to reverse the polarity of the variable supply and the meter.

Using a 10-volt peak, 60-Hz sine wave source, display the output characteristics of an *npn* and a *pnp* transistor. Use both polarities of I_B and V_{CE} for each device. From the measured characteristics determine the common emitter current gain β_F at $|V_{CE}| = 5$ volts and $|I_C| = 3$ mA in the normal operating region.

6.4 transistor curve tracer

The transistor curve tracer is an integrated oscilloscope and sweep signal source which may be used to make two- and three-terminal measurements on unknowns. Although designed for use with transistors, it is by no means limited to them. Understanding of the basic operation permits its use with many other devices. The basic operation of the Tektronix Type 575 Transistor Curve Tracer, shown in Figure 6.9, will be discussed below.

A simplified block diagram and front panel layout of the Tektronix Type 575 is shown in Figure 6.10. The horizontal and vertical amplifier controls select the signals applied to the X and Y oscilloscope axes, respectively. Controls are also provided to position the trace and to check the location of the origin of coordinates (zero check).

The sweep signal source is a rectified 60-Hz sine wave, whose amplitude and polarity are adjusted by the controls in the collector sweep block. As in the circuit of Figure 6.8, this rectified sine wave produces only one polarity

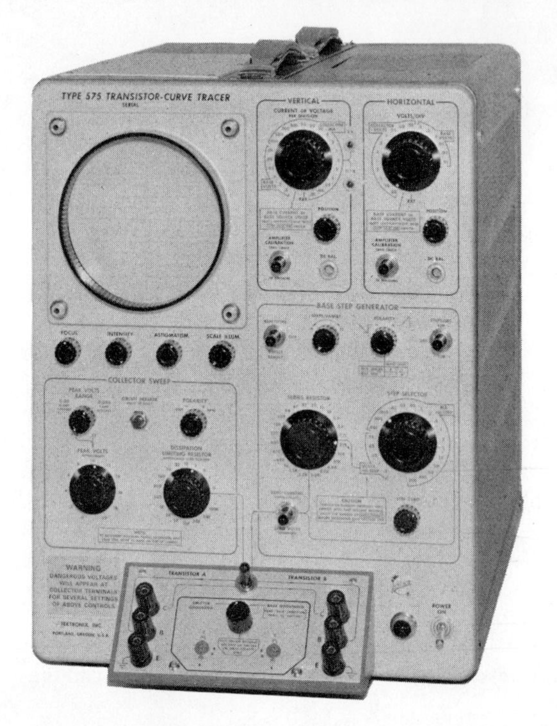

Figure 6.9. Tektronix Type 575 transistor curve tracer.

of sweep voltage at the collector terminal, selected by the polarity switch. The magnitude of the sweep voltage is adjusted by a range switch and a variable control. The dissipation limiting resistor sets the limit of the display as shown in Figure 6.3.

The transistor test panel contains two sets of binding posts for connection to the device under test. Plug-in adapters permit rapid connection of power transistors and units with long leads. A toggle switch selects which set of terminals is in use, and makes comparison between two transistors convenient.

The sweep voltage is applied to the collector and emitter terminals of the test panel. Thus, the driving-point characteristic of any device may be measured by connecting it between the collector and emitter terminals of the curve tracer.

Exercise 6.5

Use the curve tracer to repeat the previous measurements on the germanium and silicon diodes.

In measuring the three-terminal characteristics of the transistor in Exercise 6.4, only one curve was displayed on the oscilloscope at a time. This was because the base current was held constant during the sweep. Thus, if

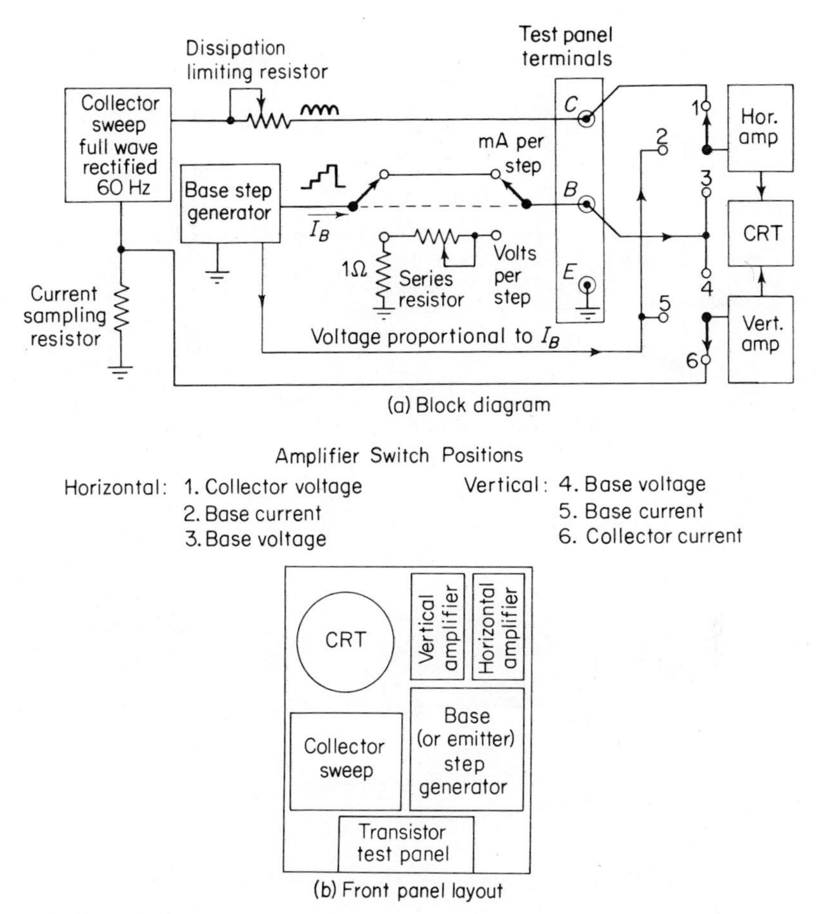

Figure 6.10. Block diagram and panel layout of Tektronix Type 575 transistor curve tracer.

an entire set of output curves were to be examined simultaneously, either a complete graph of the display must be made by hand or a multiple exposure photograph taken. The Type 575 is capable of displaying an entire set of parametric curves by changing the value of the electrical parameter at the third terminal in synchronism with the repetitive sweep voltage. Thus, although only one curve is actually present at a given instant, an apparent multiple trace display is obtained in very much the same manner that the alternate sweep provides an apparent dual-trace on a high-performance oscilloscope. (See Section 11.3.1.) To better understand this operation, consider the waveforms in Figure 6.11. The basic repetition rate is determined by the rectified 60-Hz sinewave that sweeps the collector-to-emitter voltage. A staircase current generator applies a constant base current during each collector sweep, but the value of base current is changed after each

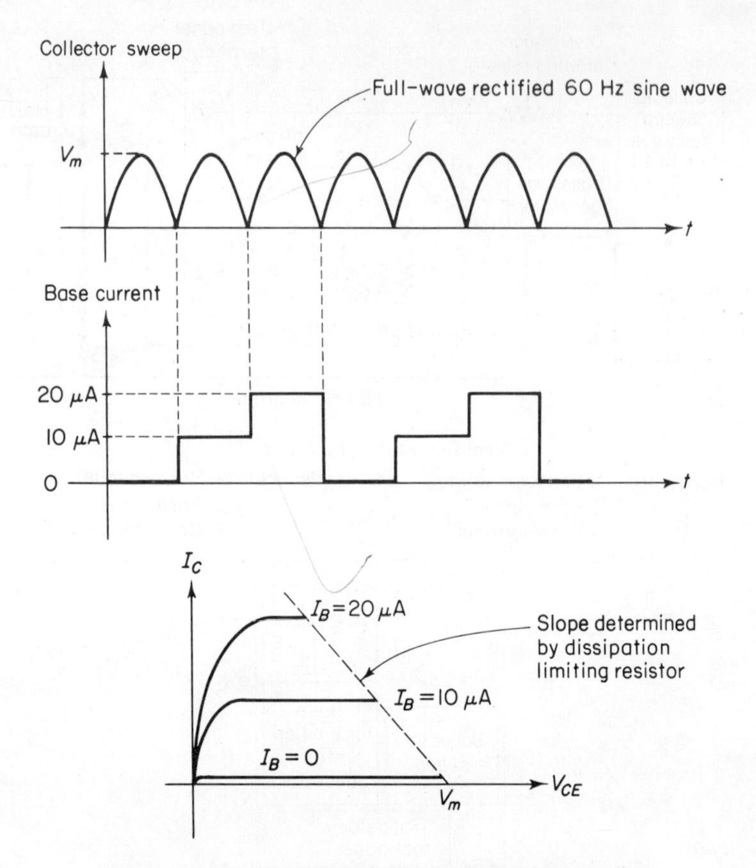

Figure 6.11. Illustration of parametric curve generation.

sweep. In Figure 6.11 three steps are shown as the complete cycle, after which the process repeats. The net result is to display on the oscilloscope three parametric curves, whose parameter values are given by the three values of the base current steps. Figure 6.11 also shows the resulting typical display, including the effect of the dissipation-limiting resistor and peak magnitude of the collector sweep voltage V_m.

The controls in the base step generator block adjust the staircase parametric variation at the base terminal. The step selector sets the step increment of base current or base-to-emitter voltage. In the voltage mode, a series resistor may be included in the base circuit to limit dissipation. The polarity of the current on voltage steps is selected by the polarity switch. The repetition and rate of steps are adjusted by the remaining controls at the top of the block. The single-family position will display one set of characteristics each time the switch is depressed. This low duty-cycle minimizes thermal effects due to power dissipation. The step zero control adjusts for drift in the DC amplifier and should be adjusted so that no change in the zero base current curve is observed when the switch at the lower left of the block is pushed toward zero current.

Additional details on the operation of the Type 575 Curve Tracer may be found in References 1 and 2.

Exercise 6.6

Use a curve tracer to display the common-emitter output characteristics of both an *npn* and a *pnp* transistor. Experiment with both base current and base-to-emitter voltage as the parametric variable.

6.5 X-Y recorders

We have seen how the characteristics of two- and three-terminal devices may be conveniently displayed on an oscilloscope. For many applications this technique is adequate, and if a permanent record is desired an oscilloscope photograph may be taken. However, in some situations more accuracy and resolution than can be obtained with an oscilloscope are desired. Then an *X-Y* recorder may be employed to produce the required graph on conventional graph paper. An example of this instrument is shown in Figure 6.12.

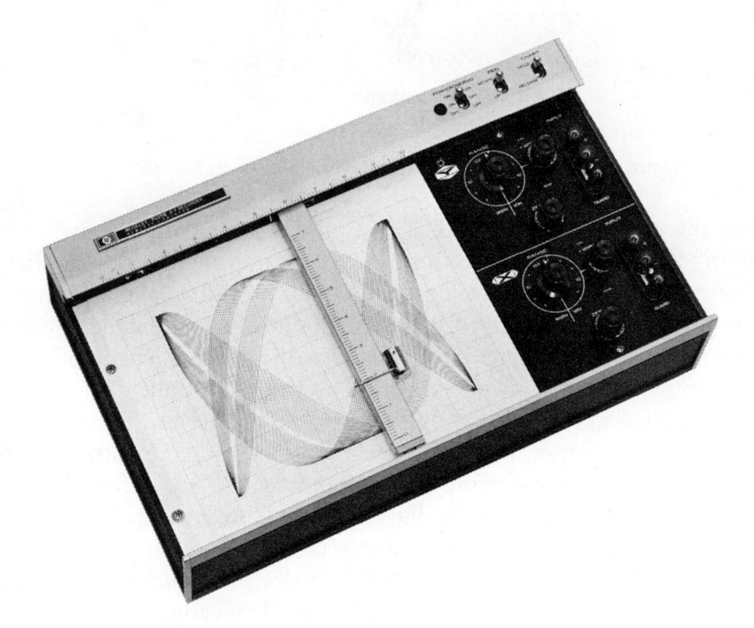

Figure 6.12. Hewlett-Packard 7035B *X-Y* recorder.

Basically, the X-Y recorder moves a pen in two orthogonal directions in response to electrical signals applied at the respective input terminals. In this sense, it is very much like an elementary oscilloscope in which the pen substitutes for the visible spot on the cathode ray tube. In many ways the functions of the controls are also similar to the elementary scope. However, there are some fundamental differences which we will discuss further.

6.5.1 X-Y recorder operation

A schematic diagram of one axis of an X-Y recorder is shown in Figure 6.13 and represents an example of a simple *servomechanism* or electromechanical feedback control system. The applied input signal is compared with a voltage derived from a reference supply and a potentiometer. The

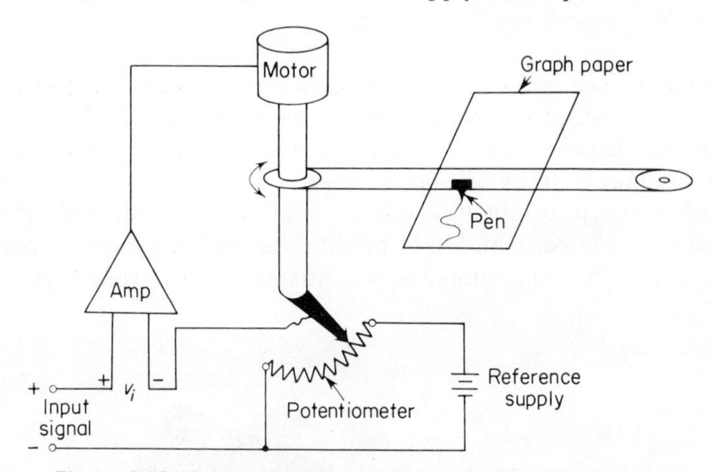

Figure 6.13. Schematic diagram for one axis of an X-Y recorder.

voltage difference v_i is amplified and applied to a motor which turns the potentiometer shaft in the direction to make v_i approach zero and thereby stop the motor. Thus, we have a potentiometer or null balancing voltmeter, such as described in Chapter 13, which is kept in constant balance by means of the amplified error signal v_i applied to the motor. The motor shaft simultaneously positions a pen through a string-and-pulley arrangement, thereby resulting in pen motion proportional to the applied input voltage.

Normally, both inputs to an X-Y recorder may be used in either differential or single-ended connections. This choice is usually accomplished by removing a ground strap from the input terminal. Because the instrument is a null potentiometer, the input impedance can be essentially infinite. However, in order to provide for variable sensitivity, the potentiometric input is only available on the most sensitive range. Typically, the less sensitive ranges have an input impedance ranging from 100 kΩ to 1 MΩ.

A calibrated, stepped sensitivity control and a variable sensitivity control adjust the deflection factor for each axis. Sensitivities as low as 100 μV/in. are available. The variable sensitivity is especially useful for fitting the recorder response from a transducer to the major divisions of the graph paper. For example, the variable sensitivity can be adjusted so that a pressure transducer coupled to the recorder will produce a graph with a scale directly in psi. Similarly, the relative response of a transducer can be plotted directly by adjusting the variable sensitivity such that the maximum output produces 100 divisions of pen motion. The basic accuracy of an *X-Y* recorder is determined by the properties of the potentiometer and the reference supply, and 0.1% is typical. Both *X* and *Y* amplifiers are also provided with a position or zeroing control which allows location of the zero voltage origin anywhere on the graph paper.

Some *X-Y* recorders, like the oscilloscopes, are provided with an internal time-base generator to provide a time axis for the recorder. On others, an external ramp generator may be used to provide a time axis. Typical sweep times range from 0.5 to 50 sec/in. Thus the *X-Y* recorder makes a useful compliment to the oscilloscopes for the display of slow waveforms.

Because the *X-Y* recorder is an electromechanical device, its frequency response is limited. The maximum velocity that the pen will move is termed the *slewing speed* and is typically 20 in./sec. When the recorded writing approaches this speed, the accuracy of the recorder becomes questionable.

The most common method of writing a trace on an *X-Y* recorder employs a pen-and-ink system. This consists of an ink reservoir connected to either a capillary-tipped or a fiber-tipped pen through a small diameter feed tube. Upon initial filling the system must be primed by forcing the ink through the tube to start the flow. Care must be used with these ink systems to prevent damage to graphs and the recorder through ink spillage. Also, care must be taken not to touch the pen tip or allow it to touch any oily or greasy substance, which would affect the ink-flow properties. Usually the recorder will incorporate an electrically operated pen lifter to raise the pen from the graph paper during adjustments and trial runs. A second switch controls the paper-holding mechanism, usually a vacuum system or an electrostatic platan.

Reliable writing with an ink system requires clean pens. Neglect of the pen system will lead to writing failure owing to dried ink and sediment in the feed tube and tip. A pen in continuous use should be thoroughly cleaned at least every two weeks to prevent the ink thickening and pen clogging that result from continuous addition of ink. Needless to say, pens should be thoroughly cleaned before storage. A pen system can be ruined by allowing ink to dry out in it.

To clean a pen system, flush it with warm water until all traces of the ink have disappeared and let it air dry. If ink has been allowed to dry in the system, the parts should be soaked in alcohol or water containing a mild detergent. Clogged capillary tubes may be opened with a fine stiff wire, usually supplied with the pen assembly.

Exercise

Use the circuit of Figure 6.14 to produce the output characteristics of an *npn* transistor in the normal operating region. The collector supply voltage is slowly varied from zero to the desired maximum value by hand to produce the sweep. This manual variation replaces the AC sweep supply in the basic configuration of Figure 6.1.

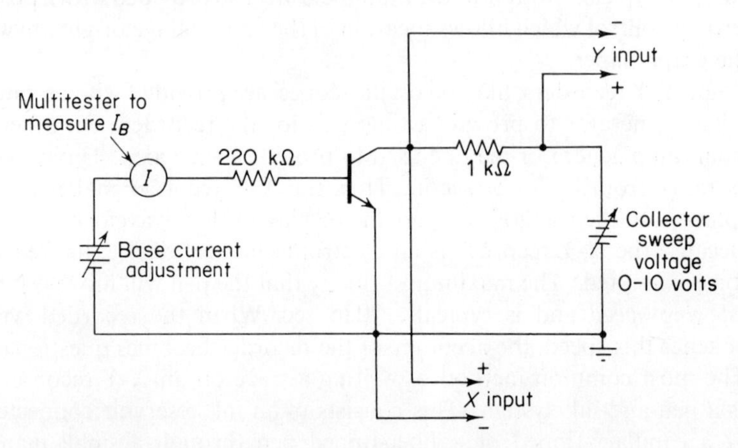

Figure 6.14. Circuit for plotting collector output characteristics with *X-Y* recorder.

6.6 references

1. Thornton et al., *Handbook of Basic Transistor Circuits and Measurements*, Sec. 7, John Wiley & Sons, Inc., New York, 1966.

2. *Tektronix Type 575 Transistor Curve Tracer Instruction Manual*, Sec. 2.

7

resistors

7.1 introduction

The ideal resistor is a two-terminal circuit element in which the instantaneous voltage between the terminals is directly proportional to the current flowing through the element. The mathematical relationship, known as Ohm's law, is simply

i R
+ v −

$$v = Ri \tag{7.1}$$

The constant of proportionality R is the *resistance* of the element, whose dimensional unit is the *ohm* (Ω).

Alternatively, Ohm's law may be expressed as

$$i = Gv \tag{7.2}$$

where $G = 1/R$ is termed the *conductance* of the circuit element. The dimensional unit of conductance is the *mho* ($\mho$).

Ohm's law states that the voltage-current relationship for a resistance element is a linear function; that is, the resistance R is independent of the current i. In practice, actual resistors can only approximate this ideal relationship; consequently Ohm's law is only an approximate mathematical description of an actual resistor. The purpose of this chapter is to explore some of the types of resistance elements commonly available, to examine their properties and limitations, and to introduce some of the common resistance measurement techniques.

7.2 resistor limitations

Before describing the various resistance elements in general use, we will consider some of the factors which affect performance as measured by the ideal description of Ohm's law.

7.2.1 resistance value

The value of resistors ranges from a small fraction of an ohm to hundreds of megohms. However, no single manufacturing method is possible for the entire range. For resistors with low values, special care must be taken that errors are not introduced by the resistance of the contacts and leads from the resistance element to the remainder of the circuit. Conversely, at high resistance values, above 100 MΩ, care must be taken that parallel-resistance leakage paths between the resistance terminals do not reduce the net resistance below the expected value. Such leakage paths are especially troublesome under conditions of high relative humidity, because moisture will have a varying effect on leakage. Leakage paths can also be produced on the resistor body by handling. Therefore, high-value resistors are usually cleaned in a solvent to remove contaminants before they are connected in a circuit. Figure 7.1 illustrates these sources of error.

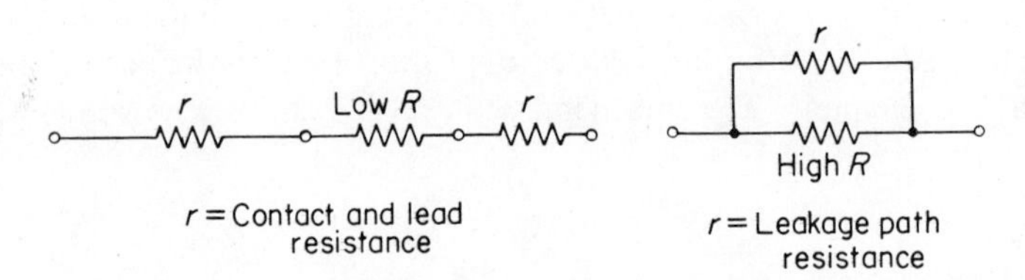

Figure 7.1. Illustration of error sources in low- and high-value resistors.

Resistance value is generally indicated on the resistor body by a color or numerical code. The common methods of marking are described in Section 2.4.1. Such resistance values, however, are only approximate. The degree of approximation is called the *tolerance* of the resistor and is expressed as a percentage of the indicated, or *nominal*, value. Thus a 1000-ohm resistor with a 5% tolerance may actually range from 950 to 1050 ohms.

Resistor manufacturers have established sets of standard resistance values with the tolerances 20%, 10%, and 5%. These values are arranged so that each successive value is approximately $(1 + 2N)$ times the preceding value, where N is the tolerance expressed as a decimal. This formula is only approximate, the round-off error being to two significant digits. This way of establishing the nominal values insures that any resistance will be within the tolerance range of one of the nominal values. This system has the obvious advantage to the manufacturer that all resistors produced are salable. Figure 7.2 illustrates the nominal values and the tolerance range for 20%, 10%, and 5% resistors. The data are presented on a logarithmic scale to emphasize that the nominal values are spaced by a constant factor.

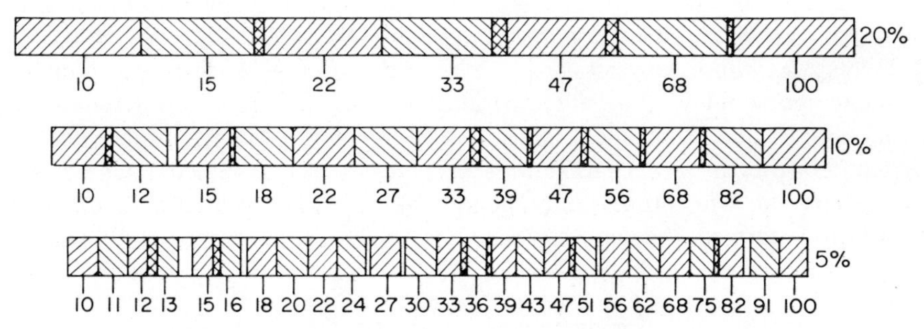

Figure 7.2. Nominal values and tolerance ranges for resistors (logarithmic scale).

7.2.2 temperature variation

All resistor values will change with temperature. In some applications, change in environmental temperature will be large. In others, a significant rise in resistance element temperature can result from the power dissipated internally by the element. Such self-heating can reach destructive proportions and can fundamentally limit performance.

The variation of resistance with temperature is specified by the *temperature coefficient* (TC). This parameter expresses the percentage change in resistance per degree Centigrade from the resistance value at 25°C. Thus, a 1000-ohm resistor with a TC of +0.1%/°C has a value of 1050 ohms at 75°C. This method of expressing resistance change with temperature implies that the change is linear in T. As with Ohm's law, this linear relation is only an approximation. Another way of expressing the temperature coefficient is in terms of parts per million (ppm) change per °C. Thus a TC of 0.001%/°C is equivalent to 10 ppm/°C.

The sign of the TC indicates the direction of resistance change. A positive coefficient indicates that R increases with increasing T; a negative TC indicates that R decreases with increasing T.

7.2.3 current and voltage limitations

As mentioned above, one of the fundamental limitations on resistor performance is the temperature rise resulting from self-heating. This is described by the power rating of the resistor, expressed in watts. Thus at

25°C, the maximum voltage and current that a resistor can handle is given by

$$V_{\max} = \sqrt{PR}, \qquad I_{\max} = \sqrt{\frac{P}{R}} \tag{7.3}$$

where P is the power rating in watts and R is the value of the resistor.

The power rating is derived from the maximum internal temperature of a resistor at which an irreversible change in resistance value will not occur. A resistor whose power rating is exceeded may not be destroyed, but its value upon cooling to room temperature may be substantially different from its initial value. Since the limitation is essentially on the maximum internal temperature, the power rating must be decreased as the ambient temperature is increased. This process of reducing maximum ratings under certain conditions is termed *derating*. A typical derating curve for carbon composition resistors is shown in Figure 7.3.

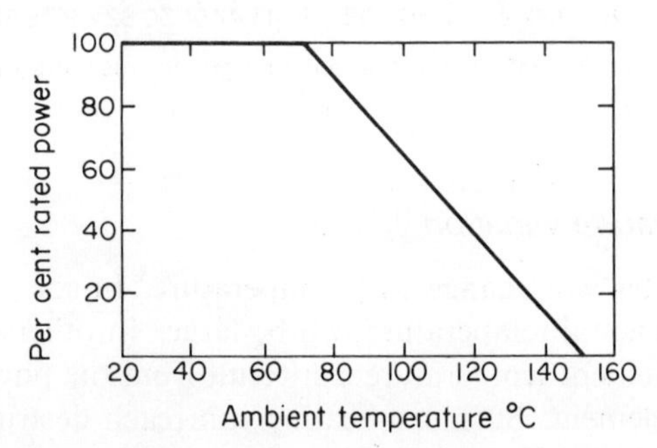

Figure 7.3. Typical power derating curve for carbon-composition resistors.

For high resistance values, at which the power dissipation may be small for large values of voltage, the high electric fields in the resistor can produce irreversible changes in value, or destruction, before the power rating is exceeded. Therefore, resistors also carry a maximum voltage rating which must be observed.

Exercise 7.1

A $\frac{1}{2}$-watt, 100-ohm carbon composition resistor has a TC of 0.1%/°C. If the temperature of the resistor increases linearly with power dissipation at a rate of 100°C/watt, how much power can the resistor dissipate if the resistance change due to self-heating is to be less than 1%?

7.2.4 Frequency limitations

At high frequencies, the performance of a resistor will depart from Ohm's law because of stray capacitance and lead inductance. Thus, the apparent resistance will change from the value measured at DC. An example of this effect is illustrated in Figure 7.4, which shows the change in effective resistance for a carbon composition resistor as a function of frequency. These high-frequency effects are discussed further in Chapter 18.

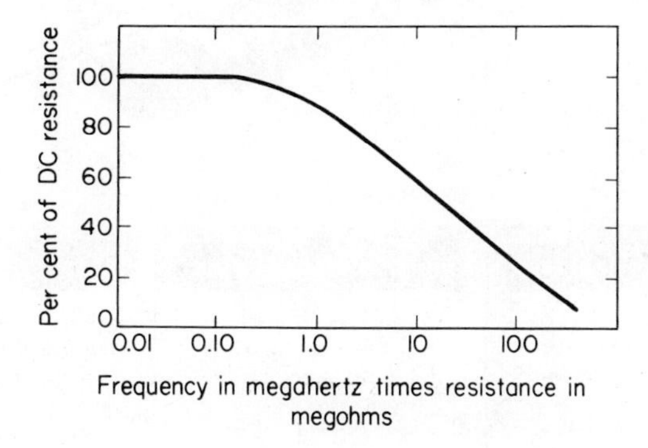

Figure 7.4. Change in resistance of a $\frac{1}{2}$-watt carbon-composition resistor as a function of frequency.

7.3 *resistor types*

Some of the most common types of resistance elements are described below. The most significant features of each are listed in Table 7.1.

Table 7.1 / Summary of Resistance Elements

Type	Available Range	Tolerance	TC	Maximum Power
Carbon composition	1 Ω to 22 MΩ	5 to 20%	0.1%/°C	2 watts
Wirewound	1 Ω to 100 kΩ	5 ppm up	5 ppm/°C up	200 watts
Metal film	0.1 Ω to 150 MΩ	50 ppm up	1 ppm/°C up	1 watt
Carbon film	10 Ω to 100 MΩ	0.5% up	−150 to −500 ppm/°C	2 watts

7.3.1 *carbon-composition resistors*

The most widely used resistor is the carbon-composition type. It is the least expensive and is highly reliable. Although Table 7.1 shows that its tolerance and TC are not so good as those of other resistors, individual units are very stable over their lifetimes. In the 0° to 60°C temperature range carbon-composition units have a very low TC, but they rapidly increase in resistance below 0° and above 60°C. Because of their low cost, reliability, and moderate power-handling capability, carbon-composition resistors are generally employed in most electronic circuit applications. Figure 7.5 illustrates the construction of a typical carbon-composition resistor.

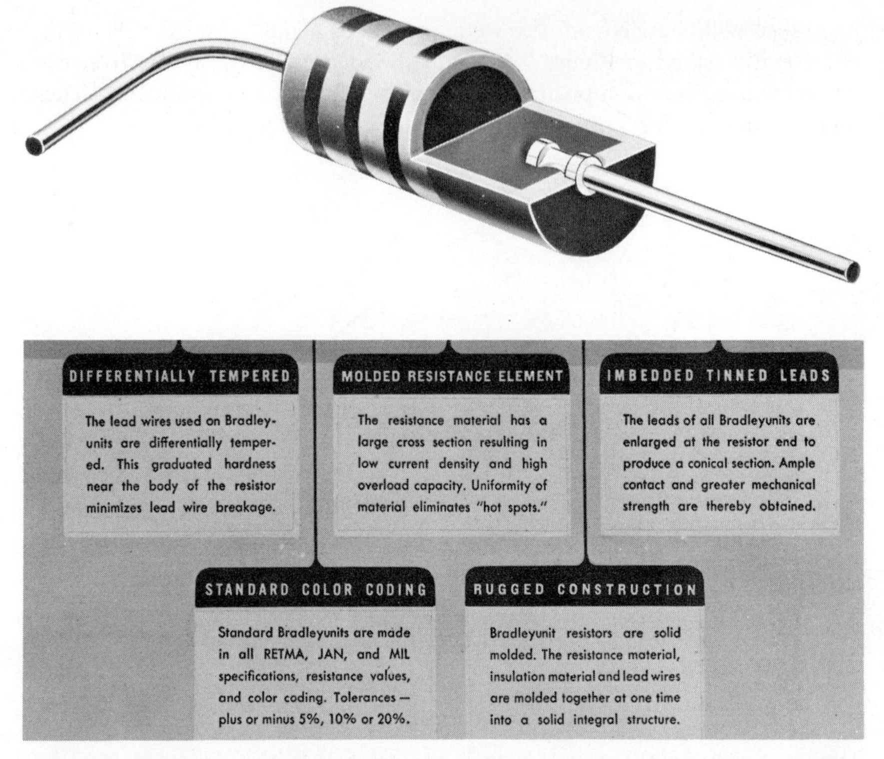

Figure 7.5. Typical construction of a carbon-composition resistor.

7.3.2 wirewound resistors

The wirewound resistor, as its name implies, is simply a length of wire whose resistance is the desired value, wound on a cylindrical core to provide a convenient package. The general range of available values is 1 ohm to 100 kΩ, the high values being limited by the long length and/or the small diameter of the wire required. Wirewound resistors are divided into three broad categories: high-power, high-accuracy, and general-purpose.

High-power wirewound resistors are generally available with 5- to 200-watt dissipation ratings and hence fill needs that carbon composition resistors cannot. The high-power wirewound resistor is designed to operate at high temperature and is usually wound on a ceramic core and covered with a vitreous enamel. Tolerances on high-power wirewound resistors are usually 5 to 10%. Examples of high-power wirewound resistors are shown in Figure 7.6.

By careful control of alloy composition, wires which exhibit a low temperature coefficient and excellent long-term stability can be produced. Such qualities are necessary for high accuracy. Although the resistance values

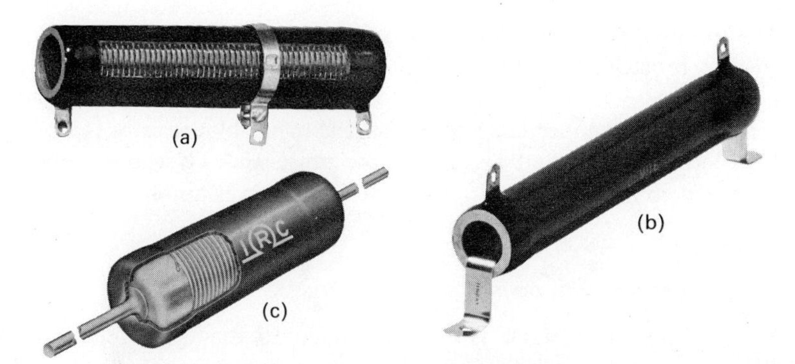

Figure 7.6. Wirewound resistors. (a) High power adjustable. (b) High power fixed. (c) General purpose. Courtesy Ohmite Manufacturing Co., Skokie, Ill., and IRC Division of TRW, Inc., Philadelphia, Pa.

are roughly the same as for high-power units, the power ratings are normally 1 watt or less and the tolerances are on the order of 5 ppm to 1%. Temperature coefficients can be as low as 5 ppm/°C. These resistors are normally cylindrical; the larger sizes have solder lugs, and units which resemble carbon-composition resistors have axial leads. This type of resistor is mainly employed in precision voltage dividers in measurement instruments, and in computer applications.

The general-purpose wirewound resistor is available with resistances from $\frac{1}{4}$ ohm to 10 kΩ and power ratings of $\frac{1}{2}$, 1, and 3 watts. An example is shown in Figure 7.6. Thus they duplicate a significant fraction of the applications of carbon-composition units. They are employed where high stability, reliability, and low TC requirements cannot be met by the composition resistor.

Since a wirewound resistor is a coil of wire, there may be a substantial inductance which is troublesome if the resistor is used at high frequencies. Some types are bifilar wound to minimize this effect and are termed *noninductive* wirewound resistors. Since the manufacturing process is more complex, they carry a premium price.

7.3.3 metal-film resistors

In order to overcome the problem associated with construction of high-resistance wirewound resistors and their inductance, the metal-film resistor was developed. This element consists of two electrodes fastened to an insulating substrate, such as ceramic or glass, with an evaporated metal film providing a conducting path between the electrodes. Since extremely thin films can be produced, resistance values as high as 150 MΩ can be obtained. Since the resistance element is metal, the tolerance and TC associated with high-accuracy wirewound resistors are also possible. The metal-film resistor is also smaller than the equivalent wirewound value. Metal-film resistors are normally supplied in a cylindrical package with axial leads.

For most performance considerations, the metal-film resistor is superior. Its long-term stability and extremely low noise and TC make it the best choice for demanding applications such as low-level amplifiers and computers.

7.3.4 carbon-film resistors

The carbon-film resistor is closely related to the metal-film unit. Its construction is essentially the same except that a deposited carbon film replaces the evaporated metal film. As shown in Table 7.1, its resistance values are higher but the tolerance is lower than those of metal-film resistors. However, one of the major attributes of the carbon-film resistor is a moderate negative TC, which makes these units suitable for many temperature-compensation networks in electronic circuits.

7.3.5 high-TC resistors

For some applications a resistance element which exhibits a large temperature coefficient is desirable. Besides the temperature-compensation applications in electronic networks, devices with a high TC may be used to measure temperature and perform a multitude of control functions based on this electronic temperature measurement.

A *thermistor* is a resistor with a high negative temperature coefficient. Thermistor elements are semiconductors generally made by sintering combinations of metallic oxides. The resistance of a thermistor as a function of temperature is given by

$$R(T) = R_0 e^{\beta[(1/T)-(1/T_0)]} \tag{7.4}$$

where

β is a constant that depends on the material.
T is the temperature in °K.
R_0 is the resistance at temperature T_0.

Usually T_0 is taken at room temperature (300°K), and values of R_0 range from a few hundred ohms to a hundred megohms. The temperature coefficient is given by

$$\text{TC} = -\frac{\beta}{T^2} \times 100\%/^\circ\text{C} \tag{7.5}$$

and may be as large as $-5\%/^\circ$C in some units. Thus the resistance will change by a factor of two for every 20°C change in temperature. It is this large TC that makes the thermistor so useful as a sensitive temperature sensor.

A second type of high-TC resistor is made from semiconducting silicon. This device, known as the Stabistor,* has a positive temperature coefficient of approximately 0.7%/°C. Although the TC of the Stabistor is much less than that of a thermistor, there are some applications in which a positive TC, which cannot be obtained with a thermistor, must be employed.

The useful temperature range of thermistors and Stabistors is on the order of $-100°$ to $+300°C$.

In the application of thermistors, care must be taken to keep the power dissipation low so that self-heating effects will not cause erroneous temperature measurements.

7.3.6 variable resistors

Frequently it is desirable to have an adjustable resistance element. Such devices, since they are often used in voltage measuring instruments (see Section 13.2), they are termed potentiometers or "pots" for short. They are available in general-purpose and precision types, examples of which are shown in Figure 7.7.

The potentiometer is simply a fixed resistor with a third terminal which can be connected to an intermediate point between the ends of the fixed resistor. The location of the third terminal is adjusted by rotation of a shaft.

The general-purpose potentiometer may have a carbon-composition or wirewound resistance element. Composition units are available in the resistance range of 100 ohms to 1 MΩ at power ratings of $\frac{1}{2}$ to 2 watts. The wirewound units cover the resistance range of 5 ohms to 50 kΩ at power ratings from 2 to 50 watts. Generally, a shaft rotation of 270 degrees will move the third terminal contact the entire length of the resistance element. The variation of resistance with shaft rotation in a potentiometer is termed the *taper*. Thus, a linear taper potentiometer will have a resistance variation with shaft position that is a linear function of the angle of rotation. For many applications, such as audio volume controls, the taper may be nonlinear. Precision potentiometers may also have composition or wirewound elements. The available resistance range is from 10 ohms to 300 kΩ in power ratings up to 5 watts. The total resistance tolerance is usually $\pm 5\%$, but the linearity of the taper is usually $\pm 0.05\%$ to $\pm 0.5\%$. Thus, although the total resistance may not be highly accurate, the variation with angular shaft position is accurate, and this feature makes the precision potentiometer useful in precision measurement applications such as discussed in Chapter 13. In order to increase the resolution of the adjustable resistance, precision

* Trademark of Texas Instruments, Inc.

Figure 7.7. Typical potentiometers. (a) General purpose. (b) Locking shaft. (c) Precision. Courtesy Clarostat Manufacturing Co., Inc., Dover, N.H., and Helipot Division of Beckman Instruments, Inc.

potentiometers usually require three or ten turns to move the slider from one end of the element to the other. Thus, as much as 3600 degrees of shaft rotation is possible.

One word of caution with respect to potentiometer power ratings should be given. The specified rating is based on uniform dissipation over the *entire* resistance element. If the circuit application results in dissipation over only a fraction of the element, the power rating must be derated by the same fraction.

7.4 resistance measurement

In the following section some of the common methods of measuring resistance will be discussed.

7.4.1 ohmmeter

The simple ohmmeter, such as employed in the multitester, is based on the voltage divider network shown in Figure 7.8. Figure 7.8 shows that if R_1 is known, then measurement of either voltage ratio V_1/V_0 or V_x/V_0 will yield the unknown resistance R_x.

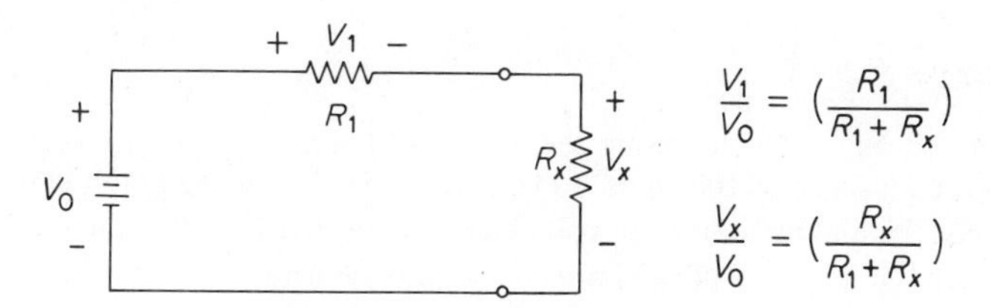

Figure 7.8. Voltage divider network.

Exercise 7.2

On Cartesian coordinates, plot the voltage ratios V_1/V_0 and V_x/V_0, defined in Figure 7.8, as a function of the resistance ratio R_x/R_1. In each case, discuss the range of resistance ratio where reasonable measurement accuracy can be expected.

Since the resistance measurement depends only on the voltage ratio and not on the measurement of absolute voltages, a variation of the d'Arsonval voltmeter may be employed to read this ratio directly. In Figure 7.9, a d'Arsonval ammeter and series resistance R_2 form an "adjustable" voltmeter used to measure the voltage ratio V_1/V_0. The resistor R_2 is adjustable, rather than fixed as in a normal voltmeter circuit, and the meter scale is calibrated linearly from 0 to 1.0. With $R_x = 0$ (ohmmeter terminals shorted), R_2 is adjusted to yield a reading of 1.0 on the meter scale. Next, the unknown resistor is connected and the meter will indicate the voltage ratio directly. With knowledge of R_1, the equations in Figure 7.8 give the unknown resistance value. Notice that by making R_2 adjustable, the measurement is independent of the absolute value of V_0. This is especially useful in the multitester where V_0 is supplied by a battery whose absolute voltage changes with age. As a final refinement, most ohmmeters carry a resistance scale rather than the 0 to 1.0 linear scale directly on the meter, eliminating the need

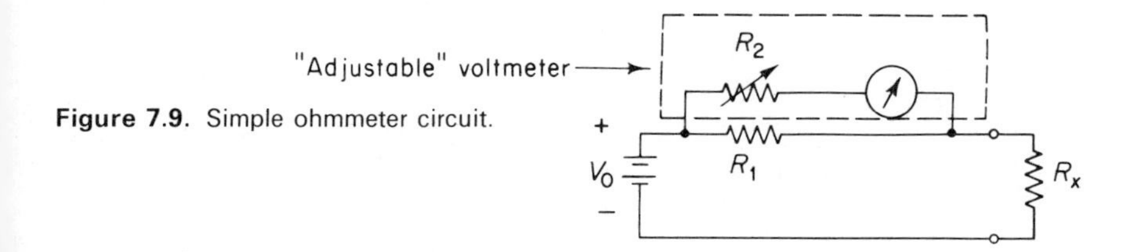

Figure 7.9. Simple ohmmeter circuit.

for calculations. In this case R_2 is adjusted to give full-scale deflection, which is equivalent to 1.0 on the voltage ratio scale, and the meter indicates the unknown resistance value directly.

If the "adjustable" voltmeter is used to measure V_x in Figure 7.8, then R_2 is adjusted for 1.0 or full-scale deflection with $R_x = \infty$ (open circuit). Otherwise the operation of the meter is the same as described above.

Exercise 7.3

Construct an ohmmeter using the circuit of Figure 7.10. Using this ohmmeter, measure resistors of 100 ohms, 1 kΩ, and 10 kΩ. Using the graph constructed in Exercise 7.2 and assuming the voltmeter is accurate to $\pm 3\%$ of full-scale reading, estimate the accuracy of each resistance measurement.

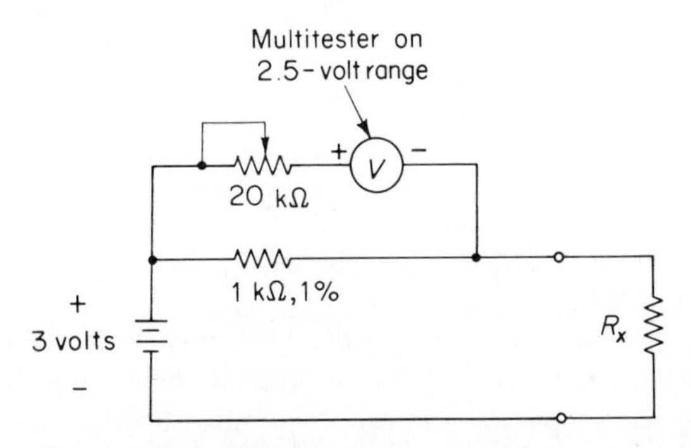

Figure 7.10. Ohmmeter circuit.

7.4.2 resistance bridge

The accuracy of the simple ohmmeter can be improved by using the meter only as a null detector in a bridge circuit. Such an arrangement is shown in Figure 7.11, in which R_x is the unknown resistance to be measured. Typically, R_1 and R_2 are precision decade-value resistors, such as 1 kΩ, 10 kΩ, etc., and R_3 is an adjustable resistor with an accuracy of 0.01%. If R_3 is adjusted for a null or zero indication on the voltmeter, then R_x is given by

$$R_x = \left(\frac{R_2}{R_1}\right) R_3 \tag{7.6}$$

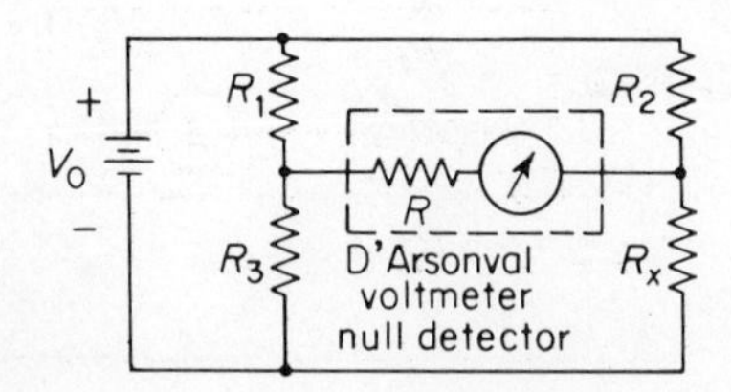

Figure 7.11. Resistance bridge.

and thus R_x is a decade multiple (R_2/R_1) times R_3. For an initial balance, the voltmeter is set on a range which includes V_0. As a balance is reached, the most sensitive voltmeter scale is employed. After balance has been achieved, care should be taken to set the voltmeter on the V_0 range before R_x is removed.

As a variation of the bridge circuit in Figure 7.11, an AC voltage source may be used in place of V_0. In this case, an AC null detector will also be required. An AC voltmeter or, particularly useful, an oscilloscope may be employed.

Exercise 7.4

Mount a rod thermistor of about 3-kΩ resistance at 25°C in intimate thermal contact with a thermometer bulb. One possible method is shown in Figure 7.12.

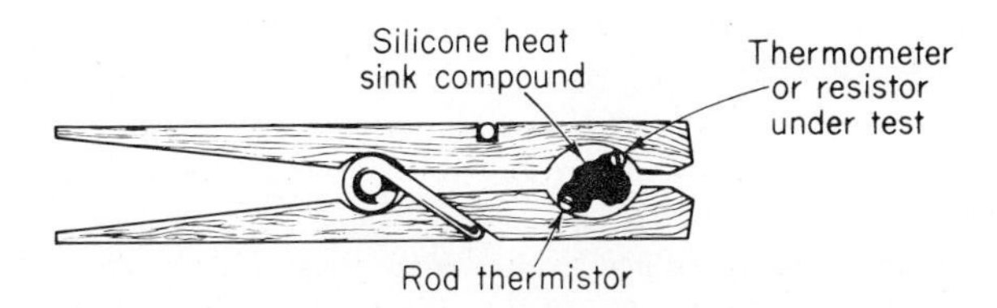

Figure 7.12. Method of mounting thermistor.

Place the thermistor in an oven and measure its resistance as a function of temperature over the range 25° to 150°C (300° to 425°K). On semilog paper plot log R vs. $1/T$ in °K^{-1} and determine the temperature constant β in Equation (7.4). Over this temperature range, R will typically decrease by a factor of 100.

The resistance measurement can be made with an ohmmeter, or, for more accuracy, a resistance bridge such as that in Figure 7.11 may be employed. Use $V_0 = 6$ volts; R_1, $R_2 = 1$ kΩ, 1%; for R_3 use an adjustable decade resistance with steps of 1 ohm.

Exercise 7.5

To obtain an appreciation of how much a carbon-composition resistor changes value with temperature and power dissipation, the following experiment is suggested.

In place of the thermometer in Figure 7.12, clamp a 100-ohm, $\frac{1}{2}$-watt resistor. Connect the resistor and thermistor in the circuit of Figure 7.13. Be especially careful of making good connections in the DC bridge circuit, since currents up to 200 mA will be flowing and a poor contact will cause erratic performance.

The power dissipated in the 100-ohm resistor is controlled by the adjustable DC voltage. The actual resistance of the 100-ohm resistor can be measured by adjusting the 0.1-ohm decade resistor until the null indicator shows that the bridge is balanced. The unknown resistance is then ten times the reading on the

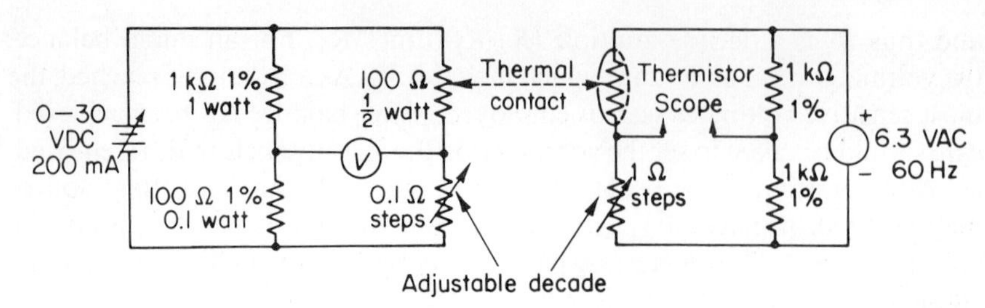

Figure 7.13. Thermal measurement of 100-ohm resistor.

decade box. The power dissipated in the resistor will be given by V^2/R, where V is the voltage across the unknown and R is the resistance of the unknown. The null voltmeter can be used to measure V by moving only one terminal. Be careful of the voltmeter range when changing the voltmeter connection.

The AC bridge is used to measure the thermistor resistance and hence temperature. In order to improve the accuracy of the temperature measurement, plot log thermistor resistance vs. degrees Centigrade over the 25° to 150°C span. This spreads the data in the useful range and aids in interpolation. The AC bridge null is observed on the oscilloscope.

Increase the applied voltage to the $\frac{1}{2}$-watt resistor gradually and measure its resistance and temperature up to 2.5 or 3 watts of dissipation. Plot R and T vs. power dissipation as the data are taken. Over this range R can be expected to nearly double and T may reach 150°C. Gradually reduce the power and continue the plot of R and T vs. power. Note the irreversible resistance change due to excessive power dissipation. When finished, *discard* the 100-ohm resistor.

7.4.3 substitution measurement

For accurate measurement of resistance, it is best to compare the unknown *directly* with an accurately calibrated resistor. One method of doing this is to first balance a resistance bridge precisely and then substitute a precision adjustable resistance for the unknown. If this precision resistance is then adjusted to restore a null reading, its value must equal the unknown value within the precision of the bridge. Thus, with a sufficiently sensitive meter, the accuracy of the measurement can be made to depend only on the accuracy of a *single* adjustable resistance.

7.4.4 commercial resistance bridges

Several common resistance bridges which operate at DC or in the audio-frequency range are available. Some examples are shown in Figure 7.14. As will be noted, these bridges are also capable of measuring inductance and capacitance. However, in this assignment we are concerned only with their resistance measurement functions. Table 7.2 lists their basic capabilities and features.

Basically, these bridges are packaged versions of the resistance bridge shown in Figure 7.11. Range switching is provided to select the ratio R_2/R_1, and the adjustable resistance is provided with a calibrated readout. Both DC and AC excitations are used; the former will eliminate any inductance or capacitance effects on the resistance measurement, whereas the AC excitation usually provides a sharper null.

Exercise 7.6

Use a multitester to measure the resistance of three resistors with values in the vicinity of 1 ohm, 1 kΩ, and 1 MΩ. Repeat the measurements on a commercial resistance bridge and compare the results. Estimate the probable accuracy for each measurement.

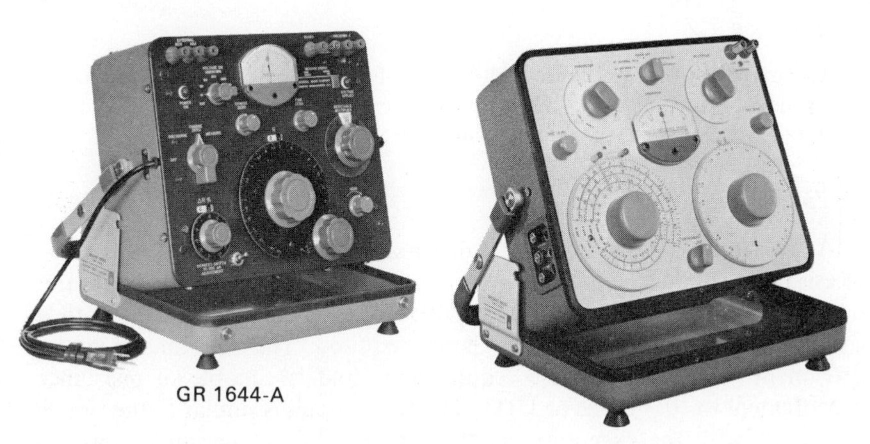

GR 1644-A

GR 1650-B

GR 1608-A

HP 4260A

Figure 7.14. Commercial resistance bridges.

Table 7.2 / Resistance Bridge Capabilities

	Range	Basic accuracy	Excitation source Internal	External	Features
HP 4260-A	10 mΩ to 10 MΩ	±1%	DC, 1 kHz	20 Hz to 20 kHz	digital readout, automatic decimal
GR 1650-B	1 mΩ to 1.1 MΩ	±1%	DC, 1 kHz	20 Hz to 120 kHz	wide range
GR 1644-A	1 kΩ to 1000 TΩ	±1%	DC	none	high resistance
GR 1608-A	50 μΩ to 20 GΩ	±0.1%	DC, 1 kHz	20 Hz to 20 kHz	high accuracy, digital readout, automatic decimal

Exercise 7.7

Using a commercial resistance bridge, repeat the measurement on the 1 kΩ resistor by the substitution method. How sensitive is the bridge to changes in the precision decade resistance? Estimate the accuracy of your measurement.

7.4.5 megohmmeter

At resistances above 10 MΩ, high voltages (500 to 1000 volts) become necessary to produce sufficient current flow to effect a resistance measurement. A voltmeter with a higher input resistance than is possible with a simple d'Arsonval meter movement is also required. The megohmmeter is an instrument which meets these requirements and is capable of resistance measurement to 10^{12} ohms or 1 TΩ. The basic circuit is similar to the simple ohmmeter, with the exceptions of the high supply voltage and improved detector mentioned above.

8

capacitors

8.1 introduction

The ideal capacitor is a two-terminal circuit element in which the current flowing through the element is proportional to the time rate of change of the terminal voltage.

$$i = C\frac{dv}{dt} \tag{8.1}$$

The constant of proportionality C is termed the *capacitance* of the element, whose dimensional unit is the *farad* (F). The ideal capacitor is a lossless element (does not dissipate energy) and is capable of energy storage. From electromagnetic field theory it can be shown that in an ideal capacitor this energy is stored in the electric field and hence is termed *electric stored energy*.

In practice, the ideal volt-ampere relationship of a capacitor can only be approximated with realizable physical structures. The purpose of this chapter is to explore some of the properties of actual capacitors, the circuit models used in their description and to introduce some instruments used in their measurement.

8.2 capacitor structure and limitations

Before describing the various types of capacitors in common use, we will consider some of the general physical factors which govern their performance as a circuit element. For the purpose of this discussion we will characterize the capacitor through the physical structure of Figure 8.1. Two parallel conducting plates of area A are separated by a distance d. The intervening volume is filled with an insulating material called the *dielectric* and characterized by a dimensionless parameter, the dielectric constant K. If electrical terminals are attached to each conducting plate, then the capacitance of the structure is given by

$$C = \frac{K\epsilon_0 A}{d} \tag{8.2}$$

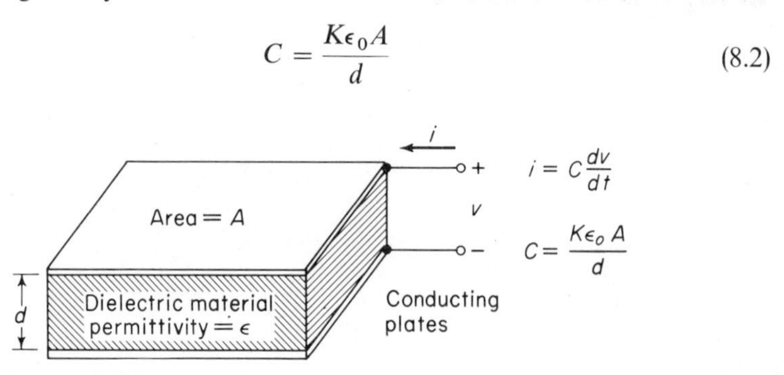

Figure 8.1. Basic capacitor structure.

where ϵ_0 is the *permittivity* of vacuum ($\epsilon_0 = 8.85 \times 10^{-11}$ farads per meter). Equation (8.2) shows that the value of a given capacitor will be a function of the dielectric constant of the material which separates the conducting plates, and of the structure geometry, increasing with plate area and inversely proportional to plate spacing.

8.2.1 properties of dielectrics

One way to reduce the physical dimensions of the structure in Figure 8.1 for a given value of capacitance is to employ a dielectric material with a large value of dielectric constant. The dielectric constants of some common materials used in capacitors are given in Table 8.1. The choice of dielectric for use in a particular application is also governed by factors such as dielectric loss, insulation resistance, and dielectric strength as discussed below.

In an ideal capacitor the dielectric material separating the conducting plates is a perfect insulator. That is, there is no energy dissipated in the dielectric as a result of an applied electric field. No present dielectric material has attained this perfection and hence there will be some energy dissipated when a voltage is applied to an actual capacitor. This dissipation, termed *dielectric loss*, will depend on the physical properties of the particular dielectric and will vary with the frequency of the applied voltage. For the special case of a DC applied voltage, the DC current that results from dielectric loss is termed the *leakage current* and the capacitor may be

Table 8.1 / Dielectric Constants of Common Capacitor Dielectrics

Material	$K = \frac{\epsilon}{\epsilon_0}$
Vacuum	1.0
Polystyrene	2.5
Mylar	3.0
Impregnated paper	4–6
Mica	6.8
Oxide films	5–25
Ceramic (low-loss)	6–20
Ceramic (high-K)	100–> 1000

characterized by a *leakage resistance.* The leakage resistance is governed by the volume resistivity of the dielectric, with polystyrene and Mylar providing the highest values of leakage resistance (lowest leakage current) of the common dielectrics. The lowest leakage resistances are found in the oxide films used in electrolytic capacitors. It should be kept in mind, however, that contributions to leakage resistance can also come from the capacitor encapsulant, surface contaminants, and high humidity as discussed in Section 7.2.1.

As the voltage applied to a capacitor is increased, a point is reached at which the electric field strength in the dielectric destroys the insulating property of the material. This effect is termed *dielectric breakdown.* The *dielectric strength* of a material is the value of electric field at which this breakdown occurs, typically on the order of 100,000 volts/cm. Thus, the product of the dielectric strength and the dielectric thickness yields the maximum voltage that may be applied to a capacitor.

In order to estimate the effect of the dielectric material on the physical size of a capacitor, it is useful to make the comparison on the basis of the stored electric energy

$$\text{Stored electric energy} = \frac{Cv^2}{2} = \left[\frac{\epsilon_0}{2}\right] \overbrace{\left[K\left(\frac{v}{d}\right)^2\right]}^{\text{dielectric properties}} \underbrace{\left[Ad\right]}_{\text{capacitor volume}} \tag{8.3}$$

Equation (8.3) shows that the maximum stored electric energy per unit volume of capacitor depends on the product of the dielectric constant and the square of the dielectric strength. Thus for a given dielectric material, increasing capacitance value and increasing voltage rating require increased volume. Conversely, to minimize the size of a required capacitor, a dielectric should be chosen to maximize the product of the dielectric constant and the square of the dielectric strength.

There are other properties of dielectric materials which will influence their application in capacitors. Some of these are the temperature coefficient of the dielectric constant and the thermal expansion, the stability of the dielectric with applied electric field and with capacitor life, and the phenomenon of *dielectric absorption*, which measures the rate at which the bound charges in the dielectric material can be displaced. The relative effects of these properties will be considered in the capacitor descriptions in Section 8.3.

8.2.2 circuit model of a lossy capacitor

It has been pointed out that imperfect dielectric materials result in energy dissipation in a capacitor. We will now consider how to characterize this loss for a given capacitor and will discuss some circuit models consisting of ideal elements which will describe the actual performance of a real capacitor.

One possible circuit model which can represent the dissipation in an actual capacitor is shown in Figure 8.2.

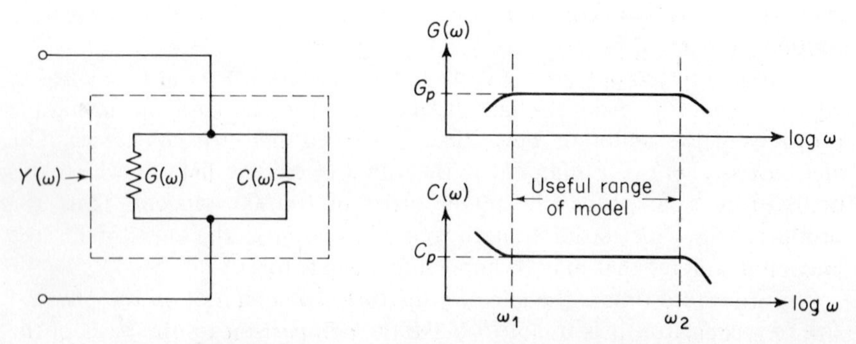

Figure 8.2. Circuit model for a lossy capacitor.

The conductance and capacitance in the model are explicitly indicated to be functions of the excitation frequency. This is because the properties of the dielectric which govern the values of $G(\omega)$ and $C(\omega)$ are, in general, frequency dependent. However, for the circuit model to be useful in analytical work, the ideal circuit elements should be frequency independent. As shown in Figure 8.2, a range of frequencies in which $G(\omega)$ and $C(\omega)$ are approximately constant usually exists. As long as we are concerned only with frequencies in this range, we may represent the lossy capacitor as the parallel combination of an ideal capacitor C_p and ideal conductance G_p which are frequency independent. Outside this frequency range we must use either the frequency-dependent conductance and capacitance in analytical expressions or construct a more complex circuit model consisting of more than two frequency-independent elements. Fortunately, most lossy capacitors can be adequately modeled with a simple parallel G_p-C_p network over the frequency range of interest. However, before assuming this model is valid, a check should be made by measuring G and C as a function of frequency to verify their frequency independence.

The admittance measured at the input terminals of a capacitor may be expressed, in general, as

$$Y(\omega) = G(\omega) + jB(\omega) \tag{8.4}$$

where $G(\omega)$ is the conductance or real part of $Y(\omega)$, and $B(\omega)$ is the susceptance or imaginary part of $Y(\omega)$. From the circuit model of Figure 8.2, we find

$$C(\omega) = \frac{B(\omega)}{\omega} \tag{8.5}$$

Although a measurement of admittance will yield the conductance term directly, this is not normally used as a measure of the loss in a capacitor. Rather, a dimensionless figure of merit called the *dissipation factor* D is employed. The dissipation factor is given by

$$D = \frac{G(\omega)}{B(\omega)} \tag{8.6}$$

In a lossless cap[illegible] = 0, and hence $D = 0$ represents the ideal case. The larger the va[illegible], the greater the loss. Typical values of D range from 10^{-4} in polystyrene units to 0.1 in electrolytic types. In general the dissipation factor will be a function of frequency. In the useful frequency range of the model of Figure 8.2, the dissipation factor becomes

$$D = \frac{G_p}{\omega C_p} \tag{8.7}$$

which shows that the performance of the capacitor increases with frequency in that particular range. Also, in specifying the dissipation factor for a capacitor, the frequency of measurement should be included for meaningful interpretation and comparisons.

Exercise 8.1

An alternate circuit model to express the loss in an actual capacitor is shown in Figure 8.3. Find $R(\omega)$ and $C(\omega)$ if this model is to be equivalent to that of Figure 8.2 over the frequency range $\omega_1 < \omega < \omega_2$. In particular, consider the special case where $D \ll 1$.

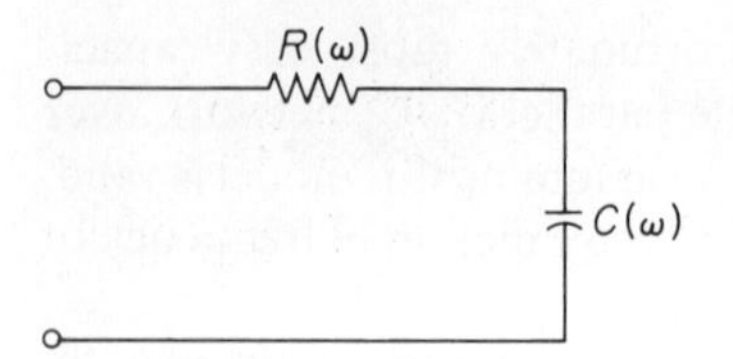

Figure 8.3. Alternate circuit model for a lossy capacitor.

8.3 capacitor types

The major characteristics of a capacitor are determined by the dielectric material from which it is constructed. For this reason capacitors are usually classified accordingly as mica, ceramic, paper, electrolytic, etc. In Section 8.2 we discussed some of the important differences between various dielectrics and proposed a circuit model which characterizes real capacitors in terms of their capacitance and dissipation factor. In order that an intelligent choice of capacitor may be made for a circuit application, we will describe some of the most common types in terms of their electrical performance.

Figure 8.4 illustrates some of the common capacitor types in general use. The method of marking for component value is discussed in Section 2.4.3. The nominal values are determined in the same manner as values are for resistors. (See Section 7.2.1.) In addition to the basic capacitance, the significant parameters from an electrical point of view are the tolerance, dissipation factor, working voltage, temperature coefficient, and useful frequency range. However, physical size and cost will also influence the capacitor choice in many applications. The general range of electrical parameters for the various capacitor types is summarized in Table 8.2. The useful frequency ranges are indicated in Figure 8.5. It should be emphasized that the data in Table 8.2 and Figure 8.5 are *representative* values and are not all inclusive. Also, not all combinations of component values indicated in Table 8.2 are normally available. For example, for a given type the larger values of capacitance may exist only at the lower voltage ratings. Similarly, in some applications, the use of a capacitor outside the indicated frequency range in Figure 8.5 may be quite acceptable.

8.3.1 paper capacitors

The general-purpose *paper capacitor* is made by rolling alternate sheets of paper and metal foil into a cylinder and filling the volume with an impregnant. The paper and impregnant form the dielectric, and the electrical terminals are made to alternate foil layers as shown in Figure 8.6a. Paper capacitors are relatively inexpensive and are widely used where tolerance and dissipation factor are not a primary consideration.

In *metallized paper capacitors*, the foil is replaced by a metallized surface on the paper dielectric. This reduces the volume for a given value of capaci-

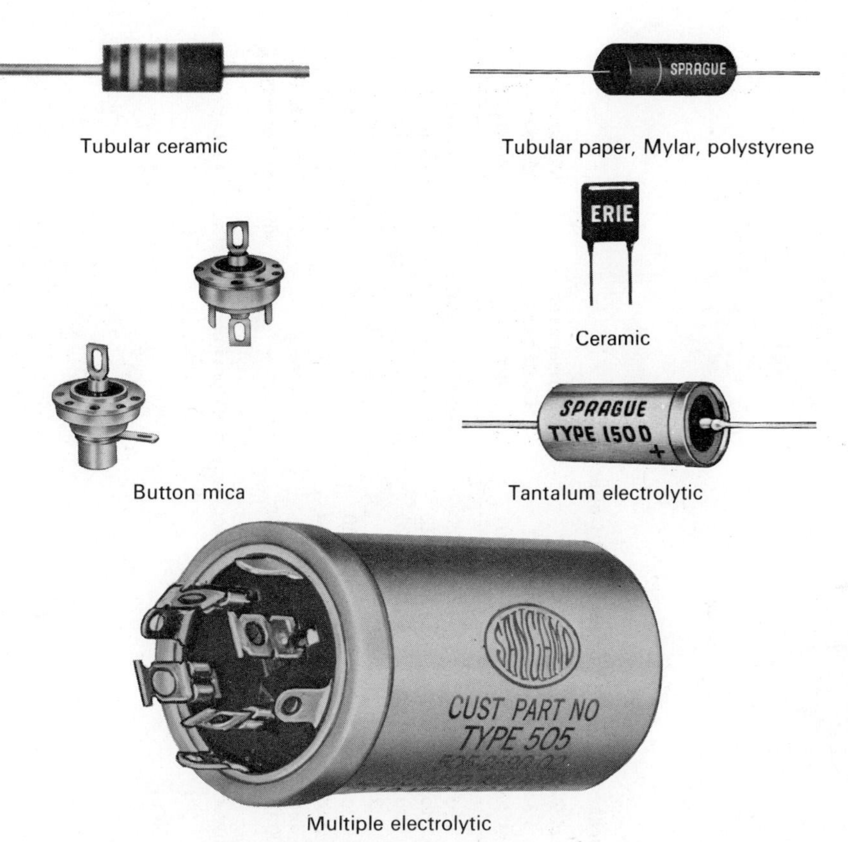

Figure 8.4. Common types of fixed capacitors. Courtesy Sprague Electric Company, North Adams, Mass.; Erie Resistor Corporation, Erie, Pa.; and Sangamo Electric Company, Springfield, Ill.

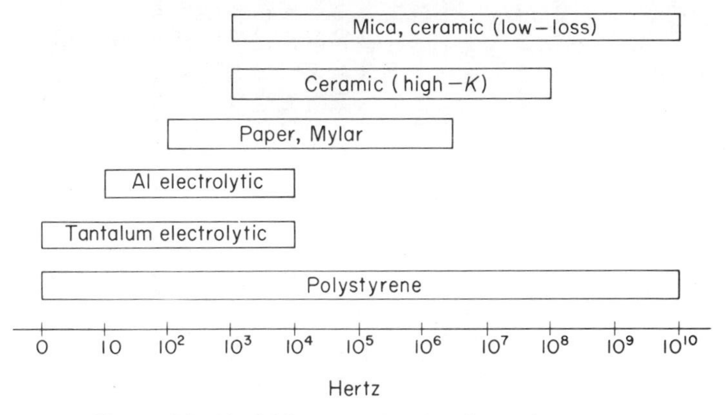

Figure 8.5. Useful frequency ranges of capacitor types.

Table 8.2 / Representative Performance of Capacitors

Type	Available Capacitance Range	Typical Dissipation Factor at 1 kHz ($\times 10^{-3}$)	Tolerance Range (%)	Temperature Coefficient (ppm/°C)	Maximum Working Voltage
Ceramic (low-loss)	1 pF to 0.001 μF	1	±5 to ±20	+100 to −750	6000 volts
Silvered mica	1 pF to 0.1 μF	0.2 to 1	±1 to ±20	±200	500 volts to 75 kV
Paper (solid dielectric)	500 pF to 10 μF	5 to 10	±10 to ±20	—	100 volts to 1.5 kV
Polystyrene	500 pF to 10 μF	0.1	±0.5	−120	1000 volts or less
Ceramic (high-*K*)	1000 pF to 0.1 μF	10 to 30	+100 to −20	—	100 volts or less
Paper (oil-filled)	1000 pF to 50 μF	2 to 3.5	±10 to ±20	—	100 volts to 100 kV
Mylar	5000 pF to 1 μF	3 to 14	±20	variable	100 volts to 600 V
Electrolytic*	1 μF to 0.1 F	100 to 300	−20 to +100	—	500 volts or less

* Electrolytic capacitors are used in DC blocking and power supply applications. Their accuracy is seldom important. The higher capacitance values are only available at low voltage ratings.

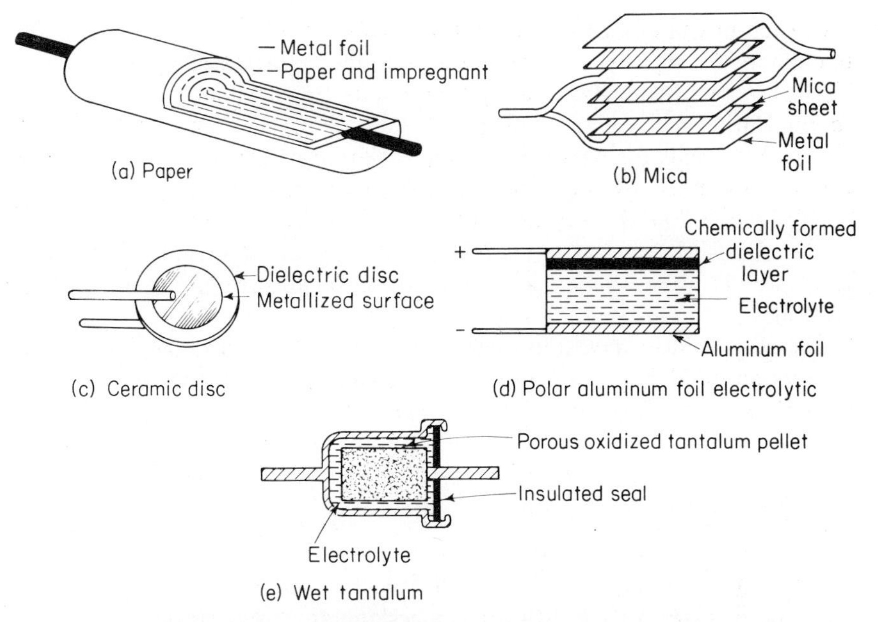

Figure 8.6. Basic construction of common capacitors.

tance by as much as one-half. Metallized units have about one-tenth the leakage current of a comparable paper unit, but are more susceptible to failure under transient voltage surges.

8.3.2 plastic film capacitors

The *plastic film capacitors* are similar to paper units, the paper dielectric being replaced by polystyrene, Mylar, polyethylene, or Teflon. The improved dielectric properties of the plastic film over the impregnated paper produce high leakage resistances and low dissipation factors, even at elevated temperatures. Plastic film capacitors are more expensive than paper types and hence are used only where their improved properties are required. Some examples are polystyrene in high-Q tuned circuits and as integrating capacitors in operational amplifiers, and Mylar and Teflon in high-temperature applications, up to 150° and 200°C, respectively.

8.3.3 mica capacitors

Mica capacitors are constructed by stacking alternate layers of mica and metal foil as shown in Figure 8.6b. These stacks are then encapsulated to provide mechanical strength and protection from moisture. They have very low dissipation factors and are very stable with life. Very high working voltages are also possible. The type of mica used and the method of construction will affect the temperature coefficient. Thus by controlling the manufacturing process, a capacitor with a predictable TC may be obtained. These units find wide application as compensating capacitors to correct for thermal drift in other circuit elements. One example is the construction of a precision oscillator whose frequency variation with ambient temperature is to be minimized by choice of a temperature compensating capacitor.

The *silvered mica* types are formed by depositing a silver layer directly to the mica sheet, similar to the metallized paper construction. This results in a lower TC than for the mica-foil construction and also improves stability with life.

Mica capacitors have a low capacitance-to-volume ratio compared with paper units and hence are normally used at high frequencies where the capacitance value is small and the low dissipation factor is important. Some examples are oscillator tuning and filter applications.

8.3.4 ceramic capacitors

Ceramic capacitors are divided into two categories: the low-dielectric constant, low-loss types and the high-dielectric constant variety. The low-loss types perform well at high frequencies and approach the quality of mica units. The TC of the dielectric constant may be controlled by the composition of the ceramic, and hence a variety of temperature compensating capacitors are available in the nominal values +100, 0, −30, −80, −150, −220, −330, −470, and −750 ppm/°C.

The high-*K* ceramic capacitors are used where a large capacitance is required in a small volume. However, the dielectric constant and dissipation factor are strong functions of temperature, applied DC voltage, and frequency. Thus they are normally used only in coupling and bypass applications where the precise capacitance value is not critical. Typical disc construction is shown in Figure 8.6c.

8.3.5 electrolytic capacitors

The most significant feature of the *electrolytic capacitor* is the large values of capacitance that can be produced in a small volume. The basic construction consists of two metal electrodes immersed in a conducting solution or *electrolyte*. A thin insulating film is chemically formed on one or both of the metal electrodes to produce the dielectric. Since the dielectric strength of these films may be as high as 10^7 volts/cm, Equation (8.3) shows that the stored energy per unit volume can be very high compared with other dielectric materials.

There are two types of electrolytics in general use. The *aluminum electrolytic* is made from aluminum foil electrodes wound in a similar fashion to a paper capacitor. One of the foil surfaces is coated with an insulating layer by anodic oxidation. The intervening volume is filled with the electrolyte. A cross section is shown in Figure 8.6d. The second and more recent type of electrolytic capacitor features a cylinder of sintered *tantalum* powder. After the sintering process, an oxide coating is produced throughout the porous pellet which provides a very large surface-to-volume ratio. Finally the pellet is mounted and filled with the electrolyte, Figure 8.6e.

Both the aluminum and tantalum electrolytic capacitors are produced in *polar* and *nonpolar* forms. The polar type, which is by far the most common, is designed to operate with only one polarity of applied voltage. These units are clearly marked with the required polarity, which must be *strictly observed.* If the reverse polarity is applied, the oxide film will not act as an insulator, and a substantial leakage current will flow. The nonpolar electrolytic has an oxide film on both electrode surfaces so that one will always be forming a dielectric layer. The disadvantage of this arrangement is that for a given volume only one-half the capacitance of a polar type can be obtained.

If a polar electrolytic has been stored unused for several months or if the wrong polarity has been momentarily applied, it may be necessary to *form* the dielectric layer. This process simply requires the application of the DC working voltage in the proper polarity for about half an hour, or until the DC leakage current has been reduced to an acceptable value.

Exercise 8.2

Using the circuit of Figure 8.7, measure the leakage current of an electrolytic capacitor. Observe the slow variation of leakage current with time. From the final value of the leakage current and the voltage applied to the capacitor, calculate the DC leakage resistance.

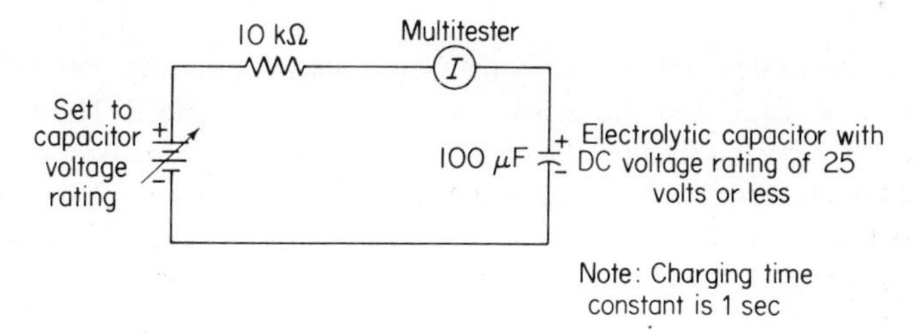

Figure 8.7. Leakage current measurement.

Set the ammeter to a high current scale (> 5 mA) and reverse the connection of the capacitor. Observe the time variation and final value of the leakage current when the incorrect polarity is applied. Reconnect the capacitor with the correct polarity and observe the leakage current change with time as the forming process takes place.

8.3.6 special-purpose capacitors

Up to now we have been considering fixed-value capacitors for general circuit applications. We will now briefly discuss some special types of capacitors which are encountered less frequently but which have important applications. These are the air-variable capacitor, the trimmer capacitor, and the feed-through capacitor. Typical units are shown in Figure 8.8.

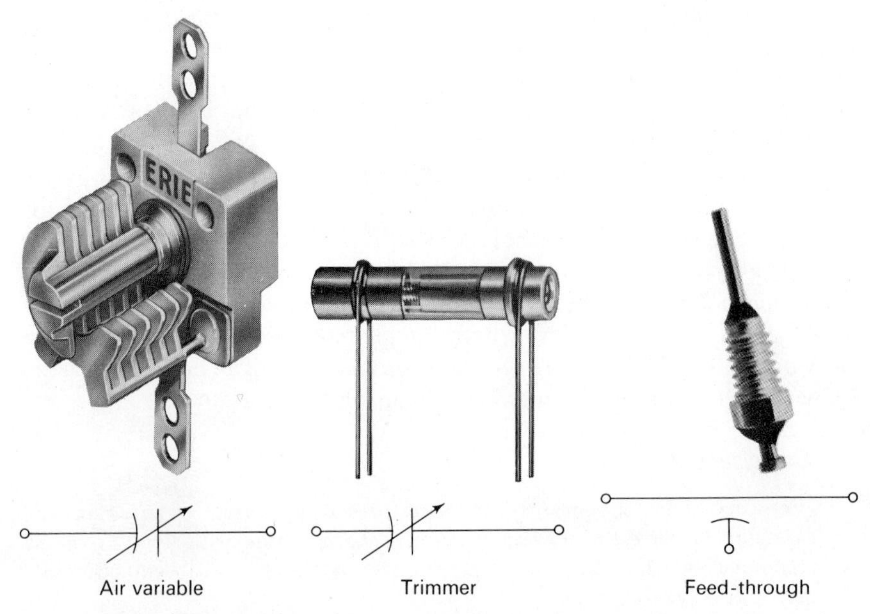

Figure 8.8. Special-purpose capacitors. Courtesy Erie Resistor Corporation, Erie, Pa.

The *air-variable capacitor* is constructed of interleaving sets of metal plates—one set fixed and the other free to rotate on a shaft. The dielectric is the air between the plates. By rotation of the shaft, the capacitor area is varied, thus changing the capacitance with shaft rotation. The range of values normally available runs from 15 pF to 300–400 pF. By shaping the moving plates properly, many different functions of capacitance vs. shaft rotation can be produced, such as linear capacitance or linear frequency tuning of a resonant circuit (square root of capacitance).

The *trimmer capacitor* is also a variable capacitor, but is used where a one-time adjustment must be made to optimize circuit performance. Some examples are precise setting of an oscillator frequency and adjusting of the bandwidth of a tuned amplifier. The range of values available runs from a few pF to several hundred pF in stages. A single unit may cover 7–45 pF or 15–130 pF typically.

The *feed-through capacitor* is used in high-frequency applications where it is desired to bypass to ground a wire passing through a shielded compartment. By the mounting of a feed-through capacitor in the wall of the compartment, terminals are provided on each side for the feed through, and a bypass capacitance to ground is obtained at the threaded body. Typical values of the bypass capacitance range from 100–5000 pF, 1000 pF being the most common.

There are many other capacitance types which have not been covered here. For discussion of these and for more details on the types covered in this chapter see Reference 1.

8.4 capacitance measurement and AC bridges

In Section 7.4.2, it was shown how a bridge circuit could be used to measure an unknown resistor. This principle can be expanded to include the measurement of the complex AC impedance of capacitors as well. Consider the bridge circuit in Figure 8.9.

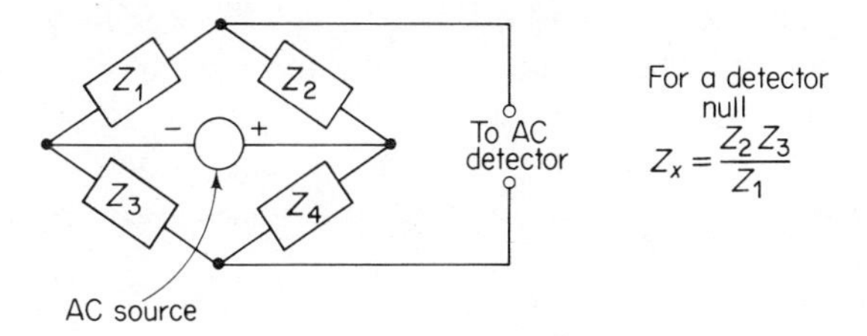

Figure 8.9. AC impedance bridge.

Because the equation for the unknown impedance

$$Z_x = R_x + jX_x = \frac{Z_2 Z_3}{Z_1} \tag{8.8}$$

is, in general, complex, two adjustments must be available to achieve bridge balance: one to satisfy the real part of (8.8) and the other to satisfy the imaginary part. Ideally, the two adjustments should be independent. That is, when the real-part adjustment is being made, there is no effect on the imaginary part of the bridge-balance equation. If this is not the case, then several successive adjustments of each control must be made in order to achieve a null output of the bridge. In some cases this process is slow to converge and special techniques must be employed.

There are many different circuit configurations for the generalized impedances Z_1, Z_2, and Z_3. The choice of a specific circuit will depend on several factors. First, the general character of the unknown impedance should be known, i.e., whether it is a low-D or high-D capacitor. This knowledge will permit selection of a circuit which will minimize the convergence problem described above. A second consideration is that it is more convenient to use variable resistors for the adjustable components rather than variable capacitors or inductors. Generally speaking, inductors in bridge circuits are to be avoided because of their limited frequency ranges of performance and their high cost and weight. The final question is how the bridge measurement is to be converted to a circuit model representation. As described above, two adjustments must be provided to balance the complex bridge equation. One of these should provide information on the real part of the unknown and the other information on the imaginary part. When all these factors have been considered, the actual bridge circuit design can be completed.

8.4.1 bridge equivalent circuits

We have seen that for a capacitor, there are two possible circuit models to account for the dissipation—the series and parallel representations. Therefore, a bridge, through the particular circuits employed, may provide the equivalent circuit information in terms of a series or parallel circuit model and may be calibrated in capacitance and resistance, conductance, or dissipation factor. In order to maximize the usefulness of this data, we present the interrelationships between the various models and parameters. These results are a concise summary of the discussion in Section 8.2.2 and are presented in Figure 8.10.

Using the relationships in Figure 8.10, a bridge measurement can be converted to the most useful form for the problem at hand. For example, if the bridge reads in C_s and D, we can calculate the equivalent C_p and R_p if that is more convenient for analysis. However, it must be reemphasized that all the parameters in Figure 8.10 are, in general, functions of frequency. Thus, the equivalent circuits are, in general, valid only at the frequency of

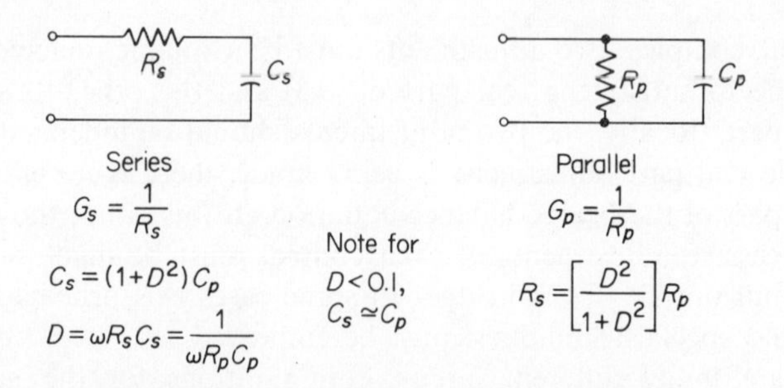

Figure 8.10. Equivalent circuits for capacitors.

measurement. The use of the equivalent resistance and capacitance values at other than the frequency of measurement is not valid unless there parameters have been checked and found independent of frequency.

Exercise 8.3

A bridge measurement at 1 kHz yields a series capacitance $C_s = 10\,\mu F$ with $D = 0.5$. Find the equivalent series and parallel circuit models valid at 1 kHz.

8.4.2 commercial capacitance bridges

AC bridges that provide capacitance measurement from 20 Hz to hundreds of megahertz are available. Some of these bridges are also capable of measuring resistance and inductance. However, in this chapter we are concerned only with their capacitance-measurement functions. Three multipurpose bridges are shown in Figure 7.14. They are the General Radio Model 1608-A and 1650-B Impedance Bridges and the Hewlett-Packard Model 4260A Universal Bridge. In Figure 8.11 there are several examples of bridges used specifically for capacitance measurement. The basic capabilities and features of these instruments are listed in Table 8.3.

The bridges which operate in the audio frequency range will have the two main controls for balancing. Range switches permit selection of the decade range of capacitance and the expected range of *D*. On some units controls for setting excitation level and detector sensitivity are provided. These should be set low while components are being connected to the bridge terminals to prevent damage to the indicator. When external generators are used, normally some conversion factor involving the change in frequency must be applied to one or both of the scales. The individual instruction manuals should be consulted for this mode of operation.

Exercise 8.4

Using a commercial capacitance bridge, measure and compare a 0.001-μF paper capacitor, a 0.001-μF high-*K* ceramic capacitor, and a 0.001-μF mica capacitor. Use the internal test frequency if one is provided.

Exercise 8.5

Review the instruction manual for your bridge to learn how to employ it over the frequency range of 100 Hz to 10 kHz and how to apply a bias voltage to a capacitor under test. Measure and compare a 1-μF paper and a 1-μF tantalum electrolytic capacitor at 0.1, 0.5, 1.0, 5.0, and 10 kHz. When measuring the tantalum unit, apply 50% of the rated voltage through the bias terminals of the bridge. Be sure the polarity is correct. From this data, determine the frequency ranges over which the models in Figure 8.10 are valid.

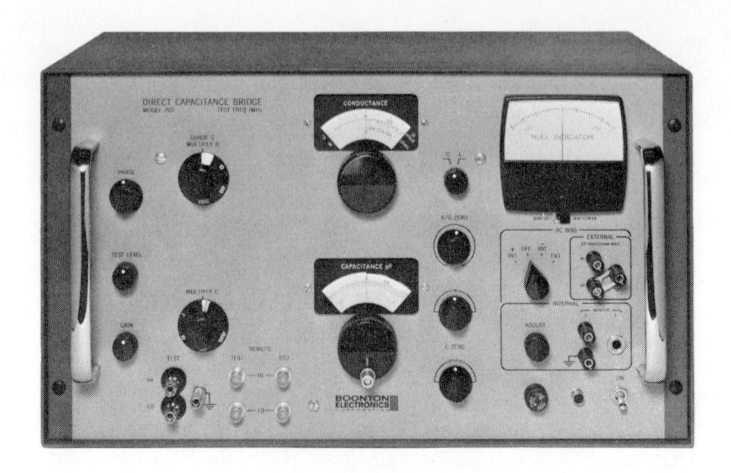

Boonton 75D

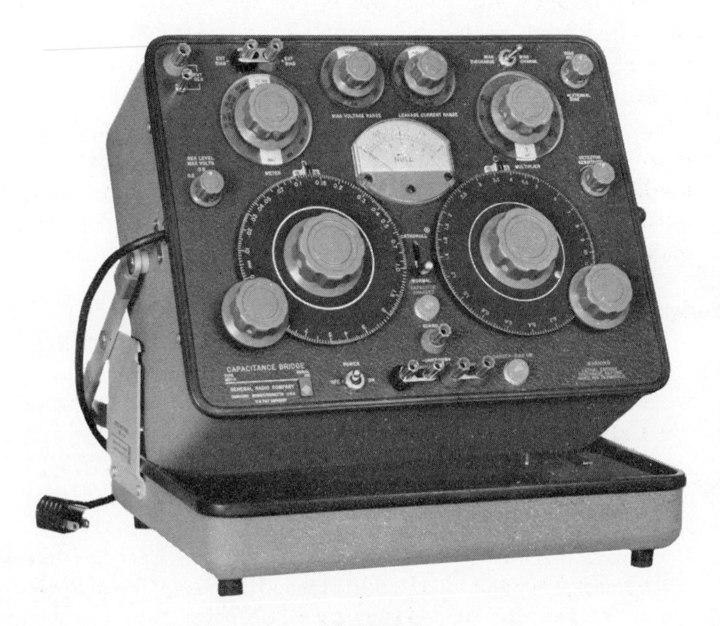

GR 1617-A

Figure 8.11. Commercial capacitance bridges.

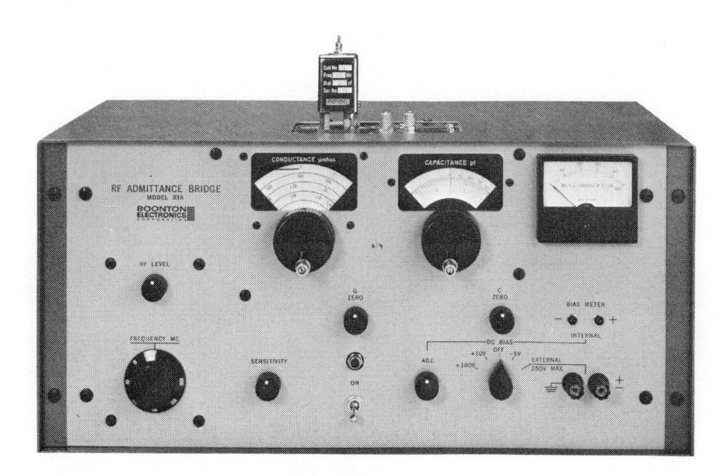

Boonton 33A

GR 716-C

GR 1680

Table 8.3 / Capacitance Bridges

Type	Capacitance Range	Basic Accuracy	Frequency	Readout	Bias Voltage	Features
HP-4260A	1 pF–1000 μF	1%	1 kHz int 20 Hz–20 kHz ext	C_p, D C_s, D	ext	electronic automatic balance
GR 1650-B	1 pF–1100 μF	1%	1 kHz int 20 Hz–20 kHz ext	C_p, D C_s, D	ext	mechanical automatic balance
GR 1608-A	0.05 pF–1100 μF	1%	1 kHz int 20 Hz–20 kHz ext	C_p, D C_s, D	ext	digital readout
GR 1617-A	2.0 pF–1.1 F	1%	120 Hz int 20 Hz–1 kHz ext	C, D	0–600	measures bias voltage and leakage current of electrolytics
GR 1680	0.01 pF–1000 μF	0.1%	120 Hz, 140 Hz 1 kHz; int	C_p, G C_p, D	ext	automatic balance, digital readout, binary-coded decimal output signal
GR 716-C	0.1 pF–1.1 μF	0.1%	30 Hz to 300 kHz ext	C, D	none	substitution measurement
Boonton 75D	0.00005 pF–1000 pF	0.25%	1 MHz int	C_p, R_p	−6 to +150 V int 400 ext	high resolution
Boonton 33A	0.02 pF–150 pF	1%	1–100 MHz int	C, G	−5 to +100 int 250 ext	wide range, high resolution

When bridges are used at frequencies above the audio range, additional controls are often provided to achieve an initial null with the unknown removed and the main balance controls set to zero. This allows a cancellation of residual parasitic effects in the bridge and improves the resolution. With this exception, high-frequency bridges are used in essentially the same fashion as the audio frequency types.

Exercise 8.6

Using a bridge capable of measurement at 1 MHz or greater, repeat the measurements on the 0.001-μF capacitors. Compare the results with the data obtained at 1 kHz.

8.5 transient measurement of capacitors

In Section 8.4 the characterization of real capacitors through their complex impedance or admittance as a function of frequency was discussed. We will now turn to some time-response measurements which present data that are more meaningful in applications such as timing of pulse circuits and analog integration.

8.5.1 self-discharge measurement

In the discussion on dielectrics, the imperfection of the dielectric material was shown to produce a steady-state leakage current in a capacitor. In Exercise 8.2 one method of measurement of the effective leakage current or leakage resistance was introduced. Here we will consider an alternate method of determining the leakage resistance.

Figure 8.12 shows a circuit model for a capacitor including the leakage resistance and the time response that results if the capacitor is initially charged to V_o volts and then disconnected from the charging circuit.

Thus, by measuring the discharge rate of the capacitor, the time constant R_LC and hence R_L can be determined. However, because R_L may be very large, in the hundreds of megohms or more for a high-quality unit, special care must be taken to avoid discharging the capacitor through the measurement circuitry.

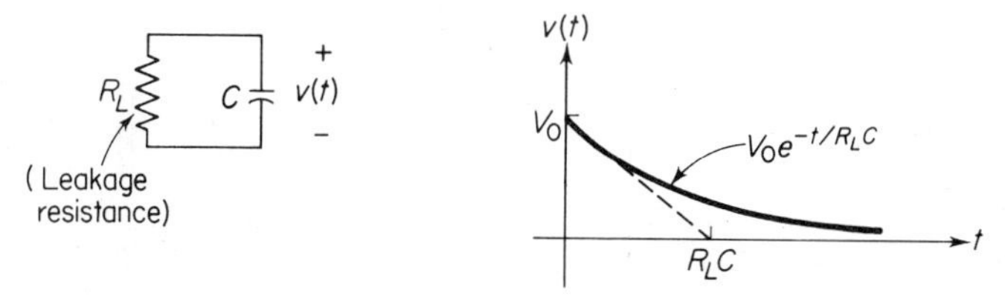

Figure 8.12. Response of a capacitor initially charged to V_0 volts.

One possible method of measuring $v(t)$ which minimizes the loading of the voltmeter is to use a sampling technique, as illustrated in Figure 8.13. In this method, a push button is momentarily closed to obtain a reading of $v(t)$ at discrete time intervals which are long compared with the measurement time. If the sample interval is t_s and the measurement interval is t_m, then the effective input impedance of the voltmeter will be increased by a factor of t_s/t_m, which may easily be 100 or more.

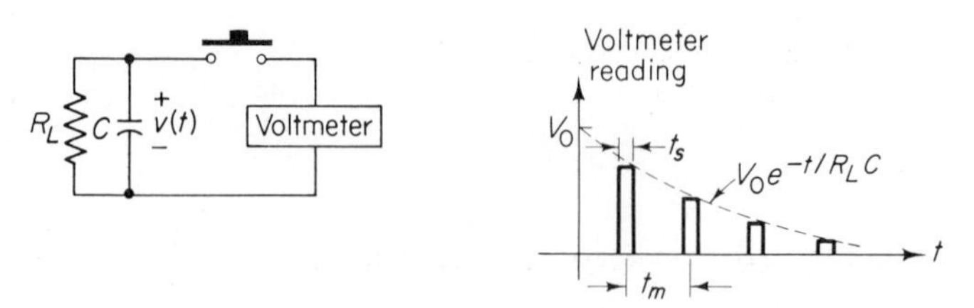

Figure 8.13. Sampling method of capacitor discharge measurement.

The choice of voltmeter is also important, since the loading is also directly proportional to its input impedance. For critical measurements, where leakage resistances are very high, the electrometer, described in Section 13.4.4, is the best choice. Moreover, if the expected leakage resistance is high, care should be taken to minimize other leakage paths, such as through the open push-button switch. In this measurement we will use an X-Y recorder with a time-base generator as a convenient method of recording both sampled voltage and sampling time. For a review of the use of an X-Y recorder, see Section 6.5.

Exercise 8.7

Construct the circuit of Figure 8.14 using an electrolytic capacitor. With S_1 in the position shown, allow the capacitor to charge to its rated voltage. Set the Y sensitivity of the recorder such that the capacitor voltage will produce at least a 75% deflection of the pen and set the time base to about 50 sec/in. Throw S_1 to the opposite position and simultaneously start the sweep. At regular intervals of about 1 in. (50 sec) depress S_2 momentarily to record $v(t)$. Keep S_2 closed only

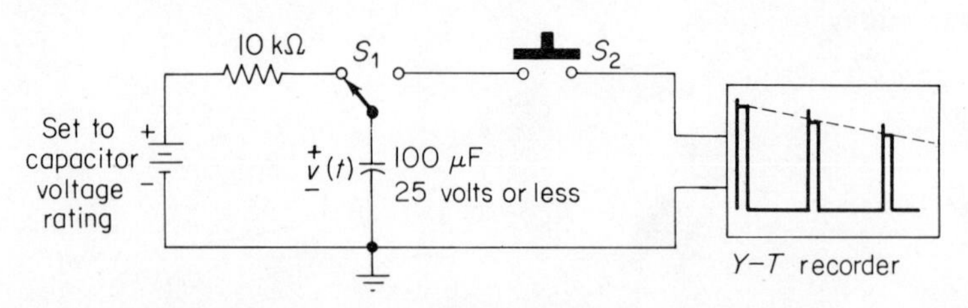

Figure 8.14. Measurement of capacitor discharge time.

long enough for the pen to reach a steady-state value (about 2 sec) so that the actual voltage $v(t)$ will be distinguished from the overshoot of the pen. From the slope of the sampled points determine the time constant $R_L C$ and calculate R_L.

8.5.2 dielectric absorption

When a fully charged capacitor is short-circuited, ideally the stored charge is completely removed. However, the time required for the stored charge in the dielectric layer to migrate to the conducting plates can be surprisingly long. The result of this migration time is that after a momentary short circuit, the voltage across an initially charged capacitor will increase from zero to as much as several percent of the initial voltage. This phenomenon is known as *dielectric absorption.*

Exercise 8.8

Construct the circuit of Figure 8.15 using an electrolytic capacitor. Set the Y-axis sensitivity to give a full-scale deflection at 10% of the initial voltage on the capacitor, and use a sweep time of 10 sec/in. With S_1 as shown, allow the capacitor to become fully charged. Throw S_1 to the opposite position, shorting the capacitor through S_2, and start the recorder sweep. Two seconds after S_1 is switched, open S_2 and record the capacitor voltage. If you wish, investigate the effect of shorter and longer short-circuited times. Note that the recorder input resistance will be continually discharging the capacitor, but for this exercise the effect may be neglected.

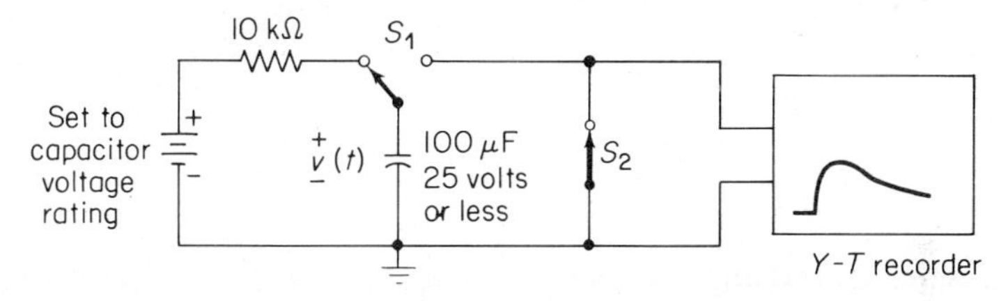

Figure 8.15. Measurement of dielectric absorption.

8.6 references

1. G. W. A. Dummer and H. M. Nordenberg, *Fixed and Variable Capacitors*, McGraw-Hill Book Company, New York, 1960.

9

inductors and transformers

9.1 introduction

The ideal inductor is a two-terminal circuit element in which the terminal voltage is proportional to the time rate of change of the current flowing through the element.

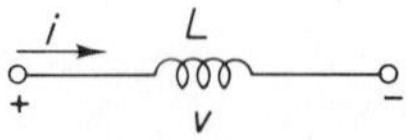

$$v = L\frac{di}{dt} \tag{9.1}$$

The constant of proportionality L is termed the *inductance* of the element, whose dimensional unit is the *henry* (H). As with the ideal capacitor, the ideal inductor is lossless and capable of energy storage. Since this energy is stored in a magnetic field, it is termed *magnetic stored energy.*

As with the resistor and capacitor, the ideal volt-ampere relationship of an inductor can be only approximated in practice, and of the three circuit elements, this approximation to the ideal element is poorest for the inductor. Circuit limitations in terms of frequency response, energy-loss mechanisms, and nonlinearities combine to restrict the performance of an actual inductor to more narrow ranges than are normally encountered for physical capacitors and resistors. The size and weight of inductors also make them by far the physically largest circuit element. Finally, inductors cannot at present be fabricated in integrated circuit form. For these reasons the inductor is often avoided in circuit applications.

When an inductor is required, a unit meeting the specifications may not be commercially available, and the engineer must then design and construct the inductor himself. In this chapter we will explore some of the basic properties of inductors and discuss some of the factors involved in their design and construction. We will also consider two magnetically coupled inductors, including the special case of a transformer.

9.2 *inductor structure*

In order to develop the factors which govern inductor performance, we will characterize the element through the simplified physical structure of Figure 9.1. A winding of N turns of wire encloses a core of magnetic material which has a toroidal shape of cross-sectional area A and mean length l. The magnetic core material is characterized by a parameter called the *permeability* μ whose dimensions are henries per meter (H/m).

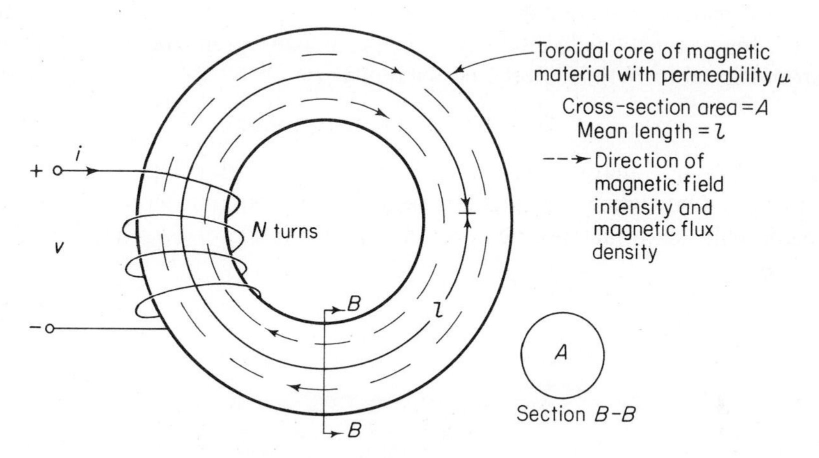

Figure 9.1. Simplified inductor structure

Assuming for the moment that the magnetic core material is linear and that the magnetic fields linking the winding are completely contained within the magnetic core, then the inductance at the terminals is given by

$$L = \frac{\mu A N^2}{l} \tag{9.2}$$

Thus, we see that the inductance of a winding depends upon the square of the number of turns, and the geometry and permeability of the magnetic core.

9.2.1 *properties of magnetic materials*

In order to describe the properties of magnetic core materials, we must first introduce two magnetic field quantities: the *magnetic field intensity* or *magnetizing force* H whose dimension is ampere-turn per meter (ampere/m), and the *magnetic flux density* or *magnetic induction* B with dimension weber per square meter (Wb/m^2). Both H and B are generally vector quantities, but for our purposes we may treat them as scalars and define them in terms of the structure of Figure 9.1 as

$$H = \frac{Ni}{l} \tag{9.3}$$

and

$$\frac{dB}{dt} = \frac{v}{NA} \tag{9.4}$$

The relationship between B and H within a magnetic material is nonlinear with the general behavior shown in Figure 9.2. At low values of current, and hence low magnetic field intensity, there is a linear region where B is directly proportional to H through the permeability μ:

$$B = \mu H \qquad \text{(linear region)} \tag{9.5}$$

By combining Equations (9.3) to (9.5), the expression for inductance (9.2) is obtained. Only when the magnetic field intensity remains within the linear region will the inductance be constant and the inductor be a linear circuit element.

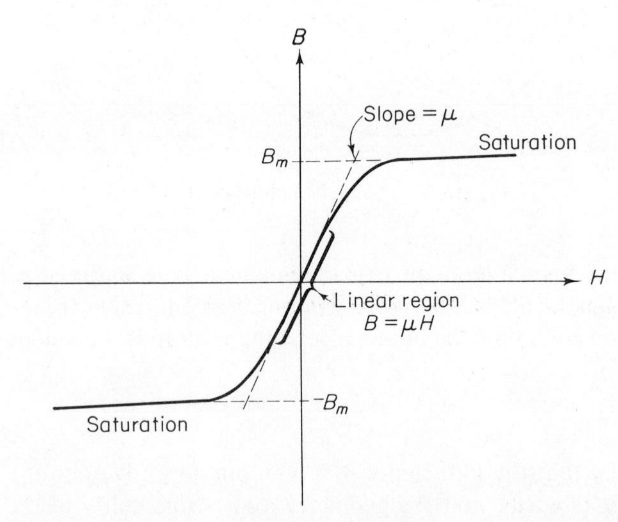

Figure 9.2. *B-H* characteristic of a magnetic core material.

As H is increased beyond the linear region, the slope of the B-H curve decreases. In this nonlinear region the inductance becomes dependent on terminal current and hence a nonlinear circuit element. Since the inductance is still related to the slope of the B-H curve, it is a decreasing function with increasing current. Finally, a point is reached where increases in magnetic field intensity produces no appreciable increase in magnetic flux density. At this point the magnetic material is termed *saturated*, with a maximum magnetic flux density B_m.

There is a wide variety of magnetic materials available for use in inductors and transformers. Selection of a specific material for a given application will depend on many factors beyond the scope of this chapter. However, we will describe the basic properties of several broad classes of materials and indicate their major applications.

Silicon steel, composed of iron with a few percent of silicon, is the most widely used magnetic material. It is characterized by a relative permeability μ_r* as high as 10^4 and a saturation flux density B_m of 1.6 Wb/m^2. Silicon steels find their major application in inductors and transformers at the 60-Hz power frequency and in AC machinery, since hysteresis loss (Section 9.3) is not excessive and low-cost material is required.

Nickel-iron alloys, containing 50% or more nickel, provide relative permeabilities as high as 10^5 while retaining a saturation flux density similar to that of silicon steel, on the order of 1.5 Wb/m^2. The higher permeability means that the saturation flux density is reached at a much lower magnetic field strength, which makes these materials desirable for applications where core saturation is a required property, i.e., magnetic amplifiers, saturable reactors, and DC-DC converters.

Powdered nickel-iron alloys are combined with a nonmagnetic binder material and formed into toroidal cores under high pressure. Because these are a mixture of magnetic and nonmagnetic material, the relative permeability is much lower, on the order of 10 to 100, and the saturation flux density is on the order of 0.6 Wb/m^2. However, these cores have excellent stability with time and temperature and useful frequency ranges to the hundreds of MHz. For these reasons they are widely used in filters where accurate inductance values are required.

Ferrites are ceramic materials composed of the oxides of iron and other magnetic metals. They are characterized by a relative permeability on the order of 10^3 and a saturation flux density of 0.3 Wb/m^2. Ferrite materials rival the powdered nickel-iron materials for application in filters and other high-quality inductor and transformer applications, particularly in the frequency range of 1 kHz to 10 MHz.

* The relative permeability μ_r is the ratio of the permeability of the material μ to the permeability of free space μ_0. Thus $\mu_r = \mu/\mu_0$, and $\mu_0 = 4\pi \times 10^{-7}$ H/m. Note that μ_r is a dimensionless quantity.

The general properties of the four classes of materials described above are intended for guidance only. Within each class there are many variations in composition, and the detailed choice must be based on the specific application. Manufacturers of magnetic materials normally apply trade names to their products, and it is not always obvious which class a particular material belongs. However, comparison of μ_r and B_m for the material with the data given above will provide some clue to the basic composition.

9.2.2 basic forms of magnetic cores

In addition to the wide variety of magnetic materials available, there is an equally wide variety of shapes and sizes. Some of the most common types are described below.

Materials fabricated in sheet form, such as silicon-steel and nickel-iron alloys, are provided in two forms: stamped laminations and tape-wound toroids. The shape of the stamped laminations is described by a similarly shaped letter—E, I, U, F, or L. Cores are then built by stacking one or more shapes to form the desired thickness. Figure 9.3a illustrates a core made from L laminations. However, when stamped laminations are assembled, there are always small air gaps interrupting the magnetic-field path. These air gaps reduce the effective permeability of the core material. Hence, in applications where this reduction in μ is undesirable, the tape-wound toroidal form of Figure 9.3b is employed. One major reason for fabricating cores from laminations rather than a solid piece of material is to reduce eddy-current losses (Section 9.3).

The powdered nickel-iron and ferrite materials may be formed in any shape desired. However, the former is normally available only as a toroid. Ferrites, on the other hand, come in a myriad of shapes, ranging from full

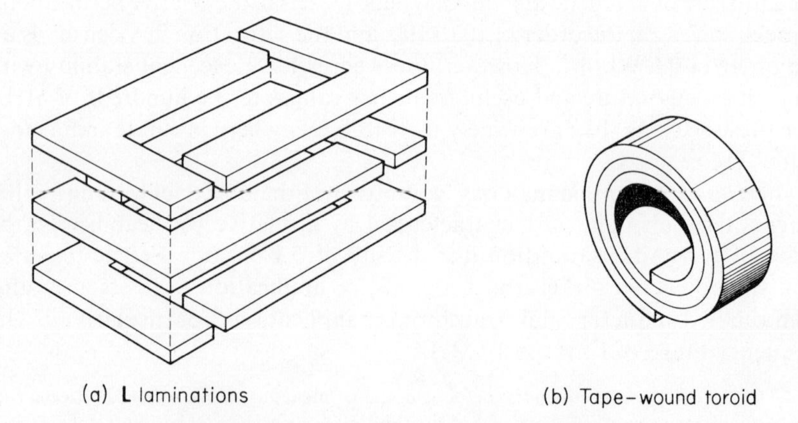

Figure 9.3. Cores made from sheet material.

Figure 9.4. Ferrite core shapes. Courtesy Ferroxcube Corporation, Saugerties, N.Y.

cross-sectional U cores, to toroids, to the cup or pot core, as illustrated in Figure 9.4.

Since these materials have a high resistivity (are poor electrical conductors), the eddy-current losses are low and hence a laminated structure is not necessary.

The pot core arrangement, shown in Figure 9.5, consists of two mating ferrite core pieces and a coil wound on a bobbin. When assembled, the coil is enclosed within the ferrite core material, which provides low magnetic flux leakage and excellent shielding from noise. Because the coil is wound on a bobbin, it is easily produced or modified, compared to the difficulty of winding coils on toroidal forms.

Figure 9.5. Pot core arrangement.

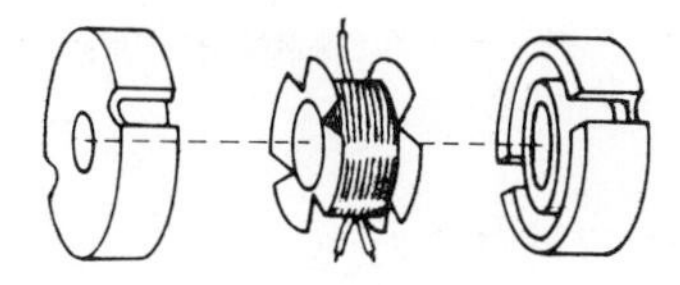

9.2.3 magnetic units and effective parameters

In discussing magnetic quantities, we have employed the rationalized MKS system of units. Manufacturers of magnetic materials, however, usually present their data in the electromagnetic CGS system. The conversion factors required to go from electromagnetic CGS units to rationalized MKS units are given in Table 9.1. Most data on permeability is given as relative permeability μ_r and conversion to MKS units is accomplished simply by using $\mu_0 = 4\pi \times 10^{-7}$ H/m. Finally, conversion of lengths and

Table 9.1 / Conversion of Magnetic Units

Quantity	To Convert	To	Multiply by
Magnetic field intensity, H	oersteds	ampere-turns/meter	79.5
Magnetic flux density, B	gauss	webers/meter^2	10^{-4}

areas to meters or meters^2 must be carried out. As familiarity with the design procedures develop, some short cuts will become obvious. However, at the initial stages it is best to convert *all* quantities to MKS units before calculating a design.

For simple core shapes, the dimensions of cross-sectional area and mean path length are straightforward. However, as the shapes become more complex, such as a pot core, these quantities are not so easily determined. Therefore, manufacturers specify the *effective area* and *effective length* of their core assemblies, and these parameters are the ones to use in electrical design formulae. Similarly, when a core assembly has a mating surface, such as in a pot core, the permeability will be affected. This effect is normally accounted for by specifying an *effective relative permeability* which should be used in design calculations.

Exercise 9.1

Using Equation (9.1) and manufacturer's specifications, determine the number of turns required to construct a 100-mH inductor on a ferrite pot core (Ferroxcube No. 2616P-L00-3B9 or equivalent). On a single-section plastic bobbin, carefully wind this number of turns using No. 28 enamel-coated copper wire. Secure the winding with a layer of electrical tape. Wind a second coil over the first, consisting of half the number of turns required to produce a 100-mH inductance. Secure this winding with a second layer of tape. The enamel insulation is easily removed from the ends of the winding by first burning it with a match and then rubbing it with steel wool. Assemble the windings and the core and provide suitable terminals for the windings.

Exercise 9.2

The inductance values of the coils wound in the previous exercise may be measured using the AC bridge circuit in Figure 9.6. *Caution:* Be careful of ground connections. Either the scope must have a differential input or the generator must have a floating output. When the decade inductor is set to produce a null indication on the oscilloscope, the value of the decade inductor and the unknown will be equal. Compare both the absolute inductance values and the ratio of the two inductances with the design values.

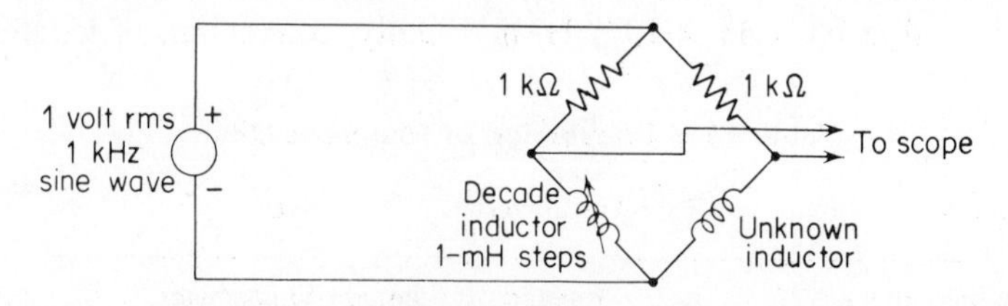

Figure 9.6. AC bridge for inductance measurement.

9.3 loss mechanisms in inductors

The performance of an actual inductor approaches, but never reaches, that of an ideal inductor. We have already seen that a nonlinear *B-H* characteristic can cause an inductance value to be a function of the current in the winding. In addition there are energy losses associated with inductors, which fall into three categories. The first of these is the *resistance loss* of the wire used in the coil. This loss mechanism is minimized by the use of the largest size wire possible consistent with the available space for the coil winding and the required number of turns.

Second are the *eddy-current* losses. Since core materials such as silicon steel and nickel iron are good electrical conductors, the changing magnetic fields in the core will induce an AC current flow within the core material. This induced current is known as an eddy current. However, laminating the core rather than making it from a solid piece of material can reduce the eddy currents. This is one reason for providing core materials as thin laminations and tape-wound toroids. Powdered metal cores and ferrites, on the other hand, are poor electrical conductors and hence have negligible eddy currents. Thus, lamination is not necessary.

The third loss mechanism involves the behavior of the *B-H* characteristic. In Figure 9.2, this characteristic was shown as a single line. Actually, on close observation it is a closed curve whose area is swept out once each cycle of applied current as shown in Figure 9.7. Since the area enclosed by the curve represents energy, this area times the frequency of operation represents power dissipation. This loss mechanism, called *hysteresis loss*, sets the

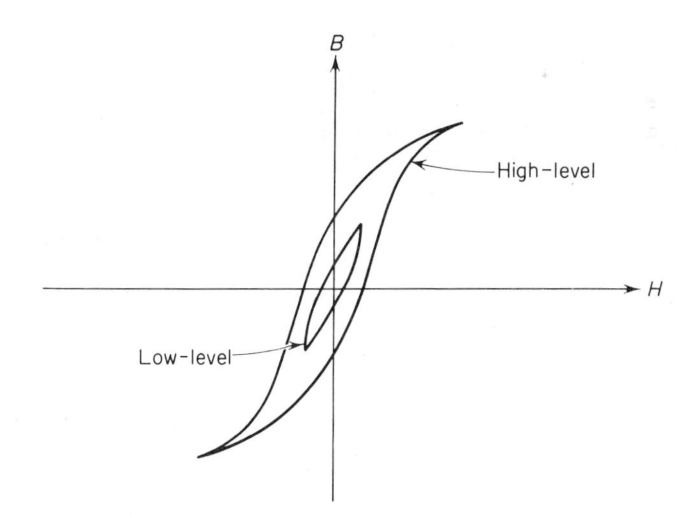

Figure 9.7. Hysteresis effect in *B-H* characteristic.

limit on frequency of operation for many materials. As shown in Figure 9.7, the area can be kept small by operating the core at low levels of magnetic-field intensity, which is also consistent with operation in the linear portion of the *B-H* characteristic.

Exercise 9.3

The *B-H* characteristic of the ferrite core constructed in Exercise 9.1 may be measured using the circuit shown in Figure 9.8. The core is excited by the 60-Hz voltage applied by the filament transformer. Since $v_x = i_1 R_I$, then

$$H = \frac{N_1 i_1}{l} = \left(\frac{N_1}{R_I l}\right) v_x \tag{9.6}$$

and thus the *X*-axis voltage yields the magnetic field intensity. Similarly, since

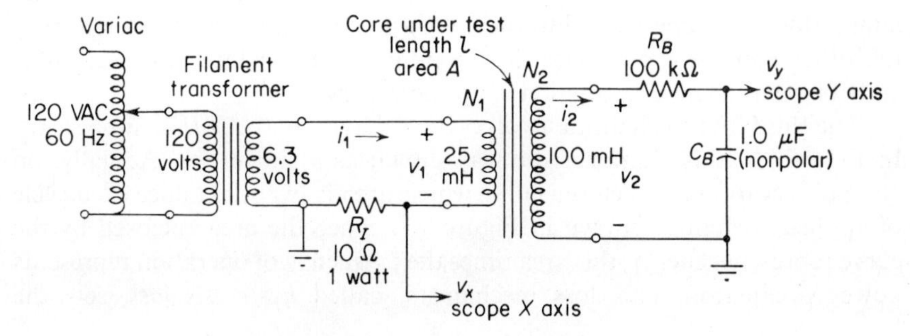

Figure 9.8. Circuit for *B-H* measurement.

$R_B \gg 1/\omega C_B$ at 60 Hz, then $i_2 \simeq v_2/R_B$ and

$$v_y = \frac{1}{C_B}\int \frac{v_2}{R_B}\,dt = \frac{N_2 A}{R_B C_B}\int \frac{dB}{dt}\,dt \tag{9.7}$$

or

$$B = \left(\frac{R_B C_B}{N_2 A}\right) v_y \tag{9.8}$$

and hence the magnetic flux density may be found from v_y.

Display the *B-H* characteristic on the oscilloscope and observe the hysteresis effect. If the display is reversed on the CRT, reverse the connections to one of the windings on the pot core. Determine B_m the saturation value of the magnetic flux density. Introduce an air gap into the magnetic path by separating the two halves of the pot core with a shim made from one thickness of paper. Compare the resulting *B-H* characteristic with that for the case of zero air gap.

9.3.1 circuit models for lossy inductors

One possible circuit model which can represent the loss mechanisms in an actual inductor is shown in Figure 9.9.

The resistance and inductance in the model are explicitly indicated to be functions of frequency. The reason is that the properties of the core material which govern $R(\omega)$ and $L(\omega)$ are, in general, frequency dependent. Moreover, to represent two or more independent loss mechanisms in an indicator by a single lumped resistive element is not likely to result in an element value that is independent of frequency. If a frequency range exists where $R(\omega)$ and $L(\omega)$ are approximately constant, as shown in Figure 9.9, then circuit

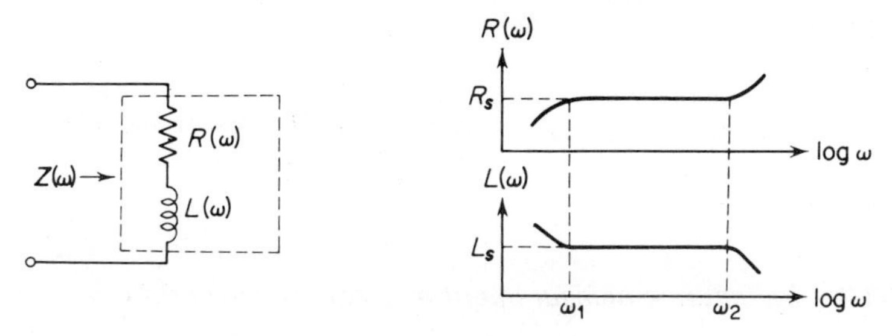

Figure 9.9. Circuit model for a lossy inductor.

calculations based on the simple model with constant element values R_s and L_s will be valid over the frequency range ω_1 to ω_2. Outside this frequency range we must either employ the frequency-dependent values of $R(\omega)$ and $L(\omega)$ in circuit calculations or construct a more complex model consisting of more than two frequency-independent elements to account for the actual behavior of the lossy inductor.

The input impedance of the inductor may be written, in general, as

$$Z(\omega) = R(\omega) + jX(\omega) \tag{9.9}$$

where $R(\omega)$ is the resistance or real part of the impedance and $X(\omega)$ is the reactance or imaginary part of $Z(\omega)$. From the circuit model of Figure 9.9 we find

$$L(\omega) = \frac{X(\omega)}{\omega} \tag{9.10}$$

Although a measurement of the impedance will yield the resistance term directly, this is not usually employed as a measure of the loss in an inductor directly. Rather, a dimensionless figure of merit called the *quality factor Q* is employed. The quality factor is given by

$$Q = \frac{X(\omega)}{R(\omega)} \tag{9.11}$$

In a lossless inductor $R(\omega) = 0$, and hence $Q = \infty$ represents the ideal case. The smaller the value of Q the more lossy the inductor. In the best inductors, the value of Q may reach 10^3. In general, the quality factor will also be a function of frequency and amplitude of excitation signal. Thus, in specifying the Q of an inductor, the amplitude and frequency of the test signal must be included for meaningful interpretations and comparisons.

Exercise 9.4

An alternate circuit model to express the loss in an actual inductor is shown in Figure 9.10. Find $G(\omega)$ and $L(\omega)$ if this model is to be equivalent to that of Figure 9.9 over the frequency range $\omega_1 < \omega < \omega_2$. In particular consider the special case where $Q \gg 1$.

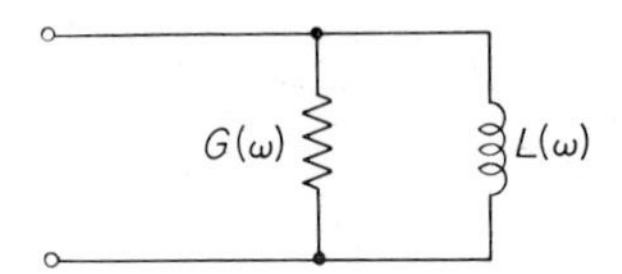

Figure 9.10. Alternate circuit model for a lossy inductor.

9.3.2 inductance measurement and commercial bridges

In Section 8.4 the method of using an AC bridge to measure a capacitor was described. The same method applies to inductors. However, the question of how the bridge measurement is to be converted to a circuit model deserves special emphasis. The two adjustments required to balance the AC bridge yield information on the real part and imaginary part of the unknown impedance. There are also two possible circuit models for the lossy inductor: the series and parallel representations. Therefore, the bridge equivalent circuit representation may be in terms of either a series or parallel model, and may be calibrated in terms of inductance and resistance, conductance or quality factor. Figure 9.11 presents the interrelationships of the various models and parameters to permit conversion to the most useful form for the problem at hand.

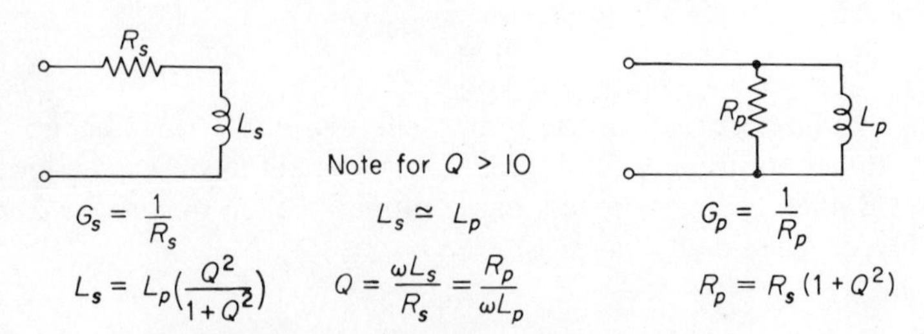

Figure 9.11. Equivalent circuits for lossy inductors.

Bridges capable of measuring inductance from 20 Hz to the hundreds of megahertz are commercially available. Some of these bridges measure capacitance and resistance also. However, in this assignment we are concerned with their inductance-measuring function only. Three multipurpose bridges capable of inductance measurements are shown in Figure 7.14. They are the General Radio Models 1608-A and 1650-B Impedance Bridges and the Hewlett-Packard Model 4260A Universal Bridge. In Figure 9.12 are some examples of bridges used specifically for inductance measurement. The basic capabilities and features of these instruments are listed in Table 9.2. Bridges and instruments designed for inductance measurements above 1 MHz (the Q meter and the R-X meter, for example) are discussed in Chapter 18.

Exercise 9.5

Using a commercial inductance bridge, measure the inductance and Q of each winding on the pot-core inductor constructed previously. Introduce an air gap with a paper shim and repeat the measurements.

9.3.3 measurement of Q

The commercial bridges described in Section 9.3.2 do not always provide good resolution of high values of Q. A circuit which permits the accurate measurement of Q as a function of frequency is shown in Figure 9.13, and forms the basis of an instrument known as the Q meter.

The operation of the Q meter is as follows. A series circuit is formed by the inductor and the capacitor C, and is driven from the variable-frequency low-impedance source. The impedances of the voltmeters are sufficiently high to be neglected. At resonance, indicated by a sharp peak in V_2, the series reactance of L_s and C cancel such that

$$|I_C| = \frac{|V_1|}{R_s} \tag{9.12}$$

Thus, at resonance we have

$$|V_2| = \frac{|I_C|}{\omega C} = \omega L_s |I_C| \tag{9.13}$$

since at resonance $\omega^2 = 1/L_s C$.

Hence at resonance

$$\frac{|V_2|}{|V_1|} = \frac{\omega L_s}{R_s} = Q \tag{9.14a}$$

and

$$L_s = \frac{1}{\omega^2 C} \tag{9.14b}$$

Boonton 63H

GR 1633-A

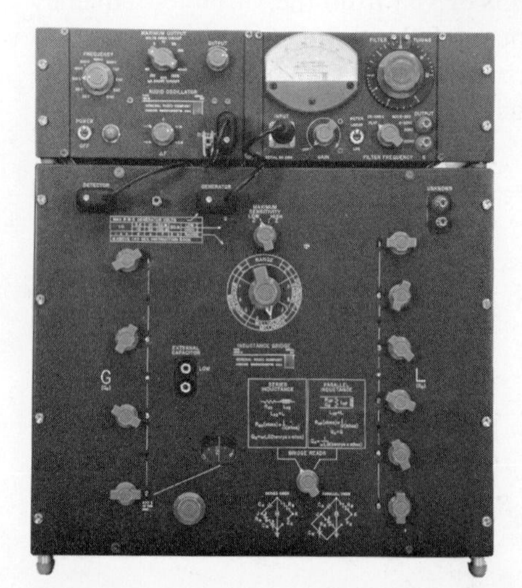

GR 1660-A

Figure 9.12. Commercial inductance bridges.

Table 9.2 / Inductance Bridges

Type	Inductance Range	Basic Accuracy	Frequency	Readout	Features
HP 4260A	1 μH–1000 H	1.0%	1 kHz int 20 Hz–20 kHz ext	L_s, Q L_p, Q	electronic automatic balance
GR 1650-B	1 μH–1100 H	1.0%	1 kHz int 20 Hz–20 kHz ext	L_s, Q L_p, Q	mechanical automatic balance
GR 1608-A	0.05 μH–1100 H	0.1%	1 kHz int 20 Hz–20 kHz ext	L_s, Q L_p, Q	digital readout
GR 1660-A	0.1 nH–1111 H	0.1%	50 Hz–10 kHz int	L_s, R_s L_p, G_p	six-digit resolution
GR 1633-A	100 μH–1000 H	1.0%	50 Hz–15 kHz ext	L_s, R_s L_s, Q	incremental measurement, applies bias
Boonton 63H	2 μH–110 mH	0.25%	5 kHz–500 kHz int	L_s, R_s	high resolution

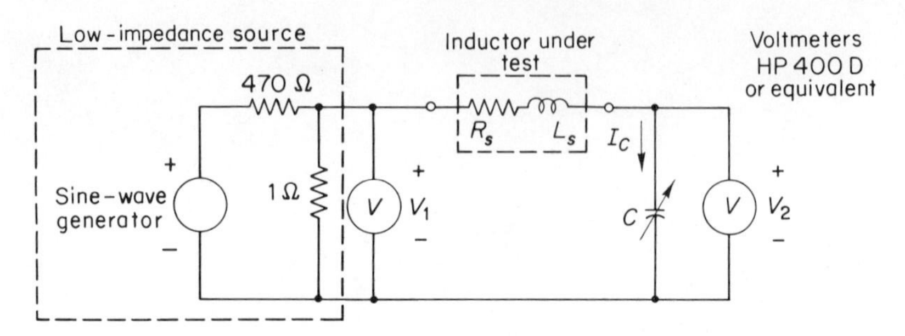

Figure 9.13. Basic Q-meter circuit.

Exercise 9.5

Construct the Q meter circuit of Figure 9.13 and use it to measure the Q of the 100-mH inductor constructed in Exercise 9.1 over the frequency range of 1 kHz to the self-resonant frequency of the coil. In practice, tuning for resonance is best accomplished by setting C and adjusting the frequency for a maximum V_2. Self-resonance is the effect of the interwinding stray capacitance forming a parallel resonant circuit with L_s and is indicated by a rapid drop in the measured Q as the test frequency is increased. At self-resonance, a minimum is observed in the reading of V_2 as the test frequency is varied. Keep the excitation signal V_1 at some convenient level, such as 10 mV rms. This permits V_2 to indicate the numerical value of Q within a power of 10. Plot the Q of the coil and the value of L_s vs. log frequency.

9.4 coupled coils

When two or more windings are incorporated on a single magnetic core, the magnetic field intensity in the core is a combined function of the currents in each winding. Such an arrangement is referred to as a set of *coupled coils.*

An arrangement of two coupled coils is shown in Figure 9.14. Note that a portion of the magnetic flux density linking one coil also links the other coil. Thus a voltage v_2 will be induced at terminal-pair 2 in response to an excitation applied at terminal-pair 1. The dot marking on the windings indicates the relative polarities of the voltages, with the convention that the dotted terminals are simultaneously positive.

If we neglect the loss mechanisms associated with the winding resistance and core properties and assume that the core is operated in the linear region, then the voltage-current equations for the coupled coils become

$$v_1 = L_1\frac{di_1}{dt} + M\frac{di_2}{dt} \tag{9.15a}$$

$$v_2 = M\frac{di_1}{dt} + L_2\frac{di_2}{dt} \tag{9.15b}$$

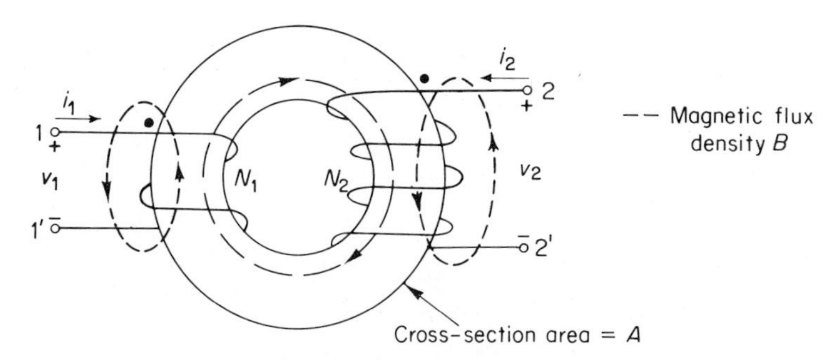

Figure 9.14. Physical arrangement of coupled cells.

where M is the *mutual inductance* between the windings. If winding 2 is open-circuited, then $di_2/dt = 0$ and (9.15a) shows that L_1 is simply the inductance of winding 1 alone. Similarly, if $i_1 = 0$, then L_2 is the inductance of winding 2 alone. For this reason L_1 and L_2 are often referred to as the *self-inductances* of the respective windings. On the other hand, if winding 2 is short-circuited, then $v_2 = 0$ and solution of (9.15a) and (9.15b) under this constraint yields

$$v_1 = \left(L_1 - \frac{M^2}{L_2}\right)\frac{di_1}{dt} \tag{9.16}$$

Thus the inductance measured at terminal-pair 1 with 2-2′ shorted is reduced from L_1 by an amount M^2/L_2. The *coupling coefficient* k is defined by

$$k = \frac{M}{\sqrt{L_1 L_2}} \leq 1 \tag{9.17}$$

and is always less than or equal to unity. Should k exceed unity, the effective inductance in (9.16) would become negative, a physical impossibility.

Exercise 9.6

Using a commercial bridge, measure the mutual inductance of the pair of coils constructed in Exercise 9.1. Compare the value of M obtained by shorting 2-2′ and measuring at 1-1′ and vice versa. (*Note:* The Q is likely to be low when one set of terminals is shorted, and a sliding null may be encountered.) From the data calculate the coupling coefficient using Equation (9.17). Use paper shims to introduce an air gap and repeat the measurements. (*Note:* When k is near unity, calculation from L_1, L_2, and M may yield poor accuracy. An alternate method is described in Section 9.4.2.)

Two circuit models which may be used to describe a set of coupled coils are shown in Figure 9.15. Although either model in Figure 9.15 may be used for AC calculations, the model in Figure 9.15a provides a DC path between terminals, whereas the model in Figure 9.15b blocks DC.

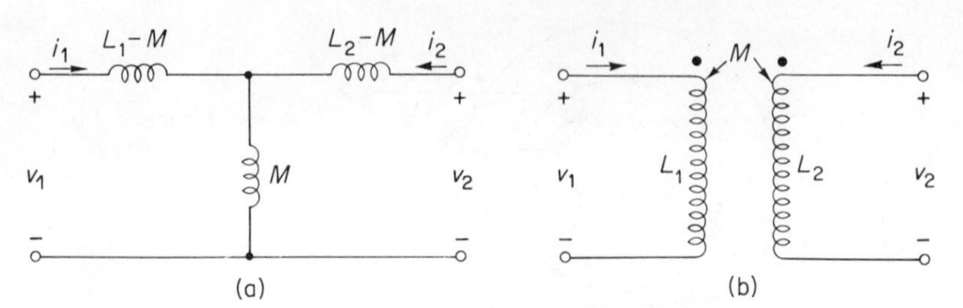

Figure 9.15. Circuit models for coupled cells.

9.4.1 ideal transformer

To illustrate how coupled coils operate as transformers, we will first describe the properties of an ideal transformer. Then, by combining the ideal transformer with a model for coupled coils, the performance of actual transformers will be derived.

In an ideal transformer all the magnetic flux produced by one winding links all the turns of the second winding. Thus in the arrangement of Figure 9.14 the value of B is the same for winding 1 and 2, and the voltages become

$$v_1 = N_1 A \frac{dB}{dt} \tag{9.18a}$$

$$v_2 = N_2 A \frac{dB}{dt} \tag{9.18b}$$

Consequently the voltage ratio is

$$\frac{v_2}{v_1} = \frac{N_2}{N_1} = n \tag{9.19}$$

where n is the *turns ratio* of the transformer. Since the ideal transformer is lossless and stores no magnetic energy the instantaneous power input is zero:

$$v_1 i_1 + v_2 i_2 = 0 \tag{9.20}$$

Combining (9.19) and (9.20) we find

$$\frac{i_2}{i_1} = -\frac{1}{n} \tag{9.21}$$

The circuit symbol for the ideal transformer is shown in Figure 9.16.

The impedance transformation properties of the ideal transformer can be demonstrated by examination of (9.19) and (9.21). If winding 2 is loaded by an impedance Z, as shown in Figure 9.17a, the complex amplitudes at the

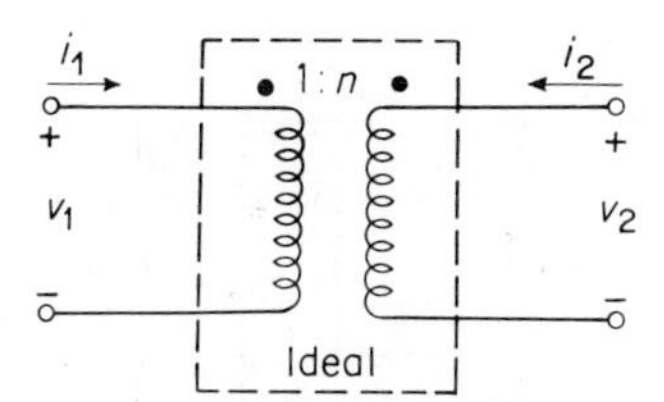

Figure 9.16. Circuit symbol for an ideal transformer.

terminals of winding 2 are related by

$$V_2 = -ZI_2 \tag{9.22a}$$

However, from Equations (9.19) and (9.21), $V_1 = V_2/n$ and $I_1 = -I_2 n$, so that the impedance seen at the terminals of winding 1 is

$$Z_1 = \frac{V_1}{I_1} = \frac{Z}{n^2} \tag{9.22b}$$

as shown in Figure 9.17b. Note that Equation (9.22b), which shows that an ideal transformer transforms impedance by the factor n^2, is valid for all values of Z and for all values of current, voltage, and frequency. An ideal transformer transforms an open circuit into an open circuit and a short circuit into a short circuit for all frequencies, including DC.

9.4.2 *lossless transformers*

Actual transformers differ from the ideal transformer in several ways. First, not all the flux produced by one winding links the second winding. This deficiency is termed *leakage*. Second, the core material stores magnetic energy. Finally there are the losses and nonlinearities associated with the core and the losses due to the resistances of the windings.

If we neglect the losses and nonlinearities for the moment, then the circuit models for the lossless coupled coils (Figure 9.15) have an equivalent representation which includes an ideal transformer as shown in Figure 9.18.

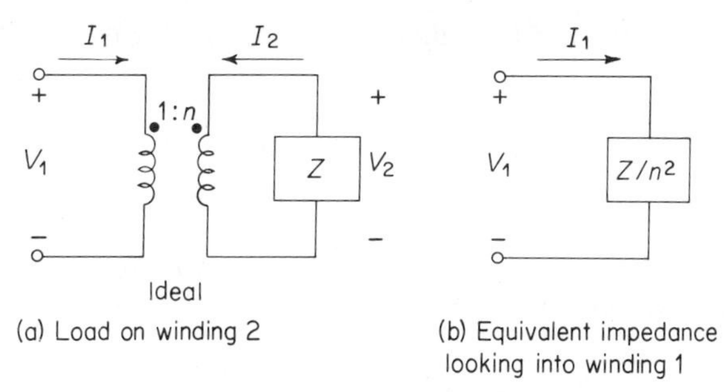

Figure 9.17. The ideal transformer as an impedance transforming device.

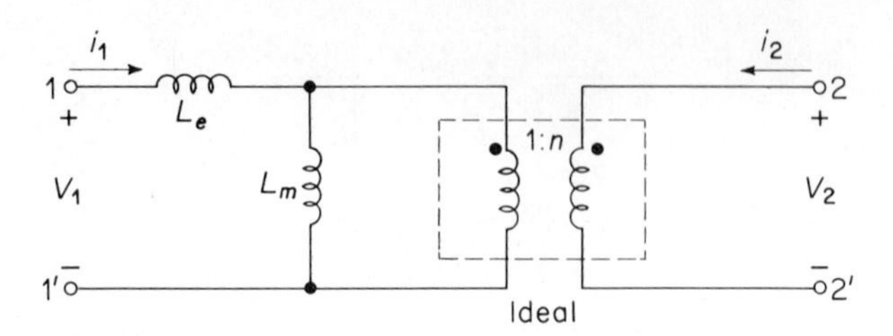

Figure 9.18. Circuit model for lossless coupled coils utilizing an ideal transformer.

By solving for the terminal voltages and currents for the model of Figure 9.18 and either model of Figure 9.15 and equating coefficients, the following relationships are found to be required for the models to be equivalent:

$$L_e = L_1(1 - k^2) \tag{9.23}$$

$$L_m = k^2 L_1 \tag{9.24}$$

$$n = \frac{L_2}{M} \tag{9.25}$$

The inductance L_e is termed the *leakage inductance*, since it is a result of the leakage flux. The *magnetizing inductance* L_m describes the energy-storage properties of the core material. In a well-designed transformer the coupling coefficient is made as near unity as possible so that L_e is normally much less than L_m.

In Section 9.4, the calculation of the coupling coefficient from measurements of L_1, L_2, and M was presented. However, when k is near unity, this calculation using Equation (9.17) gives poor accuracy. A better method is to note that L_e is the inductance measured at terminal-pair 1 with terminal-pair 2 shorted. Thus a more accurate determination of k is obtained by using Equation (9.23), since it measures the small leakage directly.

Exercise 9.7

Using the data on L_1 and L_e for the coupled coils, calculate the coupling coefficient k. Note that

$$k = \sqrt{1 - \frac{L_e}{L_1}} \simeq 1 - \frac{L_e}{2L_1} \tag{9.26}$$

for $L_e/L_1 \ll 1$.

If a resistor R is connected to terminals 2-2′ in Figure 9.18, then the impedance seen at terminals 1-1′ is

$$Z(j\omega) = j\omega L_e + \frac{(R/n^2)(j\omega L_m)}{(R/n^2) + (j\omega L_m)} \tag{9.27}$$

If these coupled coils are to provide transformer action, then for some range of frequencies $Z(j\omega)$ must be equal to R/n^2. The conditions which yield this result are

$$\omega L_e \ll \frac{R}{n^2} \tag{9.28a}$$

$$\omega L_m \gg \frac{R}{n^2} \tag{9.28b}$$

Stated in another way, the coupled coils will provide transformer action in the frequency range $\omega_l \leq \omega \leq \omega_h$, where

$$\omega_l = \frac{R}{n^2 L_m} \tag{9.29a}$$

$$\omega_h = \frac{R}{n^2 L_e} \tag{9.29b}$$

Equations (9.29) provide the bandwidth information for the transformer in terms of the load impedance R and the transformer parameters L_e, L_m, and n. Because the bandwidth depends on the load impedance, transformers are often specified in terms of an impedance transformation (25 ohms: 100 ohms) rather than a turns ratio (1:2). In this way, the manufacturer insures his bandwidth specification will be met. On the other hand, a transformer may be used at impedance levels other than those specified, provided inequalities (9.28a) and (9.28b) are satisfied.

Exercise 9.8

The coupled coils constructed in Exercise 9.1 are to be used to transform a 100-ohm load resistance to 25 ohms. From the data for the coils, calculate the upper and lower frequency limits for transformer action.

Connect a 100-ohm resistor to the coils so as to provide the required impedance transformation, and measure the input impedance on a commercial bridge at 1 kHz. Note that this impedance is primarily *resistance* if the coils are providing transformer action.

The magnitude of the input impedance of the transformer, Equation (9.27) is shown as a function of frequency in Figure 9.19. At the lower and upper frequency limits, ω_l and ω_h, the phase angle of the impedance will be 45 degrees. For frequencies between ω_l and ω_h, the phase angle will be less than 45 degrees.

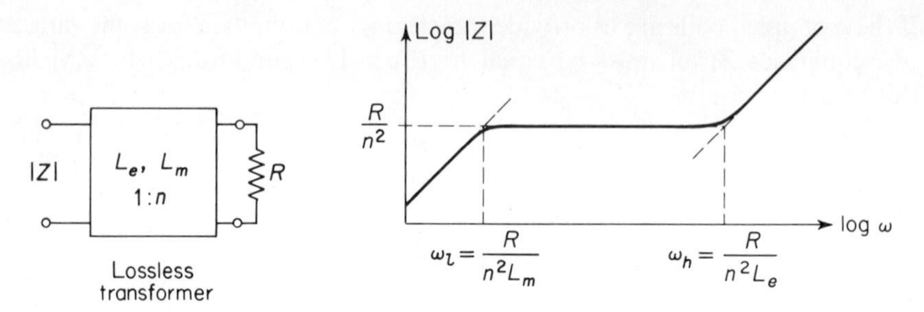

Figure 9.19. Frequency response of a lossless transformer.

Exercise 9.9

Using the circuit of Figure 9.20, measure the phase angle of the input impedance of the transformer. (See Section 4.3.4 for the method.) By finding the frequencies at which the phase angle is 45 degrees, determine ω_h and ω_l. Compare these results with the calculated values.

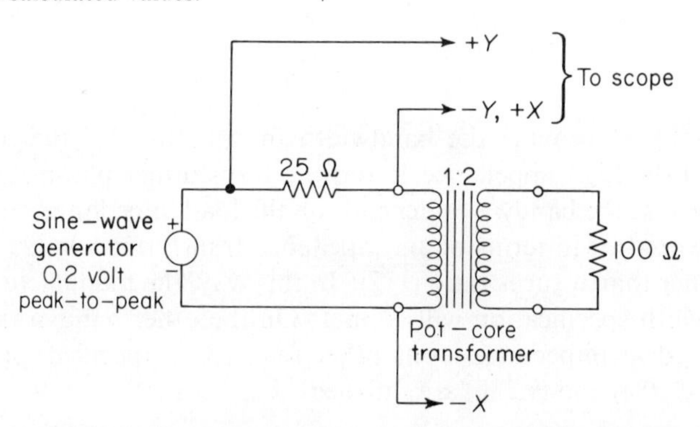

Figure 9.20. Measurement of phase angle of transformer impedance.

9.4.3 *losses in transformers*

The effects of winding and core losses on transformer performance can be included in the circuit model for a transformer by adding resistances R_{w1} and R_{w2} to represent the winding losses and R_m to represent the core loss to the model of Figure 9.18, as shown in Figure 9.21.

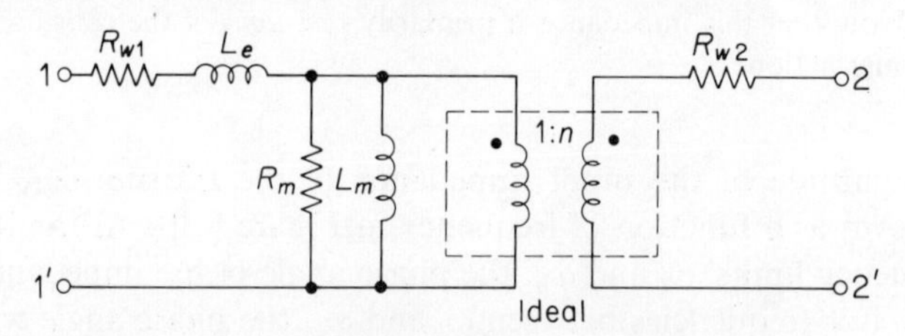

Figure 9.21. Circuit model for a lossy transformer.

The most important fact to note concerning these resistances is that they should have a small effect on the transformed resistance. Thus for a load resistance R at terminals 2-2′, then we require

$$R_{w2} \ll R \tag{9.30a}$$

$$R_m \gg \frac{R}{n^2} \tag{9.30b}$$

$$R_{w1} \ll \frac{R}{n^2} \tag{9.30c}$$

Thus, the impedance transformation ratio for a transformer not only controls the bandwidth for transformer action (9.29) but also places requirements on the transformer losses. However, by proper design, a transformer can be constructed which functions nearly as an ideal transformer over some range of frequency and for some range of load resistance. Equations (9.29) and (9.30) set these ranges and hence provide a guideline for transformer applications which can be easily checked in the laboratory.

Exercise 9.10

Using a DC ohmmeter, measure R_{w1} and R_{w2} for the transformer pot core and check the values with inequalities (9.30a) and (9.30c).

9.4.4 *transformer nonlinearity*

In Section 9.2.1 we described the saturation of the magnetic-core material at a maximum magnetic flux density B_m. This, in turn, places a limitation on the voltage that may be induced in any given winding through Equation (9.4):

$$\int v\,dt \le NAB_m \tag{9.31}$$

Equation (9.31), which is sometimes referred to as the volt-time integral, shows that if v contains a DC component, then eventually the core will saturate. For an AC waveform, the integral (9.31) is simply the area under the v vs. t curve for one half-cycle, as shown in Figure 9.22. In the case of a sinusoid of frequency f and peak amplitude V_0,

$$\int v\,dt = V_0 \int_0^{T/2} \sin 2\pi ft\,dt = \frac{V_0}{\pi f} \tag{9.32}$$

which, combined with (9.31), yields for a peak voltage to avoid saturation

$$V_0 \le \pi fNAB_m \tag{9.33}$$

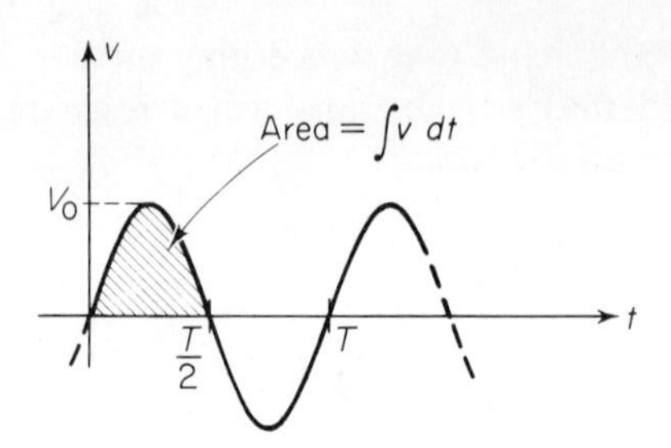

Figure 9.22. Evaluation of volt-time integral.

Exercise 9.11

Using the measured value of B_m for the pot core from Exercise 9.3, calculate the peak voltage that may be induced in the 100-mH winding without saturation. Using the circuit of Figure 9.8, observe v_2 as a function of time as the amplitude and frequency of v_1 are varied. Compare the maximum observed value of v_2 with the value calculated from Equation (9.33).

9.5 references

1. Reuben, Lee, *Electronic Transformers and Circuits*, 2nd ed., John Wiley & Sons, Inc., New York, 1955.

2. Most manufacturers of core materials provide design manuals for use with their products. Some of these are excellent references for inductor and transformer design. A partial list of manufacturers is:

   Arnold Engineering Co., Marengo, Ill.
   Ferroxcube Corp., Saugerties, N.Y.
   Magnetics, Inc., Butler, Pa.
   Magnetic Metals Co., Camden, N.J.
   U.S. Steel Co., Pittsburgh, Pa.
   Indiana General Corp., Keasbey, N.J.

10

DC power sources

10.1 introduction

This chapter introduces some of the elements used to supply DC power to electronic circuits and indicates how some of their important characteristics are defined and measured. Power sources are a universally used circuit element, since *every* circuit must include at least one power source if it is to exhibit any interesting behavior.

The two types of power sources discussed in detail in this chapter are batteries and DC power supplies operated from the AC power line. These types of supplies are the ones most frequently used in typical experimental situations.

Several important properties of solar cells are also discussed. This type of power source is used almost exclusively in space electronic systems since it can provide continuing service for many years. Solar cells are also used in rural telephone systems, and at least one manufacturer has made a portable radio powered by solar cells.

10.2 batteries

When electronics was in its infancy, virtually all power was supplied by batteries of one form or another. As alternative sources of power became available, batteries were rapidly replaced as power sources because of their relatively large size, limited life, and high cost. However, batteries are regaining favor as power sources for two reasons. First, the almost universal switch from vacuum tube to transistor circuits has resulted in a significant decrease in the amount of power that a particular circuit requires for operation. Second, newer types of batteries are now available with improved characteristics, such as higher capacity per unit volume. These two developments, singly or in combination, have led to the use of batteries in a list of applications which ranges from portable power tools to oscilloscopes.

10.2.1 types of cells

A *battery* is made up of a series-parallel combination of individual elements called *cells*. Cells are subdivided into two categories, wet and dry. A *wet cell* is formed by placing two dissimilar plates in an *electrolyte* solution. The most familiar example is the lead-acid cell, used in automobile batteries, which employs plates of lead and lead oxide immersed in a sulfuric-acid solution. Wet cells are not normally used to power portable electronic equipment and are not discussed further in this chapter.

The term *dry cell* is a misnomer, in that all cells require some type of liquid electrolyte. In dry cells, this electrolyte is combined with other substances to produce a gelatinous material. This electrolyte material is placed between two dissimilar plates such as zinc and carbon.

Cells can be further classified as primary and secondary. *Primary cells* are electrochemical power sources which cannot be recharged any significant amount. *Secondary cells* are designed to be rechargeable.

Figure 10.1 compares the internal structure of a lead-acid wet cell with that of a zinc-carbon dry cell.

10.2.2 common characteristics of cells

The unloaded or open-circuit voltage of a new cell depends only on its chemical composition. This voltage is typically 1 to 1.6 volts in dry cells, and may be somewhat higher for wet cells. However, the cell capacity, or the total amount of energy that a cell can deliver over its life, depends on the physical size of the cell.

A cell's *capacity* is normally specified as the number of ampere-hours that it can deliver before its terminal voltage is reduced below some prescribed level. Cell capacity is also strongly dependent on the way the cell is discharged. For example, a D-size zinc-carbon cell (the common flashlight cell) will deliver 30 mA for a total of 175 hours, or 5.25 ampere-hours, before its voltage drops below 1 volt if it is discharged intermittently at a rate of 30 mA for 2 hours per day. Its capacity drops to 3.45 ampere-hours if the load drawing 30 mA is applied continuously, and to 1.05 ampere-hours if a 300-mA load is applied continuously. The capacity of a cell is further influenced by factors such as temperature and the length of storage prior to use.

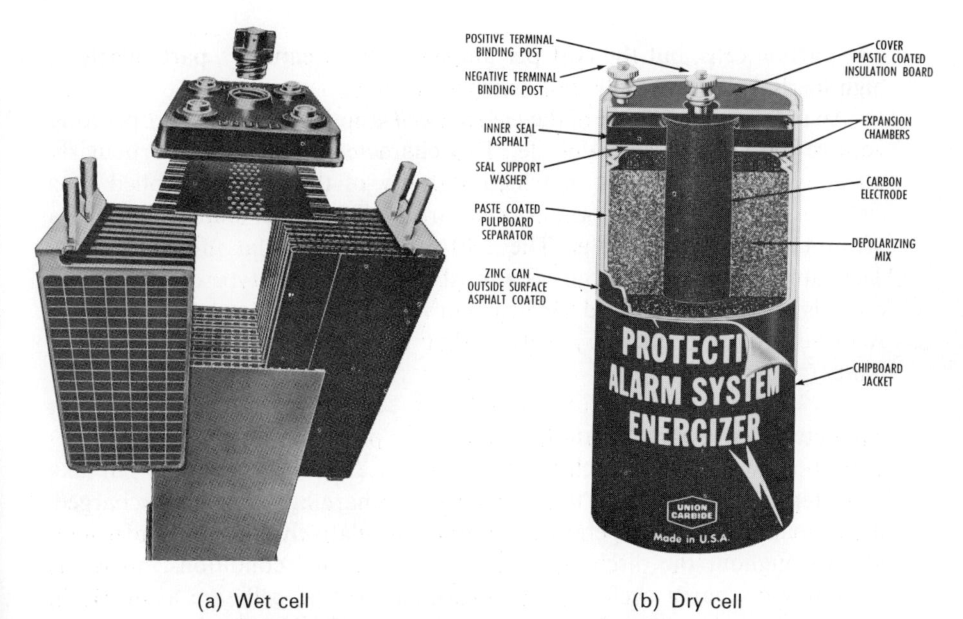

(a) Wet cell (b) Dry cell

Figure 10.1. Examples of cell construction.

A more meaningful measure of capacity in many applications may be the total energy a cell can deliver before its terminal voltage drops below a given value. Although the voltage during discharge is not constant, the delivered energy in joules can be approximated as the average of the initial and final output voltages multiplied by the ampere-seconds it delivers. Thus, the D cell can deliver approximately 2×10^4 joules under light load conditions. An automobile storage battery can deliver approximately 2×10^6 joules if discharged over a 10-hour interval.

10.2.3 characteristics of dry cells

The most widely used type of dry cell, the *zinc-carbon cell*, was developed in 1868 by Georges Leclanche. This primary cell produces an initial open-circuit voltage of 1.55 volts. This cell has the advantages of low cost and wide availability. Also, the zinc-carbon cell has recuperative powers in that its unloaded voltage will recover somewhat after its load is removed. On the other hand, the capacity of this type of cell is more dependent on operating schedule than that of most other types, and has limited temperature tolerance.

The *alkaline-manganese cell* produces an initial voltage of 1.50 volts and is available in both primary and secondary versions. The primary type is the more widely used and is discussed here. The nature of the load has considerably less effect on the ampere-hour capacity of alkaline-manganese cells. The cost of alkaline-manganese cells is somewhat greater than that of zinc-carbon cells, but the cost per ampere-hour of capacity, particularly at high loads, is lower for alkaline-manganese units.

An outstanding feature of the *mercury cell* is an almost constant operating voltage throughout its useful life. This characteristic makes it particularly attractive in applications in which stability of the voltage applied to a circuit is required. These primary cells are available with unloaded voltages of either 1.35 or 1.40 volts. The 1.40-volt version is far more common. The capacity in ampere-hours is almost independent of type of service and exceeds that of equal-sized zinc-carbon cells by at least a factor of three. A disadvantage of mercury cells is that their capacity drops severely at temperatures below 40°F.

Nickel-cadmium (or Ni-Cad) cells are secondary cells with excellent characteristics. Ni-Cad batteries provide the power for numerous rechargeable appliances. The fully charged voltage is 1.25 volts. These cells can take considerable abuse, and are unharmed by overcharging, being left discharged, heavy loads, or low temperatures. Voltage is relatively constant under load and throughout the discharge cycle. Under normal conditions, hundreds of charge-discharge cycles can be expected. Ni-Cad cells are available in many sizes and shapes. A set of D-size cells and a simple charger can be used to provide a flashlight with virtually infinite battery life. Although the initial cost of these cells is quite high, the cost per hour of use drops to insignificant levels after repeated recharging.

The discharge characteristics of the four different types of cells described above are compared in Figure 10.2. This figure illustrates the voltage change during discharge and the life to a particular final voltage for D-size cells under the indicated load conditions.

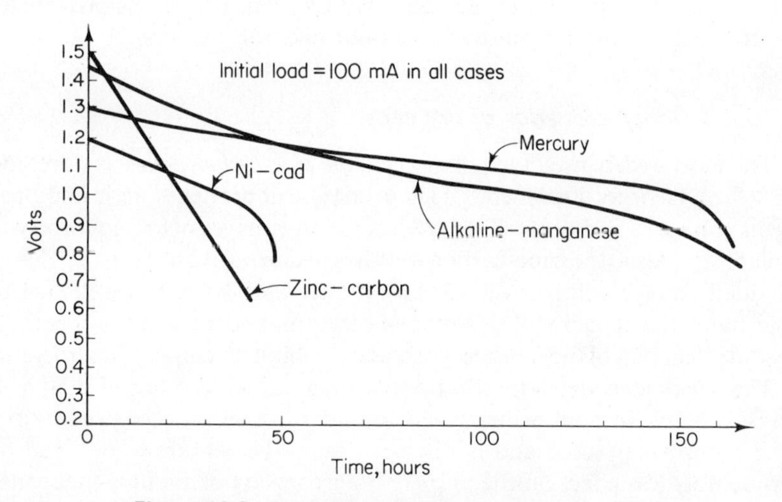

Figure 10.2. Discharge characteristics of D-size cells.

10.2.4 *internal resistance*

Measurements made on cells indicate that the decrease in terminal voltage with increase in load is approximately linearly related to load current. This implies that the cell can be represented from the terminals as shown in Figure 10.3. The voltage drop across the internal resistance results in the difference between open-circuit and loaded terminal voltage.

Note that the elements in the model of Figure 10.3 are a function of many factors such as battery age and state of discharge. However, we can measure the element values if we complete the measurement over a short time interval.

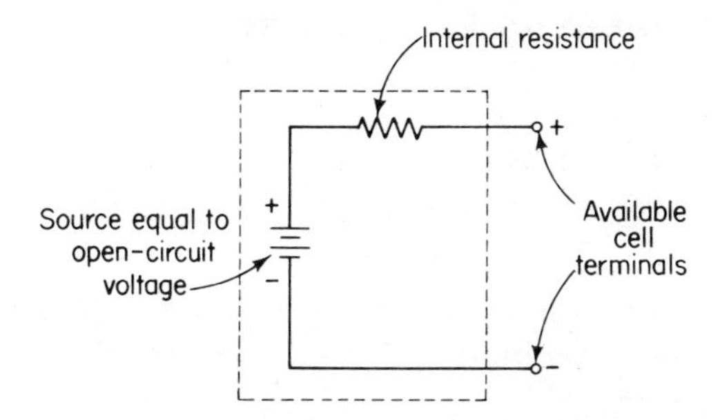

Figure 10.3. Model which accounts for changes in voltage as a function of load current.

Exercise 10.1

Connect one AA-size (penlight) cell as shown in Figure 10.4 and measure the unloaded voltage of the cell (the switch is open for the measurement) with a digital voltmeter. Close the switch, quickly measure the change in terminal voltage, and then reopen the switch. Compute the internal resistance of your cell from your measurements and the known value of the loading resistor.

If a digital voltmeter is not available, any meter which can measure small changes about a nominal value of 1.5 volts can be used, although there may be some loss of accuracy.

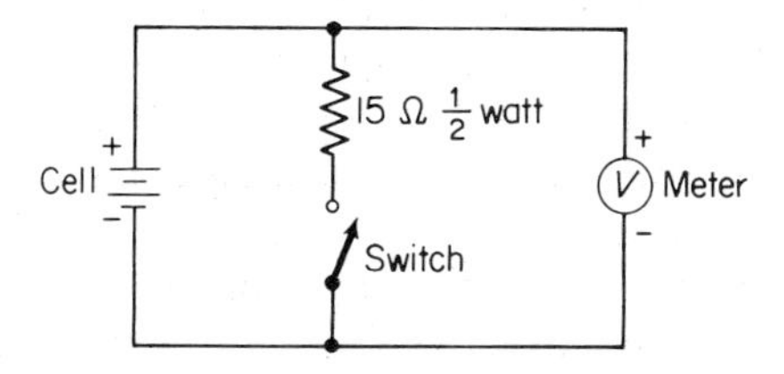

Figure 10.4. Connection for measuring internal resistance and load characteristics.

Exercise 10.2

Using the circuit shown in Figure 10.4, load the cell with the 15-ohm resistor for 1 hour, remove the load for 1 hour, and then reapply the load for 1 hour. Plot the terminal voltage as a function of time. Record points at least once every 15 minutes during the 3-hour interval.

Compute the number of ampere-hours and the total energy in joules the cell delivers during its initial discharge period until a voltage of 1.2 volts is reached. Contrast these figures with the 0.5 ampere-hour and 2.3×10^3 joules which would be obtained if the cell were discharged at a 2-mA rate for 2 hours a day.

10.3 line-operated power supplies

The power supply which converts the AC line voltage to a DC voltage compatible with electronic circuits is standard equipment in virtually every electronics laboratory. For such a supply to be useful it must perform at least three functions:

1. It must *rectify*, or convert the AC line voltage to the DC voltage normally required for circuit operation.
2. It must change voltage levels if the output voltage requirement differs from that which is conveniently obtained directly from the line.
3. It must *regulate* or provide a relatively constant output voltage in spite of changing load conditions or changing line voltage.

In addition to these basic functions, we will see that several additional features such as overload protection are frequently included to enhance the versatility of a power supply.

A block diagram of the elements common to most power supplies is shown in Figure 10.5. The input voltage from the AC power line supplies a transformer which is used to change voltage levels. The transformer turns ratio (which sets the transformer output voltage) is selected to produce the desired output voltage range. The rectifier and filter convert the AC voltage at the transformer output to a DC voltage, V_1.

The regulating portion of the supply consists of a voltage reference, a differential amplifier, and a series regulating element. This connection is an example of a *feedback control system.* In order to understand its operation, it is convenient to consider the series regulating element as a resistor whose value is controlled by the output of the differential amplifier. The differential amplifier, in turn, produces an output voltage proportional to the difference between the supply output voltage V_0 and the reference voltage V_r. Polarities are chosen so that if the reference voltage exceeds the output voltage, the resistance of the series regulating element decreases, thereby

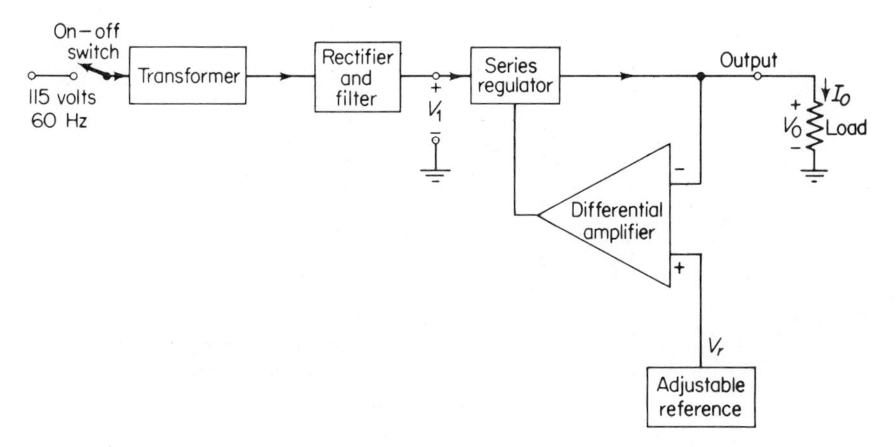

Figure 10.5. Elements of typical power supply.

increasing the output voltage. Similarly, if the output exceeds the reference, the regulating element's resistance increases. It is assumed that there is always some current flow through the regulating element, and that $V_1 > V_0$.

As a result of the feedback through the series regulator, the output voltage should track the value of the reference voltage. Thus, by making the reference voltage variable it is possible to adjust the output level.

10.3.1 safe operating areas and current limiting

Power supplies of the type described in this section are designed to be constant-voltage sources. However, there is some limitation of the maximum power that any particular supply can deliver. The reason for this limitation is evident from Figure 10.5. The voltage drop across the series regulating element is equal to $(V_1 - V_0)$. The current through this element is the load current I_L and, therefore, the power dissipated in the series regulating element is $(V_1 - V_0)I_L$. If the power dissipated in the series element exceeds its capacity, the regulating element will be destroyed. Figure 10.6 shows a curve of load current vs. output voltage which results in constant power dissipation in the regulating element. This curve indicates a maximum safe value of load current at $V_0 = 0$ of $I_L = P_d/V_1$. The allowable current increases as V_0 approaches V_1.

Some supplies incorporate electronic *current limiters.* These elements sense the load current and override the voltage regulator if necessary to keep the load current at a safe value. Usually these limiters are constructed to provide a voltage-independent limit current. The maximum current is chosen to insure protection regardless of output voltage as shown in Figure 10.6.

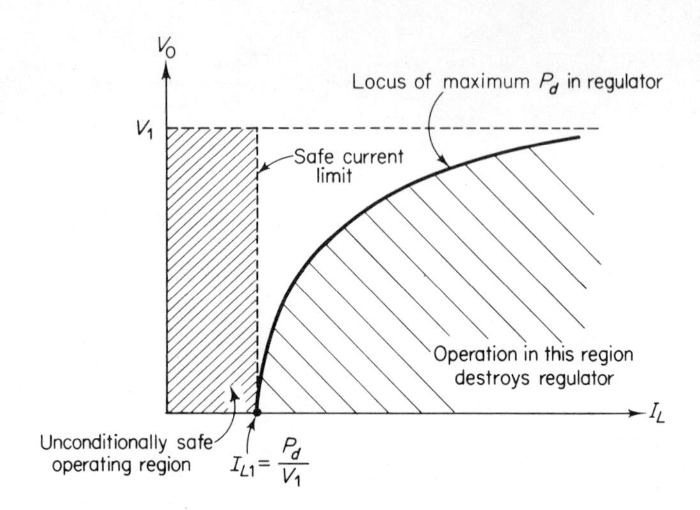

Figure 10.6. Safe region of power supply operation.

In many power supplies it is possible to adjust the current limit to values below the maximum level. This feature can be used to protect circuits which may be operated from the supply. It is also possible to use such a supply as a current source by adjusting the current limit to a desired value and setting the voltage adjustment to its maximum value. The supply will then function as a current source as long as the selected value of current multiplied by the load resistance connected to the supply is less than the maximum voltage capability of the supply.

Exercise 10.3

Obtain a supply which has an adjustable current limit. Set the current limit and the voltage adjustment to convenient values such that their product does not exceed 1 watt. (This power limitation is required to protect the measuring circuitry.) Connect the supply as shown in Figure 10.7. Measure and plot the output voltage-current (V-I) relationship by changing the load resistor R.

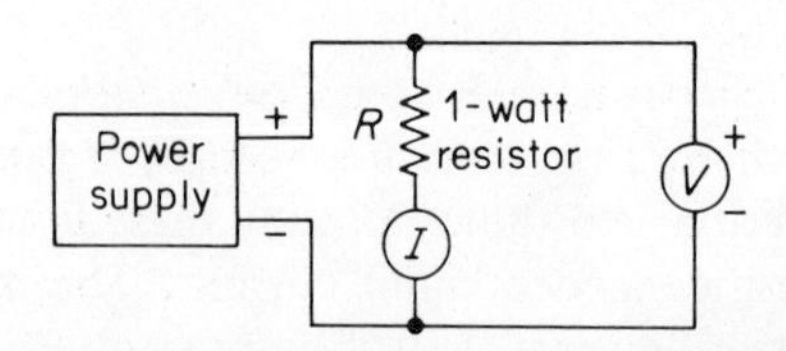

Figure 10.7. Connections for measuring V-I characteristics.

10.3.2 regulation

The degree to which a supply maintains a fixed output voltage independent of disturbances such as changes in line voltage, load current, or temperature is called *regulation.*

The regulation against changes in line voltage is usually specified as either the maximum amount the output voltage will change as the line voltage is varied over some range (for example, 105 to 125 volts rms) or as the percentage change in output voltage for a 1% change in input voltage.

Exercise 10.4

Connect a power supply as shown in Figure 10.8. The variable transformer is used to change the magnitude of AC voltage applied to the supply.

Measure the regulation of this supply as the transformer output is varied from 105 to 125 volts rms. Do this for nominal output voltages of 5 volts and 20 volts. Perform measurements at each voltage for $R = 1\ \text{k}\Omega$, $\frac{1}{2}$ watt and $R = 100$ ohms, 5 watts. Express your results both in terms of maximum change in output voltage over the input range used and as a percentage output change for a 1% change in input voltage.

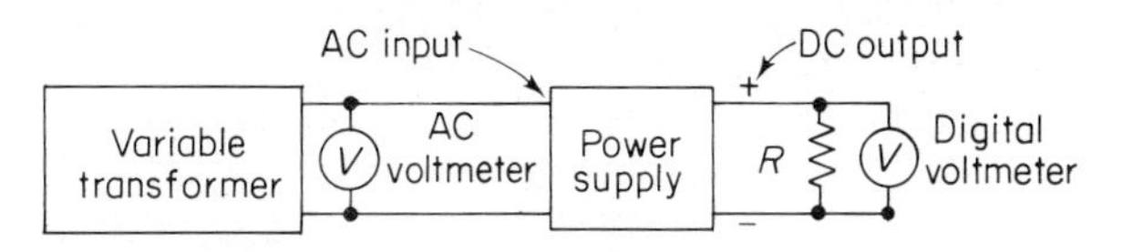

Figure 10.8. Connections for measuring line regulation.

Exercise 10.5

Measure the regulation of the supply against changes in load current. Connect the supply to the AC line and adjust the unloaded output for a nominal value of 5 volts. Load the supply with a 100-ohm, 5-watt resistor and note the change in output voltage. (Use a digital voltmeter for these measurements.) Assume that we can model the supply as an ideal voltage source in series with an output resistance. (See Figure 10.3 and the associated discussion.) Determine the internal resistance of the supply. Repeat the measurement and calculation for a nominal output voltage of 20 volts.

10.3.3 output ripple

The output of a line-operated power supply contains an AC component in addition to the desired DC output. This AC component is called *ripple.*

Exercise 10.6

Measure the peak-to-peak value of output ripple of the supply with the output voltage set to 20 volts. Use an oscilloscope with an AC input coupling for this measurement. Observe the change in ripple magnitude as the load is changed from an open circuit to a 100-ohm, 5-watt resistor. What is the frequency of the largest component of ripple?

10.3.4 additional features

The basic function of a line-operated power supply is to provide a well-regulated output voltage. Additionally, many supplies include an adjustable current limit. Other functions are included in some higher-priced designs. Two of the most common functions are described below.

Remote sensing is a technique which permits control of the voltage delivered to a load independent of the length of the leads which connect the power supply to the load. The principle is illustrated in Figure 10.9.

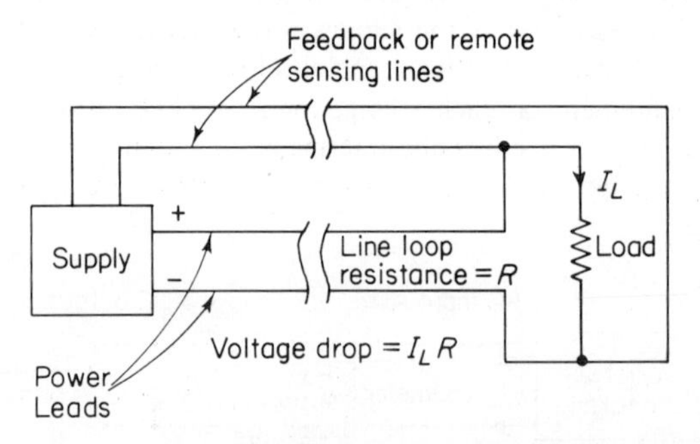

Figure 10.9. Remote sensing.

The voltage delivered to a load will not be identical with that available at the supply terminals because the load current produces a voltage drop along the leads connecting the supply to the load. Although the voltage at the supply terminals may be well regulated, it is not uncommon to have voltages on the order of 0.1 to 1 volt along the leads in high-current applications. In order to circumvent this problem, some supplies are constructed so that the feedback signal can be obtained at a pair of terminals different from the output terminals. In practice, the voltage *at the load* is measured and returned to the supply on a separate pair of wires. Since the current necessary for sensing is negligible, no *IR* drop error is introduced in these signal leads. The regulator can then be arranged to keep the voltage at the load, rather than at the supply output terminals, constant.

Programming offers a method for remote control of the output voltage of a power supply. In some cases this takes the form of an external voltage reference for the regulator. This is particularly useful when two related voltages are required. For example, it is possible to program two supplies to provide equal magnitude but opposite polarity outputs by using a single reference or control voltage.

In other cases, the supply output voltage is made proportional to the value of a resistance connected to a pair of terminals. This permits remote voltage control with a passive element.

10.4 solar cells

The solar cell is an example of an energy converter since it provides direct conversion of light energy into electrical energy. These devices can be used either as light detectors or as power sources. Examples of the former application include photographic light meters, which utilize the output of a solar cell to deflect a calibrated d'Arsonval meter, and certain types of sunfinders (used for navigation in deep space) which combine solar cells with optical equipment to determine the location of the sun. As mentioned earlier, solar cells are used as power sources in space systems, rural telephone systems, and portable radios, among other applications.

10.4.1 solar cell V-I characteristics

Two types of *V-I* characteristics have been introduced in this assignment. One is a linear terminal-current–terminal-voltage relationship which can be obtained from a circuit as shown in Figure 10.3. The other is the *V-I* characteristic of a power supply with current limiting, which can be represented by two intersecting straight lines in the *V-I* plane.

A solar cell is basically a *pn* semiconductor junction, and therefore exhibits a *V-I* characteristic related to that of diodes. Because diode *V-I* characteristics are nonlinear, they cannot be represented as straight lines.

Exercise 10.7

Illuminate a solar cell with a 150 watt reflector-spot lamp located 1 foot away from the cell. Measure the *V-I* characteristics of the cell using the connection shown in Figure 10.7, with the power supply replaced by the solar cell. Repeat your measurements with the lamp located 3 feet from the solar cell. *Caution:* Solar-cell characteristics are temperature dependent. In order to minimize cell heating and possible damage to the cell, keep the lamp off when not making measurements. Under no circumstances should the lamp remain lit more than 5 seconds at a time.

10.4.2 power output

The power available from a solar cell is a function of the load resistor connected to the cell. For example, the cell delivers maximum current into a short circuit, yet supplies no power under these conditions since its terminal voltage is zero. Similarly, open-circuit operation results in maximum voltage but again no power is supplied. The power delivered by the solar cell can be maximized by proper choice of load resistor. Determine the optimum value of load resistor at both levels of illumination from the data obtained from your solar cell. Calculate power output with the optimum load.

10.4.3 efficiency

The efficiency of the solar cell, defined as the ratio of its electrical power output to the incident light power, is not particularly high. The purpose of Exercise 10.8 is to make a rough approximation of the efficiency of your solar cell.

Exercise 10.8

Determine (from information printed on the bulb) the wattage of the reflector-spot lamp you used to measure *V-I* characteristics. Assume that 10% of the electrical power supplied to the lamp is converted into light power, which is typical for a tungsten lamp. Measure the diameter of the illuminated area when the lamp is held 1 and 3 feet from a surface. (It is recognized that there is some uncertainty about this figure because there is no sharp dividing line between illuminated and nonilluminated areas.) Compute the density of light power striking the surface in watts per unit area. Then calculate the amount of power which strikes the solar cell by multiplying its area by the power density. Estimate the efficiency of your solar cell from the light power it intercepts and its output power when connected to an optimum-value load resistor. Typical efficiencies of commercial solar cells are on the order of 6–8%.

11

advanced oscilloscopes

11.1 introduction

This chapter will explain the operation of more complex oscilloscopes. Although the refinements and additional features found on these instruments extend their usefulness considerably, their basic mode of operation is essentially the same as in the elementary scope. For this reason, Chapter 4 should be reviewed before proceeding with this chapter.

In addition to the advanced oscilloscopes, the oscilloscope camera will be introduced as a tool for recording scope displays.

11.2 CRT controls

The controls for adjusting the spot on the CRT face are essentially the same as on an elementary oscilloscope. Usually in addition to the intensity and focus controls, an astigmatism adjustment is provided to improve the overall sharpness of the trace.

On some scopes, beam-position indicator lamps are located above the CRT face. These show the direction in which the electron beam has been deflected when it is not on the screen and thus aid in adjusting the horizontal and vertical position controls. On other types, a trace-finder control is provided to locate an off-screen display. With the trace finder activated, the horizontal and vertical position controls are used to center the trace on the screen. Then, with the trace finder turned off the display will remain on the screen.

11.3 vertical amplifiers

In most simple oscilloscopes the response of the vertical amplifier is from DC to the neighborhood of 1 MHz. Since this is inadequate for many applications, vertical amplifiers have been developed which extend the upper-frequency range to approximately 150 MHz. Beyond 150 MHz, sampling oscilloscopes can display periodic waveforms up to 12 GHz. Sampling oscilloscopes will be discussed in Chapter 12; here we will be concerned with the more conventional, direct-writing oscilloscopes.

Since the largest application of oscilloscopes is to display electrical signals as a function of time, the advanced oscilloscopes usually feature high performance of the vertical amplifier and flexible time-base operation. Because of the wide variety of practical situations, many high-performance scopes have interchangeable or "plug-in" vertical amplifier sections. In this way numerous vertical amplifiers with different performance characteristics are available, and selection is possible according to the particular situation at hand. Plug-in versatility adds to the cost of an oscilloscope and therefore some high-performance scopes are available without plug-in capability. Examples of both types of high-performance oscilloscopes are shown in Figure 11.1.

Figure 11.2 shows some of the wide variety of plug-in units available for oscilloscopes. They include units that provide one, two, or four simultaneous traces; sensitivities as high as 10 μV/cm with variable bandwidth for noise rejection; differential input capability; high-frequency performance to 50 MHz; and conversion to sampling oscilloscope operation.

11.3.1 high-performance, dual-trace amplifier

For most general laboratory work, a vertical amplifier with high gain (5 mV/cm or more), high-frequency response (50 to 150 MHz), and dual-trace capability will serve very well. As well as being available as a plug-in unit, amplifiers with these general performance specifications are available in non-plug-in scopes. For the remainder of this chapter we will be concerned specifically with this type of vertical amplifier.

The dual-trace vertical amplifier consists of two identical amplifier channels. Each channel is normally provided with a single-ended input, input selector switch, variable and calibrated gain controls, balance adjustment, and vertical position control. The function of each of these controls is fully discussed in Chapter 4.

The *mode* of operation of the vertical amplifier is controlled by a selector switch. In single-trace operation either channel may be displayed independently, or the outputs of both channels can be combined to form the algebraic sum or difference according to the setting of the polarity or trace inversion

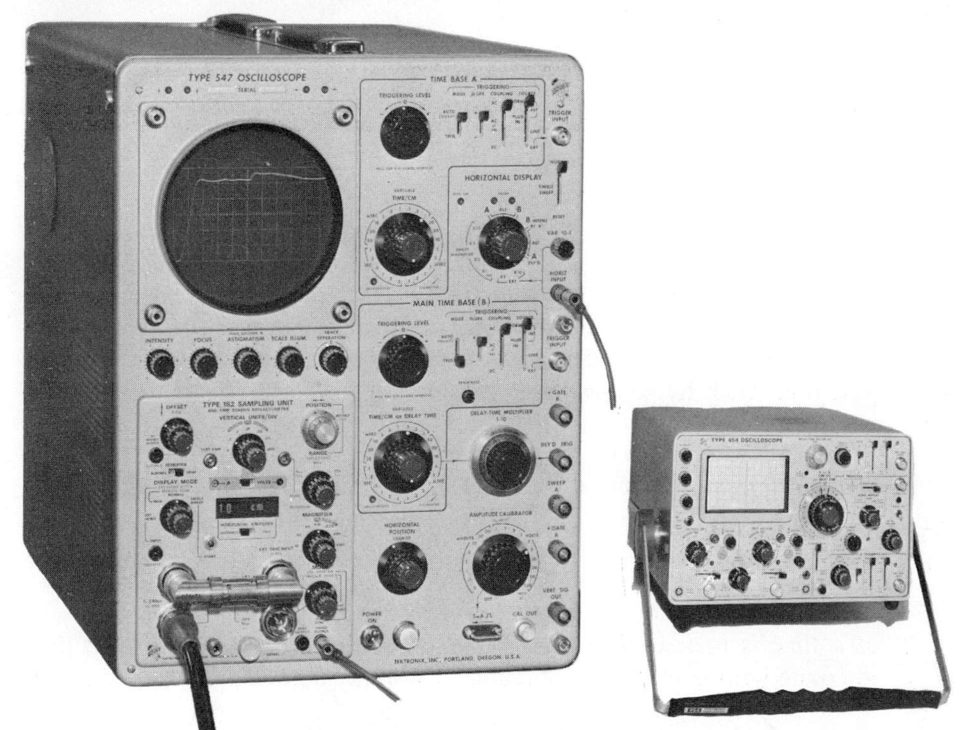

Figure 11.1. High-performance oscilloscopes.

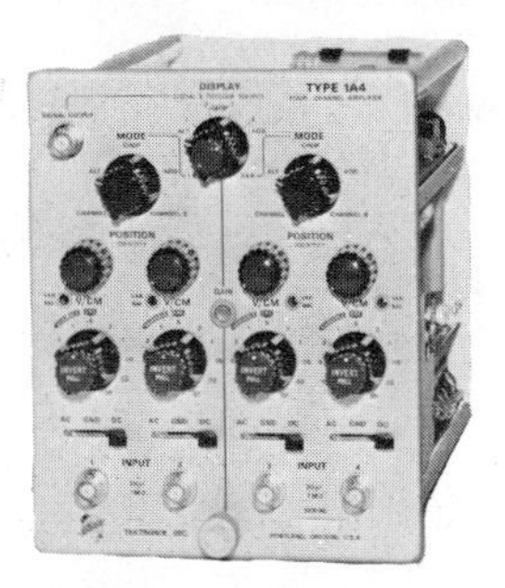

Four-channel

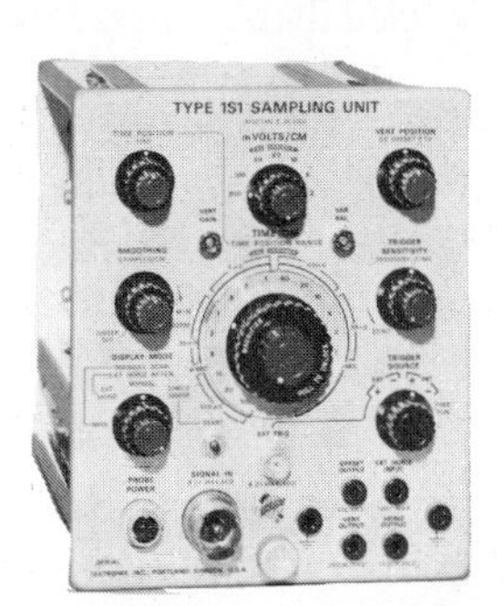

High-frequency sampling

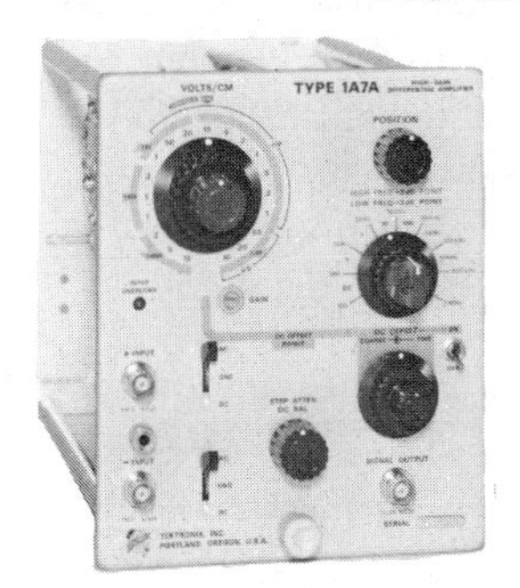

Variable bandwidth differential

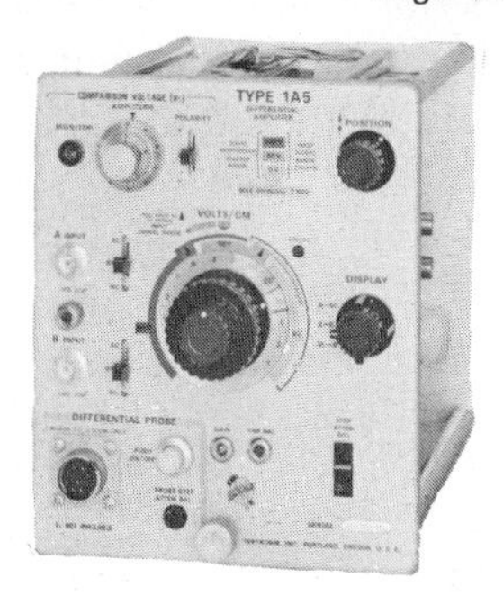

High-gain differential

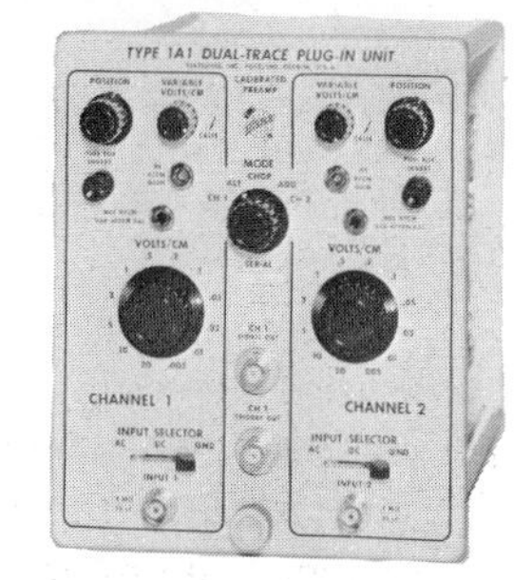

Wide-band dual-trace

Figure 11.2. Examples of vertical plug-in amplifiers.

switches in one or both channels. This feature permits differential-input operation.

Dual-trace operation is provided by electronically switching between the outputs of each channel and applying the resulting signal to a single set of CRT deflection plates. This should not be confused with a dual-beam oscilloscope which has a CRT with two independent electron guns, each connected to an independent vertical amplifier channel. The *mode* selector switch provides for two methods of switching the vertical-channel outputs, alternate and chopped.

In the *alternate mode*, the deflection plates are connected to a single channel for one complete sweep cycle, and the connection is alternated between channel 1 and channel 2. Thus, every other sweep displays channel 1, the channel-2 display being interlaced on the remaining sweeps. The image-retention characteristics of the CRT phosphor screen and the human eye result in an apparent dual-trace image, although in reality only one trace at a time is present. However, if the sweep time is sufficiently long, the alternate writing becomes noticeable, first as an annoying flicker and then, at the slowest times, as clearly alternate displays.

The problems of alternate sweep at long sweep times can be overcome by using a *chopped* display presentation. In the chopped mode the signal to the deflection plates is electronically switched at approximately 1 MHz during every sweep. At slow sweep speeds this switching occurs so rapidly that the "gaps" in each waveform are imperceptible. However, at fast sweep times, the waveform chopping is clearly evident. As a rule of thumb, a sweep time of 1 ms/cm is the dividing line for choosing between alternate or chopped dual-trace displays. Figure 11.3 summarizes the differences between the two display modes.

11.3.2 frequency response

The high frequency performance of a vertical amplifier is usually specified as the frequency at which the indicated voltage drops to 0.707 of the true voltage applied at the input. This deviation of the indicated voltage from the true value actually begins at a frequency of about half the specified cutoff frequency. Conversely, by correcting for the error, the amplifier may be used beyond the cutoff frequency.

The cutoff frequency of a vertical amplifier will generally depend on the sensitivity setting at the higher gains (typically 20 mV/cm and higher), decreasing with increasing gain. With plug-in units the frequency response will also depend on the main oscilloscope. Figure 11.4 illustrates the frequency response of a vertical amplifier and its dependence on gain.

An alternative measure of amplifier performance is the rise time of the display in response to an ideal voltage step applied to the input. The definition

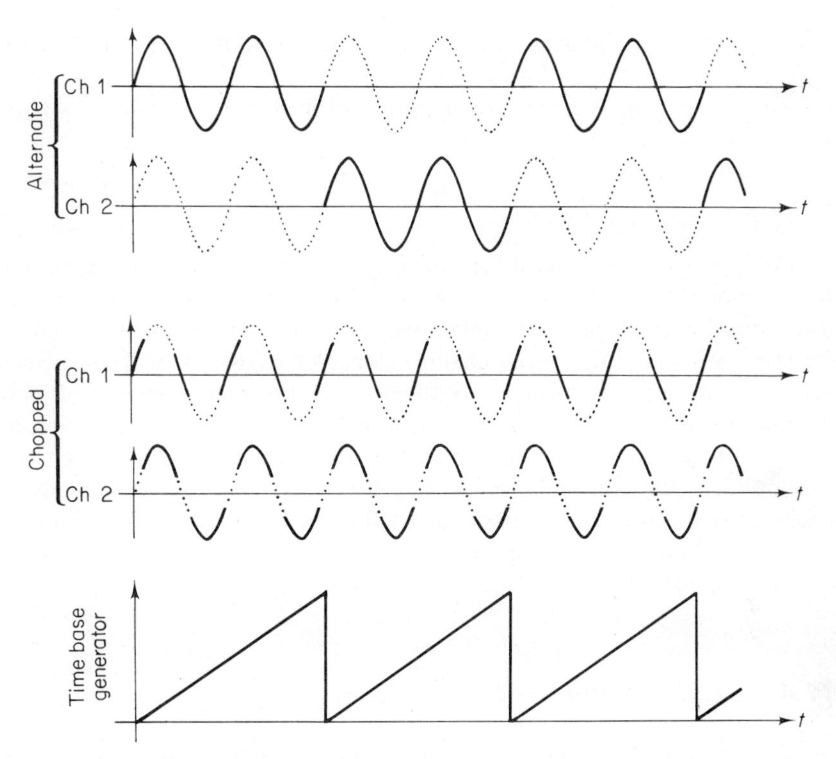

Figure 11.3. Real-time illustrations of alternate and chopped dual-trace displays.

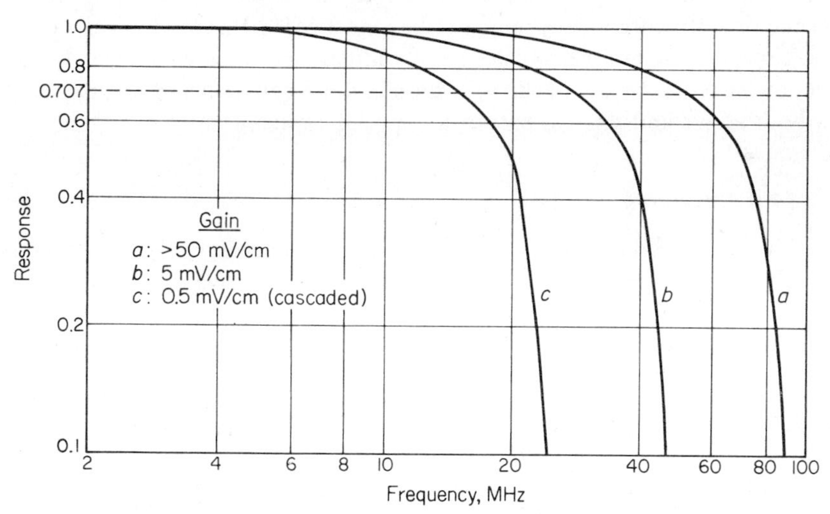

Figure 11.4. Dependence of frequency response on sensitivity for a Tektronix 1A1 plug-in mounted in a 547 main frame (logarithmic scales).

of rise time and its interpretation is discussed more fully in Section 14.3.1. Typically, the rise time of an amplifier is approximately $1/3f_c$, where f_c is the cutoff frequency. Thus, the rise time of a 50-MHz amplifier will be about 7 ns.

11.3.3 cascaded operation

Most dual-channel amplifiers have provision for cascaded operation. In this mode, one channel is used as a preamplifier for the other, resulting in a single-trace display with a sensitivity about ten times greater than could be obtained with a single channel alone. One penalty for the increased gain is a further reduction in bandwidth, as shown in Figure 11.4. The method of cascading may be by means of an external cable connection or by a switch. In the case of the former, the oscilloscope can also be used as a preamplifier for a voltmeter or other signal-processing device. Usually in cascaded operation there are specified settings for the individual sensitivity controls and limitations on input levels. The instruction manual should be consulted before this type of operation is attempted.

11.4 oscilloscope probes

In Chapter 4, the use of a 1 × scope probe was introduced as a method of reducing noise pickup. Also, in that chapter the effects of loading by the probe and vertical amplifier on the circuit under test were discussed. As the operating frequency increases beyond 1 MHz, this loading, especially the capacitive component, can severely affect circuit performance. In this section we will consider probe operation in more detail, including discussion of both active and passive probes for special applications.

11.4.1 voltage probes

The simplest method of reducing the circuit loading would be to insert a high value of series resistance in the probe. This arrangement is shown in Figure 11.5a.

The parallel combination R_2C_2 represents the effective input impedance of the vertical amplifier. The input impedance at the probe tip is

$$Z_1(j\omega) = R_1 + \frac{R_2}{1 + j\omega R_2 C_2} \tag{11.1}$$

Thus the input impedance is never less than R_1, which may be made large, by perhaps ten times, compared with R_2. If, however, we calculate the

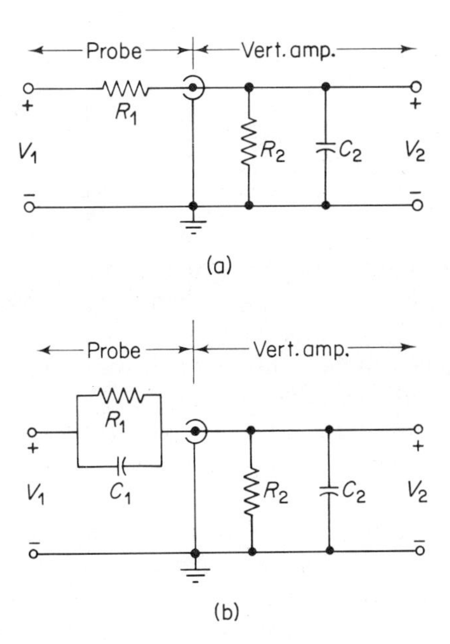

Figure 11.5. Voltage probe compensation.

voltage transfer ratio

$$\frac{V_2(j\omega)}{V_1(j\omega)} = \underbrace{\left[\frac{R_2}{R_1 + R_2}\right]}_{\text{attenuation}} \underbrace{\left[\frac{1}{1 + j\omega[R_1R_2/(R_1 + R_2)]C_2}\right]}_{\text{frequency dependence}} \tag{11.2}$$

we find there is an attenuation factor which is always less than unity multiplied by a frequency-dependent term which approaches unity at low frequencies. Although the attenuation term can be corrected by increasing the gain of the amplifier, the frequency-dependent term will cause considerable distortion of high-frequency waveforms.

The arrangement of Figure 11.5b provides a method of compensating for the frequency-dependent term in the voltage transfer ratio. The voltage transfer ratio for the circuit of Figure 11.5b is

$$\frac{V_2(j\omega)}{V_1(j\omega)} = \frac{R_2}{R_2 + R_1[(1 + j\omega R_2C_2)/(1 + j\omega R_1C_1)]} \tag{11.3}$$

If $R_1C_1 = R_2C_2$, then the brackets reduce to unity *at all frequencies* and the resulting frequency-*independent* voltage transfer becomes

$$\frac{V_2}{V_1} = \frac{R_2}{R_1 + R_2} \tag{11.4}$$

Under the condition $R_1C_1 = R_2C_2$, the input impedance in Figure 11.5b is

$$Z_1(j\omega) = \underbrace{\left[\frac{R_1 + R_2}{R_2}\right]}_{\text{multiplying factor}} \underbrace{\left[\frac{R_2}{1 + j\omega R_2 C_2}\right]}_{\text{original impedance}} \tag{11.5}$$

which is the original amplifier input impedance with no compensation (R_2 and C_2 alone) multiplied by a factor greater than unity. As an example, suppose $R_1 = 9R_2$. Then the input impedance will be increased by a factor of 10. On the other hand, the voltage transfer ratio becomes 0.1 or 10 × attenuation. Thus by sacrificing some sensitivity we can reduce the loading on the circuit under test. In this example the loading becomes $10R_2$ in parallel with $0.1C_2$ to ground, compared with R_2 in parallel with C_2 to ground without the probe compensation.

The construction of oscilloscope probes varies with manufacturer. The adjustment of C_1 for proper compensation is usually by a recessed screw on the probe barrel or at the input connector. Some probes provide this adjustment by rotation of a sleeve on the probe barrel. Most probes used in high-frequency applications are 10 × and carry that designation on the probe barrel. For situations in which extremely low capacitance is paramount, a 100 × probe is sometimes used. However, owing to stray capacitance, these probes do not provide a factor of 100 reduction in loading.

Just as vertical amplifiers have frequency limitations, so do oscilloscope probes. In choosing a probe for a particular application, care must be taken to insure that the probe is designed to operate with the particular vertical amplifier in use. Generally speaking, a probe designed for an amplifier with a given high frequency response will work satisfactorily with amplifiers with a lower cutoff frequency. The safest course in critical applications, however, is to obtain the proper probe for use with the vertical amplifier as specified by the manufacturer.

Exercise 11.1

Connect a 10 × probe to a simple oscilloscope and use it to display the square-wave calibration signal available on the front panel. Adjust the compensation to produce a square-wave display as shown in Figure 11.6.

11.4.2 current probes

One common method of measuring a current waveform is to measure the voltage across a resistor. If the resistor has neither node connected to ground, a scope with a differential input or a dual-channel vertical amplifier in the subtract mode can be used. The latter uses both channels and consequently eliminates the dual-trace feature. If the current does not pass

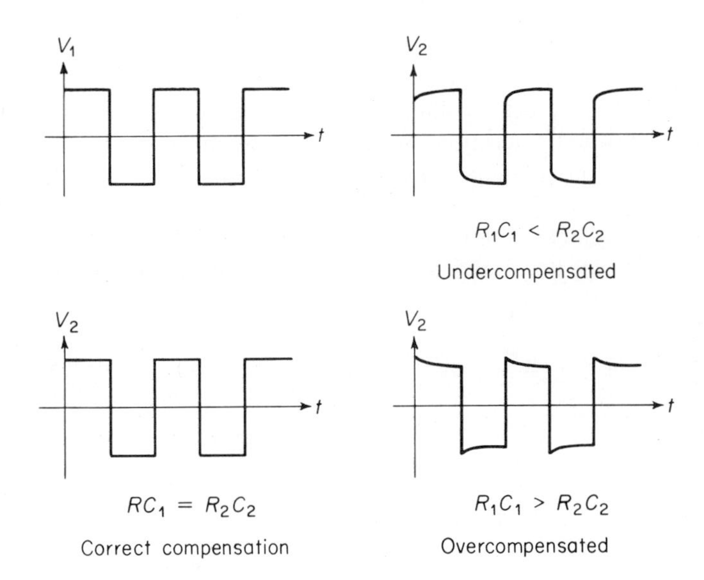

Figure 11.6. Waveforms for probe compensation adjustment.

through a resistive element, a small sampling resistor can be inserted in the current loop to produce the required voltage drop. However, this sampling resistor can adversely affect circuit performance.

An alternative method of measuring current is to use a current probe. A current probe, shown in Figure 11.7, is designed to be clipped around a current-carrying conductor and to produce a CRT display proportional to the current flowing in the conductor. There are two basic methods by which this may be accomplished: by transformer coupling and by the use of the Hall effect.*

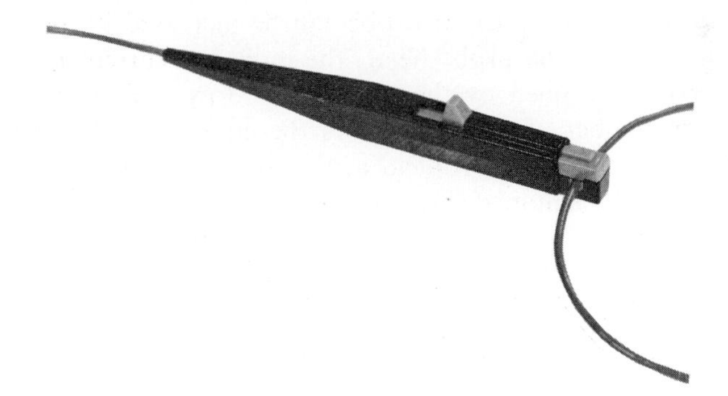

Figure 11.7. Current probe.

* See for example, *Introduction to Semiconductor Physics* by Adler, Smith, and Longini, pp. 55 and 205, John Wiley and Sons, New York, 1964.

The transformer system is illustrated in Figure 11.8. The conductor carrying the current to be measured $i(t)$ forms a one-turn primary of a transformer. The induced voltage in the secondary is then applied to the oscilloscope through a termination. However, because a transformer is used to produce the voltage signal, this method can be used only for AC currents. Typically, such a probe can operate from a few hundred hertz to 50 MHz. By using an auxiliary amplifier the low frequency response of the transformer current probe can be extended to about 10 Hz; its sensitivity is increased as well.

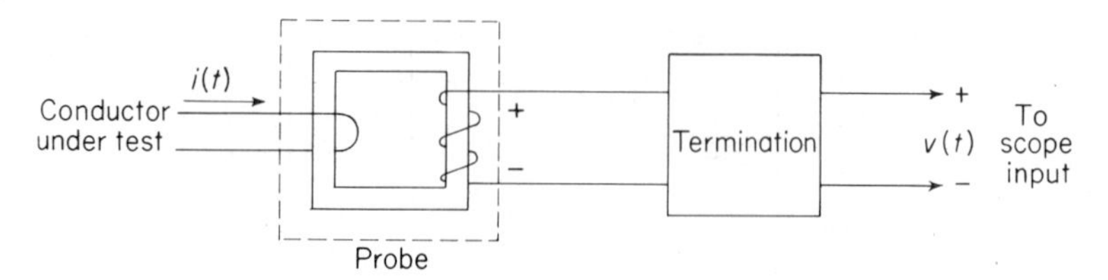

Figure 11.8. Transformer-type current probe.

Because the Hall effect produces a voltage proportional to an applied magnetic field, the field surrounding a current-carrying conductor can be used as an indication of the current. Since the probe output does not depend on induction of a voltage, these probes have a response to DC. However, they do require auxiliary equipment to supply a bias current. A typical Hall-effect current probe has a frequency response from DC to 50 MHz and a sensitivity of 1 mA/division on the CRT screen.

Although a current probe does not make a direct electrical contact to the circuit under test, the magnetic-field coupling does result in circuit loading. In most applications, however, this is negligible.

The sensitivity of a current probe can be increased by placing two or more turns through the probe head; the indicated current then being the actual current multiplied by the number of turns through the probe. Increasing the number of turns increases the circuit loading, and the loop inductance may also be a problem at the higher frequencies. The current probe may also be employed in a differential mode by inserting two wires carrying different currents through the probe simultaneously. The indicated current will then be either the sum or difference of the individual currents, depending on the current directions in the two conductors.

Exercise 11.2

In the circuit of Figure 11.9, measure the current $i(t)$ both with a current probe and by measuring the voltage $v(t)$. Compare the results over the frequency range of 20 Hz to 100 kHz.

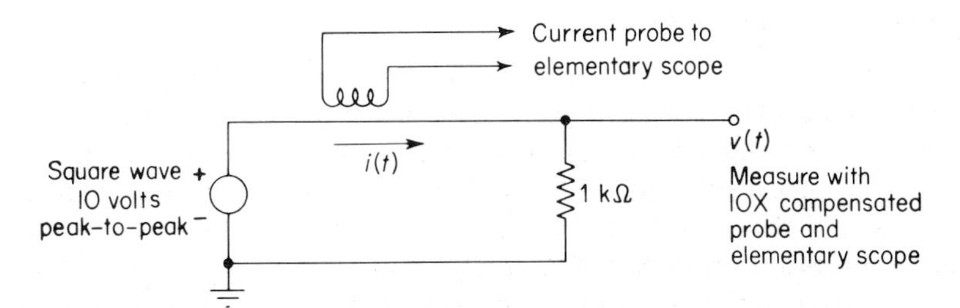

Figure 11.9. Comparison of current probe and current-sampling resistor.

11.4.3 active probes and preamplifiers

In some small-signal applications the attenuation of a passive probe reduces the amplitude to a level that is difficult to measure. On the other hand, a 1× probe will often result in excessive loading on the circuit under test. This problem can be solved by using a preamplifier or an active probe. A *preamplifier*, as its name suggests, is an auxiliary signal amplifier which may be inserted between the probe and the input to the scope. Typical units have a gain of 50 to 100, resulting in an overall gain of 5 to 10 when used with a 10× passive probe. In high frequency applications, the bandwidth or rise time of the preamplifier must be considered when interpreting data.

An *active probe* is one which has the preamplifier built into the probe itself. One unit, using a field effect transistor, results in a 1× probe with a DC to 230-MHz bandwidth and 10-MΩ input resistance. Such a probe is capable of critical measurements with little signal degradation when used with any presently available, direct-writing oscilloscope.

11.5 time base

Chapter 4 showed how the time-base generator and trigger circuits provide a stable display of signal waveform as a function of time. In the high-performance oscilloscope two separate time-base generators and trigger controls are provided. By using these separate time-base generators in various combinations, the flexibility of waveform display is greatly increased.

11.5.1 single time base

In scopes which have more than one time-base generator, the horizontal display switch will permit selection of one time-base generator and will disable the second. In this mode of operation the function of the selected time-base generator is identical with the time-base generator in the elementary oscilloscope.

11.5.2 trigger controls

As in the elementary oscilloscope, the function of the trigger controls is to synchronize the horizontal axis sawtooth waveform with the signal waveform. Selector switches for the source, coupling, and slope, plus a continuous-level adjustment, perform in the same fashion they do in the elementary oscilloscope. However, in the high-performance oscilloscope there are some additional features which are frequently useful.

In addition to the AC and DC coupling, the advanced scope also provides two additional coupling modes: low-frequency rejection (LF-REJ) and high-frequency rejection (HF-REJ). In the LF-REJ mode, the trigger is AC coupled, the attenuation of the trigger signal beginning in the range of 1 to 10 kHz and increasing with decreasing frequency. This mode is useful when observing a high-frequency signal with a significant low-frequency component, such as 60-Hz noise. Use of the LF-REJ coupling greatly simplifies triggering on the high-frequency signal.

In the HF-REJ mode, the trigger signal is AC coupled and also attenuates high-frequency signals above 100 kHz. Thus if a low-frequency signal with a significant high-frequency component is to be observed, the HF-REJ mode simplifies triggering on the low-frequency component. Figure 11.10 illustrates the bandwidth limitations of these various coupling modes.

The level and slope controls select the particular point on the triggering waveform at which a sawtooth waveform begins. In the automatic mode, a baseline sweep is displayed in the absence of a triggering signal. However, unlike the elementary scope in the automatic mode, the *level is not automatically selected*, and proper adjustment of the level control must be made. In the normal trigger mode, no baseline sweep is displayed when the triggering signal is removed.

Provision is also made on the advanced scope for a single-sweep display. This is extremely useful for observing nonrepetitive phenomena, such as contact bounce in a relay. In the single-sweep mode, the sweep is triggered by a signal according to the settings of the trigger controls, but does not recur until the sweep has been reset by a control on the scope panel. Convenient observation of the single sweep can be made on CRT which has a long-persistance screen, by making a photograph of the single trace, or by using a storage scope.

As in the elementary scope, the trigger signal source may be selected from the internal signal of the vertical amplifier, an external signal, or the 60-Hz line frequency. On many advanced scopes, provision is made for a factor of ten reduction in the external trigger signal amplitude by means of a switch. This extends the range of useful triggering signals that may be applied to the external trigger input.

In selecting internal triggering with a dual-trace vertical amplifier, some additional factors must be considered. In Section 11.3.1 the methods of

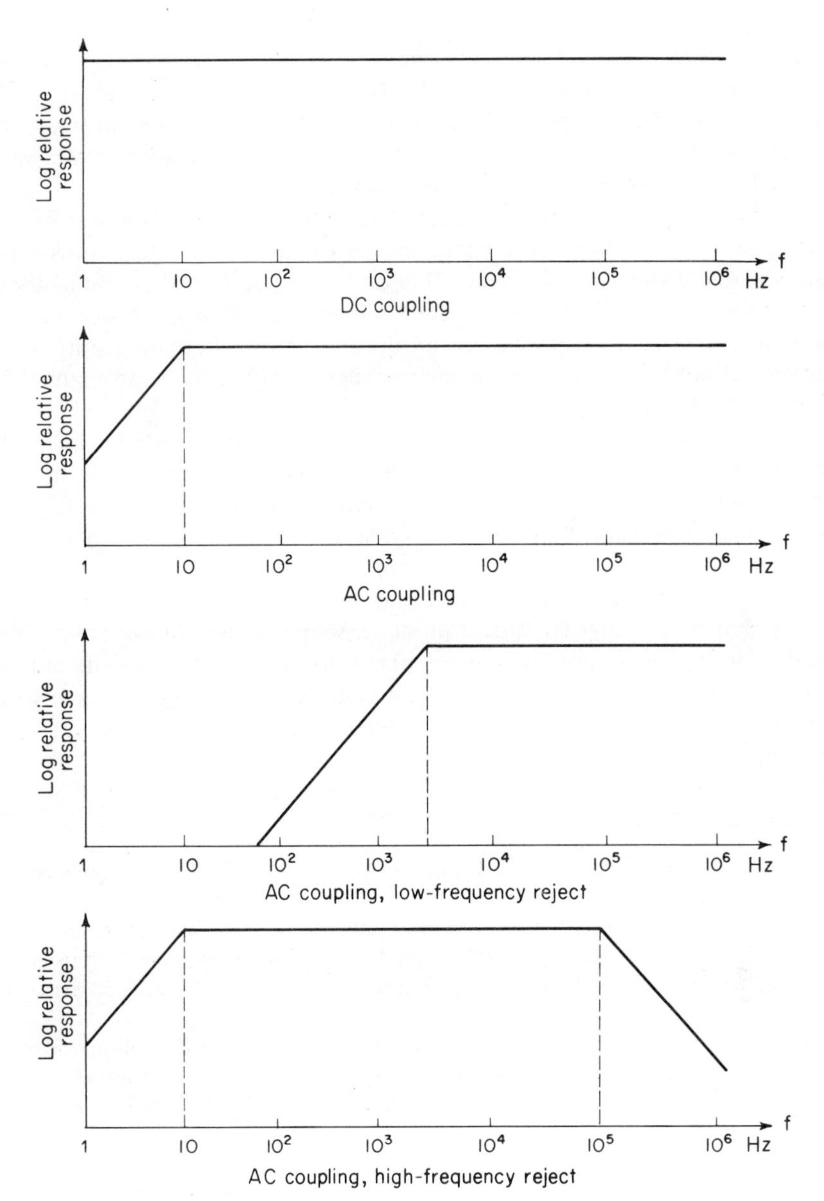

Figure 11.10. Bandwidth limitations of various trigger coupling modes.

producing a dual-trace display were described. Since the signal at the output of the vertical amplifier is the source for internal triggering, there are two possible situations. In the alternate mode of operation, the normal internal trigger signal will be derived alternately from channel-1 and channel-2 signals. Although in some cases this operation will result in a stable display, it requires that both channel-1 and channel-2 signals have a waveform that will simultaneously satisfy the settings of the trigger coupling, slope, and level controls. If this is not the case, a stable display cannot be obtained on both channels. Furthermore, since each channel triggers its own sweep, any time reference between the two signals is lost.

In the chopped mode of dual-trace display, the use of the normal internal trigger source will cause false triggering on the chopping signal transients and a stable display will be impossible to obtain. Both of these difficulties can be overcome by the use of an external trigger signal in dual-trace operation. In this way, the trigger does not depend on the alternating output of the vertical amplifier and the time reference between channel-1 and channel-2 signals is also preserved.

On most dual-trace scopes, however, provision is also made for internal triggering on channel 1 only. This control may be located either in the vertical amplifier section or the trigger section. In this mode, the trigger signal is derived from the channel-1 signal and operates the sweep similar to an external trigger signal. Also, on some dual-trace plug-in amplifiers, a trigger signal derived from channel 1 is available at the front panel for connection to the external trigger input on scopes which do not have provision for channel-1-only triggering. Thus for most operation in either dual-trace mode, use of channel 1 only or an external trigger source will provide the most useful display. In single-channel operation, usually the normal trigger mode is employed.

Exercise 11.3

Connect a high-performance oscilloscope with a dual-trace vertical amplifier to the circuit of Figure 11.11.

Use the horizontal display control to obtain single time-base operation with normal internal triggering. Obtain a stable display in the channel 1, channel 2, add, and subtract modes of operation. The trigger level may require adjustment in each case.

Reset the amplifier to channel 1, the trigger source to channel 1 only and adjust the trigger controls for a stable display. Now set the amplifier to the alternate mode and obtain a dual trace. Observe the effect of returning the trigger source to

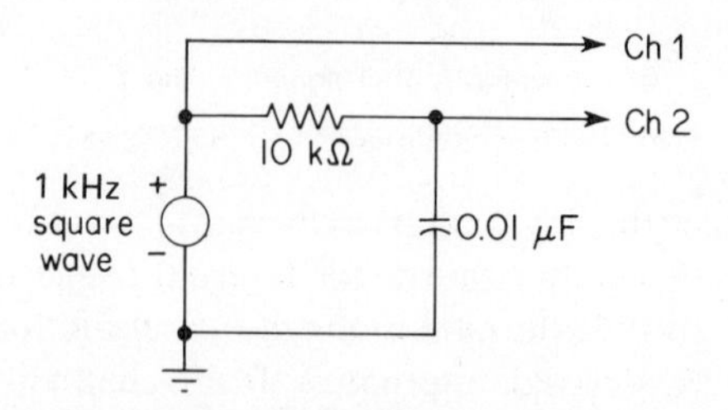

Figure 11.11. Circuit for dual-trace display.

normal internal triggering, and attempt to obtain a stable display. Adjustment of the vertical position controls to cause the traces to overlap will be helpful.

With a stable display in the alternate mode and normal internal triggering, change the amplifier to the chopped mode. Try to obtain a stable display. Reset the trigger to channel 1 only and obtain a stable display.

With the amplifier in the chopped mode, increase the square-wave frequency to about 100 kHz and obtain a stable display. Slowly vary the frequency of the square wave until the chopped trace becomes visible. What is the approximate chopping frequency?

11.5.3 delayed sweep

The delayed sweep feature of a high-performance scope serves a similar function as the horizontal magnifier on the elementary scope. However, the delayed sweep technique permits increased accuracy of time measurements and more versatile magnification. This is achieved through the use of a second time-base generator and trigger section. For the purpose of this discussion we will refer to the main time base as A and the secondary or delayed time base as B. However, this nomenclature is not universal and A and B have the opposite meaning on some scopes. We have already explained the use of the A time base alone. Next we shall discuss the use of the A and B time bases together to obtain a magnified display.

Figure 11.12 is a simplified block diagram which illustrates how a delayed sweep is produced. The waveform diagram located at the right of each block represents the output waveform of that block, $t = 0$ being chosen as the time of the A trigger pulse. With the horizontal display switch in the A sweep position, single time-base operation is obtained as described in Section 11.5.1. However, the output of the A trigger is also applied to a delay circuit which allows an adjustable time delay of the A trigger pulse. This adjustable delay is controlled by a ten-turn dial on the scope panel and the A sweep time. The amount of the time delay is found by multiplying the A sweep time by the setting of the delay-time dial. Thus, if the A sweep is set at 50 μs/cm and the delay-time dial reads 450, the actual delay time will be $50 \times 4.50 = 225$ μs. This calculation can be easily remembered if it is kept in mind that the maximum delay is just the total length of the A sweep (500 μs in the above example).

With the B triggering mode set to start the B sawtooth after the delay time (in some scopes this is automatic when the delayed sweep is selected by the horizontal display switch), the delayed trigger pulse is applied to the B time base and will produce the B sawtooth starting at the end of the delay time and increasing at a rate controlled by the B sweep time setting, which should be shorter than the A sweep time.

If the horizontal display is now set at the A intensified position, the horizontal deflection remains controlled by the A time base. However,

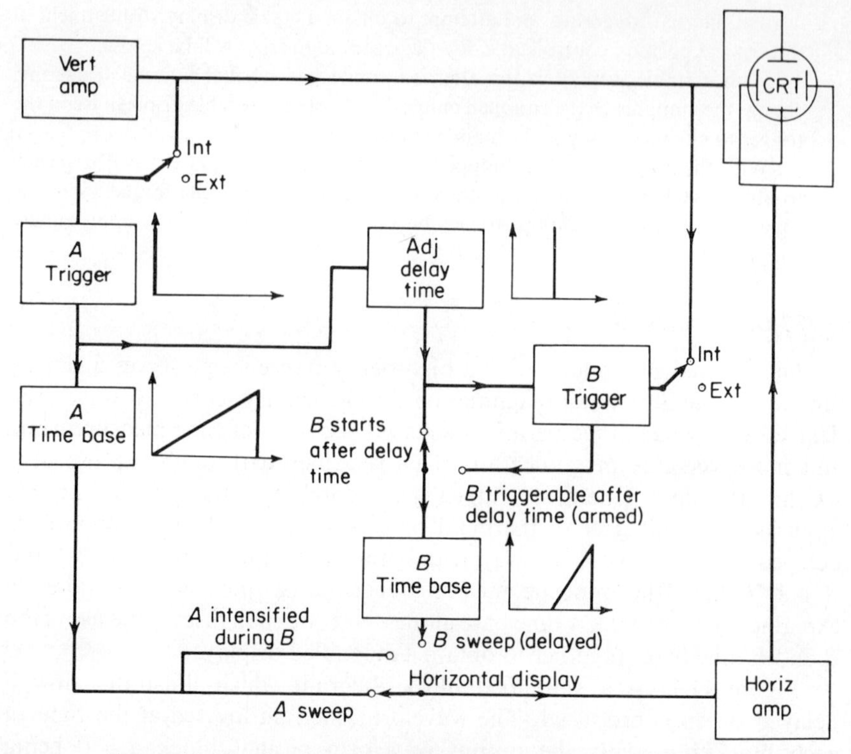

Figure 11.12. Simplified block diagram illustrating delayed sweep.

additional circuitry, not shown in Figure 11.12, increases the trace intensity during the *B* sweep. Since the intensified portion occurs only during the *B* sweep, its starting point will be controlled by the delay-time dial and its length by the *B* sweep time.

Finally by setting the horizontal display switch to *B* sweep or delayed sweep, the horizontal deflection will be controlled by the *B* sawtooth which produces a faster sweep and hence a magnified trace. These three cases are illustrated in Figure 11.13.

Exercise 11.4

Using the circuit of Figure 11.11, obtain a stable display of the channel-1 input only with the *A* (main) sweep. Set the delay-time control to zero and the *B* sweep speed five times faster than the *A* sweep speed. Set the *B* trigger mode so that the *B* sweep starts after the delay time. Using the horizontal display switch, select the *A* intensified mode and explore the effect of the delay-time and *B* sweep time controls. (*Note:* it may be necessary to readjust the CRT intensity control to obtain a

Horizontal Display Switch — Scope Trace

A sweep

0.2 msec/div.

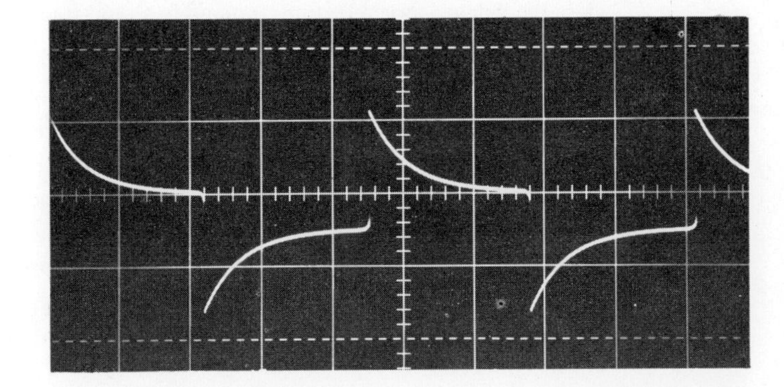

A intensified during *B*

0.2 msec/div.

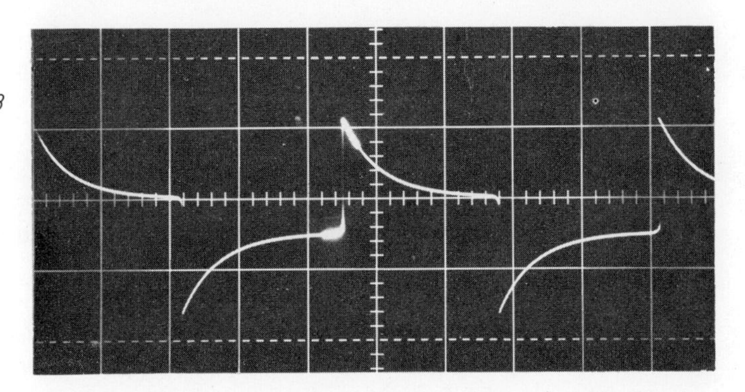

$B = 10\ \mu$sec/div.
Delay = 0.2 X 4.30 = 8.6 msec

B (delayed) sweep

10 μsec/div.

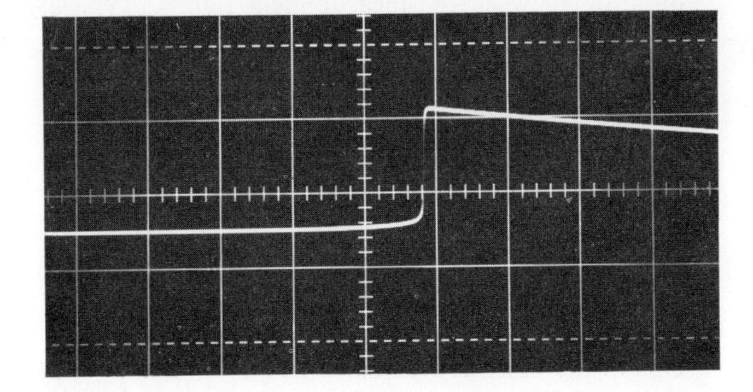

Figure 11.13. Oscilloscope waveforms illustrating the delayed sweep.

proper intensified trace.) Using the horizontal display switch, obtain the delayed sweep trace.

Use the delayed sweep feature to magnify a dual-trace display from the circuit of Figure 11.11.

In some instances, it is desirable to delay the B sweep but to trigger it rather than start it at the end of the delay time. This is called *triggered delayed sweep* or *armed* delayed sweep. In Figure 11.12 the B trigger is shown to be controlled in part by the output of the delay time control. When the B time base derives its input from the B trigger, the B trigger can produce an output pulse only when the trigger conditions are satisfied (level, slope, etc.) *and* the delayed pulse from the A trigger has been received. Thus in the armed mode, the B sweep is triggerable only after the delay time.

Exercise 11.5

Obtain a delayed sweep from the circuit of Figure 11.11 using the triggered delayed sweep feature. Since individual scope models are different, it may be necessary to consult the instruction manual to determine exact setup for triggered delayed sweep.

Exercise 11.6

Use the horizontal magnifier to further magnify a delayed sweep trace.

The delayed sweep may also be used to make more accurate time measurements than is possible by using a single time-base display. To accomplish this, the starting point on the delayed sweep waveform is set to a specific point on the CRT screen and the delay-time multiplier reading noted. Next the end point of the time measurement is set at the same point on the CRT screen by adjusting the delay-time multiplier. The difference between the delay-time multiplier readings times the A sweep time yields the time duration between the measurement points. Typically this reading will be accurate to 1% ± 2 minor divisions of the delay-time multiplier.

Exercise 11.7

Use the delay-time control and the delayed sweep to carefully measure the period of the channel-1 waveform of Figure 11.11.

11.6 oscilloscope photography

One of the most convenient methods of making a permanent record of an oscilloscope display is photographing the trace. A photograph is also more convenient to use in attempting to obtain detailed data from a scope trace.

Finally, in observing fast single-trace or "one-shot" phenomena a photograph is a useful way to obtain permanent data.

There are a variety of camera arrangements for use with oscilloscopes. In this section some general information is given to aid in making scope photographs, but it is expected that the particular instruction manual will be consulted for detailed instructions such as camera mounting, film loading, and focus adjustments.

11.6.1 film

Polaroid film which produces immediate pictures is the most widely used film for scope photography. In some cases, such as when a large number of pictures must be taken very rapidly or without film changing, conventional film is used. However, for general laboratory work the Polaroid film is most convenient.

The sensitivity of film to light is specified by the ASA number or "speed." The higher the ASA number, the more sensitive the film. For most work, Polaroid film with an ASA rating of 3000 is the best choice since it is the most sensitive and hence results in short exposure times.

In using the Polaroid film, care must be taken to properly coat the pictures after development. This coating fixes the image and renders the print permanent. If the print curls substantially, draw it across a sharp surface, such as a bench edge, to flatten it before applying the coating. At all times care must be taken not to touch the image surface, because fingerprint impressions will result.

11.6.2 exposure control

The exposure of a film is controlled by two parameters: the exposure time and the lens aperture or diameter of the lens opening. The exposure time is controlled by the shutter speed, which is adjustable. The speed is usually specified in fractions of a second, 1/10, 1/125, etc. However, on the shutter speed control, the 1/ is usually omitted and only the denominator, 10, 125, etc., is shown. Two additional positions, B (bulb) and T (time), are also provided for longer time exposures. With B, the shutter will remain open as long as the cable release is depressed. With T, the shutter opens on the first depression of the cable release and *remains open* until the cable release is depressed a second time.

The lens aperture is specified by the $f/$ setting, which ranges from 1.9 to 16 in most scope cameras. The light passing through the lens is maximum at the smallest $f/$ number, and decreases by a factor of two as the $f/$ number increases by $\sqrt{2}$. Thus at $f/2.8$, half as much light is incident on the film plane as at $f/2$.

11.6.3 focusing

When attaching a camera to an oscilloscope it is good practice to check the focus. A frosted-glass attachment is placed at the film plane instead of the film holder, and the image observed as the distance from the lens to the CRT is adjusted. The shutter should be opened by means of the *T* position, and the widest aperture used for the most critical focus.

11.6.4 taking the picture

Experience plays a sizable role in making good oscilloscope pictures. Some guidelines are given below as a starting point for making a picture.

1. With ASA 3000 film, a good starting point for exposure is $f/5.6$ at 1/5 sec.
2. Keep the intensity of the trace low. This increases sharpness and reduces background lighting on the film. As a rule of thumb, the trace is too bright if it makes the background of the CRT visible through the viewing hood.
3. Keep the graticule illumination low, using the same principles outlined in (2).
4. In some cases, separate exposures of trace and graticule will be necessary, each made with the other turned off.
5. Do not be afraid of making a bold adjustment in exposure. A factor of two in exposure makes only a modest difference in the final print. If your initial try is very far off, try a change of a factor of four, eight, or even sixteen.

Exercise 11.8

Photograph a dual-trace display from the circuit of Figure 11.11.

Exercise 11.9

Using the single sweep display, photograph the waveform $v(t)$ as the switch in the circuit of Figure 11.14 is closed. This shows the contact bounce as a function of time in the switch. Use of *B* shutter setting will insure the shutter is open when the switch is closed.

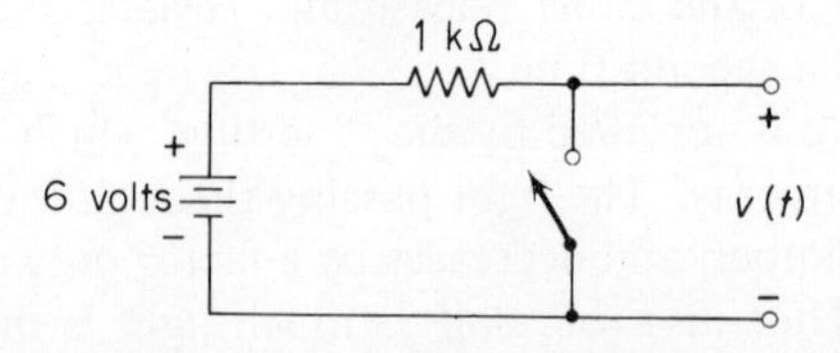

Figure 11.14. Circuit for measurement of switch contact bounce.

12

storage and sampling oscilloscopes

12.1 introduction

The basic operation and performance of oscilloscopes have been examined in Chapter 4 and Chapter 11. This chapter introduces two special-purpose oscilloscopes, the storage oscilloscope and the sampling oscilloscope. Although these types are not used as frequently as conventional units, there are many measurement problems which are either greatly simplified or are only made possible by such equipment. The storage oscilloscope allows convenient observation of very slow or infrequently occurring signals, and the sampling oscilloscope permits observation of events which last for less than 1 ns, providing such events occur periodically.

12.2 storage oscilloscopes

A conventional oscilloscope is normally used to display repetitive signals. If the repetition rate of the displayed waveform is less than approximately 30 times/sec, the display flickers. On the other hand, if a waveform to be observed occurs only once, then it can be displayed only during the single occurrence and observation on the cathode ray tube (CRT) screen for a prolonged period to permit detailed evaluation of the signal is not possible. If a longer-lasting display of infrequent events is required, it is necessary to photograph the trace as it occurs.

A *storage* cathode ray tube, however, has the capability of retaining the image of a single-trace signal display for an hour or more after it has occurred. Thus, the use of a storage oscilloscope offers an attractive alternative to photography in almost any application in which a single-trace waveform is to be observed but a permanent record is not required.

Except for the storage CRT, the storage oscilloscope is identical with a conventional oscilloscope. The operation of the horizontal and vertical amplifiers and the time base generators are unchanged. Differences in electrical performance may arise, however, because of differences between the storage CRT and the conventional CRT.

12.2.1 the bistable, direct-view storage CRT

Storage cathode ray tubes may be divided into two categories: *bistable* and *halftone*. The stored display on a bistable tube has only one level of intensity, while the trace of a halftone tube may have a range of intensities. Storage cathode ray tubes may also be classified as *direct-viewing* or *electrical-readout* tubes. Direct-viewing tubes provide a visual display as do conventional cathode ray tubes, while electrical-readout tubes provide an electrical output rather than a visual signal as a function of the stored information. This section briefly outlines the operation of the direct-view bistable storage tube used in many commercially available storage oscilloscopes. A more detailed description of storage circuitry is provided in Reference 1.

The operation of the bistable storage tube depends on the process of *secondary emission.* If a beam of electrons with the proper energy strikes certain types of surfaces, electrons will be emitted from the surface in response to the incident electron beam. These emitted electrons are called *secondary electrons.* If one secondary electron is emitted for each incident electron, then the average charge on the surface does not change. If the rate of emission is less than the rate of incidence, then the surface will become negatively charged. Under some conditions it is possible for the number of secondary electrons emitted to exceed the number supplied by the incident beam, and in these cases the surface loses rather than gains electrons when it is struck by the beam and becomes positively charged. The average number of secondary electrons emitted per incident electron is called the *secondary emission ratio.* Figure 12.1 shows a curve of the secondary emission ratio ρ vs. energy of the incident electron beam.

Figure 12.2 shows the major elements of a storage cathode ray tube. The inside of the screen is covered with a pattern of tiny phosphor dots, which are electrically isolated from one another. These phosphor dots will emit light if they are struck by electrons of sufficient energy, and are also capable of secondary emission.

Assume that the entire phosphor-dot screen is initially near ground potential. (The screen is returned to this state whenever it is *erased.*) The

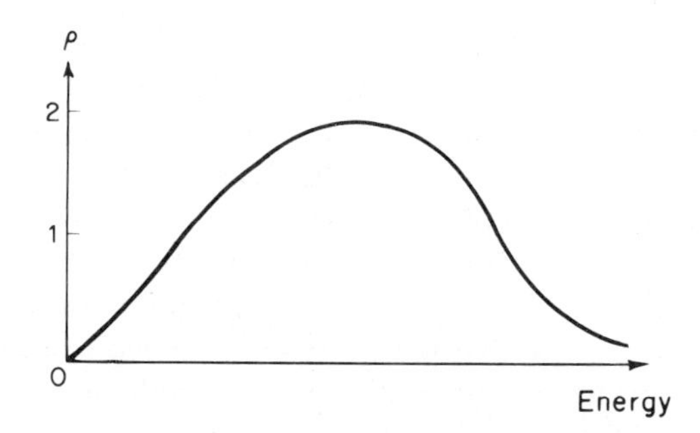

Figure 12.1. Secondary emission ratio vs. energy of incident electrons.

flood guns (which are on continuously) provide a stream of low-energy electrons toward the screen, since the flood guns are held at ground potential. The flood electrons which strike the phosphor dots have insufficient energy to light them or to cause significant secondary emission. Therefore, as the screen collects the flood-gun electrons, it becomes negatively charged until it reaches a potential where it repels all further flood-gun electrons.

The potential of the writing gun is much more negative than the potential of the screen when it has reached an equilibrium with the flood guns. Thus, when the writing gun is turned on, its electrons strike the screen with high energy, causing the phosphor dots hit by these writing electrons to emit light, and, because of secondary emission with $\rho > 1$, become positively charged. Since adjacent phosphor dots are electrically isolated, a positively charged dot maintains its charge even when the writing beam is removed. However,

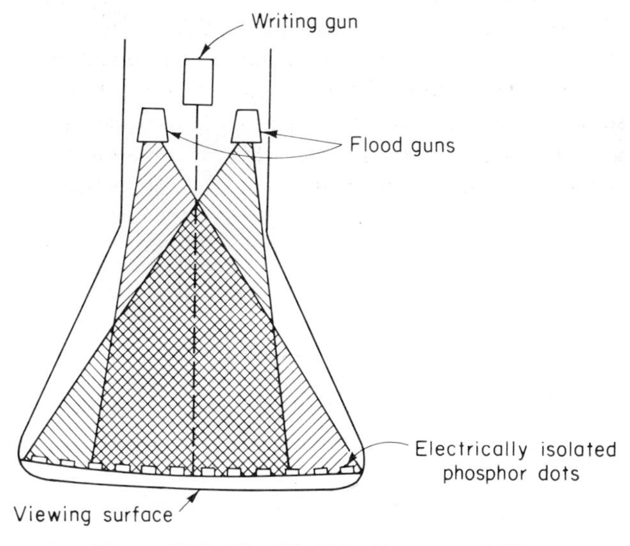

Figure 12.2. Simplified bistable storage CRT.

positively charged portions of the screen attract and accelerate flood-gun electrons to the point at which their energy is sufficient to maintain illumination. Secondary emission insures that the positively charged areas remain positively charged as they continue to be struck by flood-gun electrons.

It can be seen that when the entire screen is flooded with electrons, only those portions which have been previously excited by the writing gun will emit light. This selective light emission, dependent on past history, provides the storage mechanism. Stored images tend to spread somewhat with time owing to charge leakage to adjacent dots, but usable displays can easily be maintained for one hour or more. Erasing an image is accomplished when desired by applying a negative potential to a mesh located close to the screen, thereby removing the positive stored charge from the previously excited phosphor dots.

12.2.2 performance features and limitations

The engineering compromises necessary in the design of a storage tube degrade certain performance characteristics compared with conventional cathode ray tubes. The most important of these limitations are discussed in this section along with some special features offered by various manufacturers.

Maximum writing speed or writing rate is defined as the maximum spot velocity which will still produce a stored display. This quantity is normally specified in centimeters per second. Note that it is the total velocity of the spot, not simply the horizontal component, that is important, as illustrated in Figure 12.3. The writing rate of some available instruments exceeds 0.1 cm/μs. Some storage tubes permit the buildup of a fast signal if the event is repetitive, and therefore under certain conditions it may be possible to store a signal which could not be observed on a conventional oscilloscope. It is important to realize that the writing speed of some storage tubes deteriorates as the tube ages, with the result that the manufacturer's original specifications may not be met over the lifetime of the instrument.

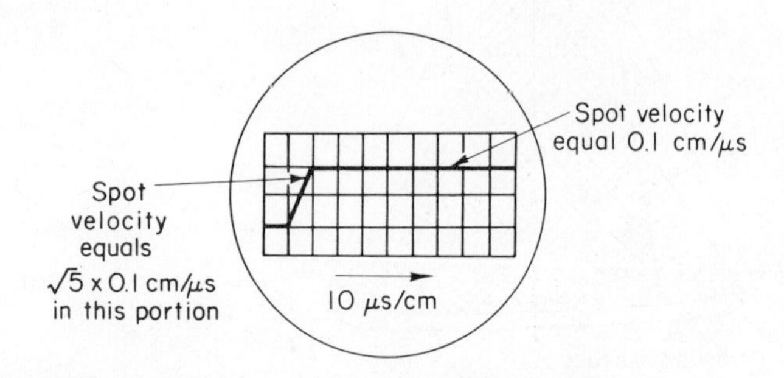

Figure 12.3. Illustration of spot velocity.

Storage time is simply the time the display can be retained on the storage-tube screen. As mentioned earlier, many storage tubes have a storage time in excess of 1 hour. One type can store a display for a week or more, and this can be done with the instrument turned off. One manufacturer offers a storage oscilloscope with a variable persistance feature, i.e., the stored display fades with time with a rate which can be controlled from the front panel of the oscilloscope. This feature is attractive for viewing repetitive, low frequency signals.

Some storage scopes provide a *split-screen* capability. Either the top or the bottom half of such a screen can be independently operated in either the store or nonstore modes.

The storage phosphor is more easily damaged than conventional CRT phosphors. A repetitive sweep over the same area may damage the phosphor dots. Also, it is important to use only the *minimum* intensity setting required to produce a well-defined display. Even with these precautions, the lifetime of a storage CRT used in the storage mode is limited to approximately 1000 hours. Some storage tubes can be switched into a nonstorage mode to provide a conventional type display. Conventional operation of such a tube does not deteriorate the storage screen and should be used whenever possible.

12.2.3 storage oscilloscope operation

The exercises in this section have been developed for use with the Tektronix Type 549 storage oscilloscope. The basic techniques can be used with other storage oscilloscopes, but the nomenclature and certain operating features may be somewhat different. *Caution*: The following rules, which are summarized from Section 2 of the instruction manual for the Type 549 storage oscilloscope, must be observed to prevent storage-tube damage.

1. A repetitive sweep over the same area for extended periods should be avoided.
2. Use only the intensity level required to write a well-defined display.
3. Turn the INTENSITY control to its minimum setting during warm-up time and before changing to slower sweep speeds or switching the HORIZONTAL DISPLAY selector.
4. Turn the INTENSITY control counterclockwise and the POWER switch to OFF when changing plug-in units.
5. Avoid continued use of one target area. This causes differential aging of the storage target and may result in differential light-emitting qualities over the target area.

6. Avoid leaving stored display on the screen longer than required. Operation in the ready-to-write (fully erased) state will give longer target life.
7. When storage operation is no longer required, shift to conventional mode operation. Conventional operation does not deteriorate the storage screens; however, avoid prolonged display periods of a repetitive waveform or a free-running sweep.

Most controls for the Type 549 storage oscilloscope are similar to those described in Chapter 11. The main difference is in the store controls located to the top right of the front panel. These controls include the fast-writing controls, the split-screen storage controls, and the programmed erase controls.

Exercise 12.1

Before turning on power, set the store controls as follows:

ENHANCE MODE	OFF
UPPER and LOWER SCREEN STORAGE	OFF
AUTO ERASE	OFF
SCREEN SELECTOR VIEWING TIME	ANY POSITION

These control positions place the oscilloscope in the conventional mode of operation. Turn on the oscilloscope and obtain a stable display using a signal from the calibrator as an input. At this point the display should be similar to that obtained on any conventional oscilloscope. Using a low-power magnifying glass, observe the display carefully to see if the phosphor dots are distinguishable.

Position the trace so that it is entirely in the upper half of the screen and set the oscilloscope for operation in the single-sweep mode. Depress the store button of UPPER SCREEN STORAGE ERASE button. *Note:* It may be necessary to adjust the triggering a single sweep. The stored display is removed from the screen with the UPPER SCREEN STORAGE ERASE button. *Note:* It may be necessary to adjust the CRT controls (intensity, focus, and astigmatism) slightly to obtain the sharpest stored image. Turn off the upper screen storage. Reposition the trace to the lower half of the screen, and store the display. Return the oscilloscope to the nonstore mode and increase vertical amplifier gain so that the trace covers 5 cm vertically, and center this trace. Store this display by depressing both store buttons simultaneously.

Notice that the screen is automatically erased and the trigger circuit reset simultaneously if the ERASE AND RESET button is pushed. Also notice that the vertical transitions of the calibrator square wave exceed the writing rate of the storage tube. Insert the filter shown in Figure 12.4 between the calibrator and the vertical input. Note that this network slows the transitions sufficiently to permit storage. Turn off both the screen storage controls after completing this exercise.

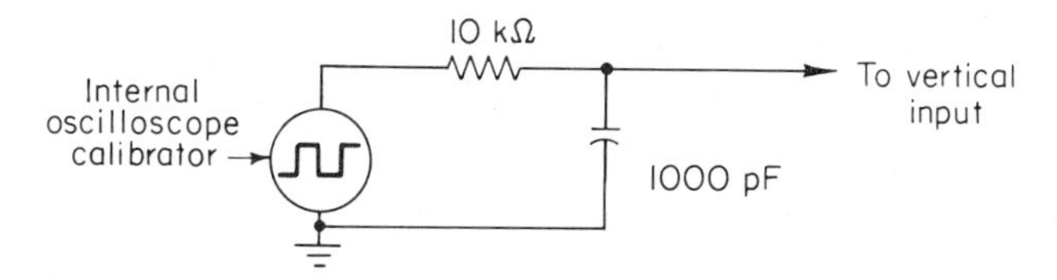

Figure 12.4. Filter network.

Certain types of pushbutton switches exhibit a phenomenon known as *switch contact bounce.* This process was photographed in Exercise 11.9. It is difficult to operate a pushbutton switch at a fast enough rate for convenient display on a conventional oscilloscope. Furthermore, the exact bounce pattern changes slightly from operation to operation. This is an excellent example of the type of signal best observed with a storage oscilloscope.

Exercise 12.2

Construct the circuit shown in Figure 12.5. When the switch is open, +6 volts is applied to the oscilloscope input, and its input is shorted when the switch is closed. With the oscilloscope in a nonstore mode, set the trigger controls to trigger normally on the negative transition of a 5-volt calibrator signal. Remove the calibrator signal and attach the pushbutton network to the vertical input. The oscilloscope should trigger whenever the button is pushed. Store a contact bounce transient. How long does the switch continue to bounce? Approximately how many times does it open and close? Return the oscilloscope to the nonstore mode after completing this exercise.

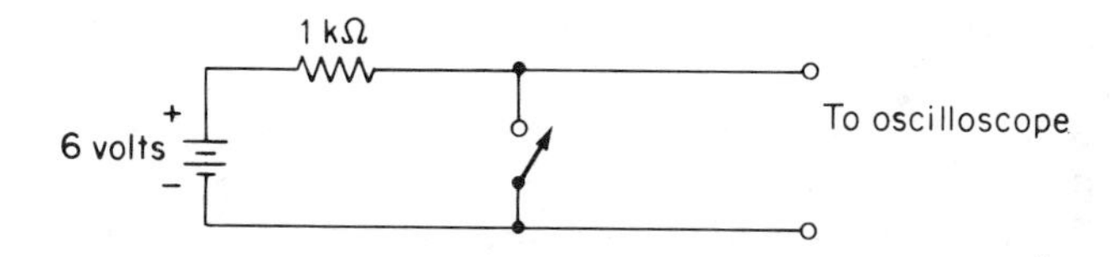

Figure 12.5. Circuit for displaying switch contact bounce.

The ERASE PROGRAM controls provide a means for periodically erasing the stored display.

Exercise 12.3

Adjust the oscilloscope to display a calibrator waveform of 5-cm amplitude. Adjust the trigger controls for single sweep. Turn the SCREEN SELECTOR control to FULL, and the AUTO ERASE switch to periodic. Depress the STORE buttons for both halves of the screen. With these settings, the screen is erased periodically. Rotate the red VIEWING TIME knob and note how the storage period changes. Set the VIEWING TIME knob to maximum. A trace will be stored and displayed until the end of the viewing time period whenever the trigger circuit is reset by depressing the HORIZONTAL DISPLAY MODE control lever. Switch the AUTO ERASE lever to AFTER SWEEP. In this position the trigger circuitry is reset automatically by the erase pulse.

Return the AUTO ERASE lever and the storage controls to OFF.

The FAST WRITING controls can be used to increase the writing speed of the instrument under certain conditions.

Exercise 12.4

Display a 5-cm calibrator waveform at 1 ms/cm. Set the HORIZONTAL DISPLAY lever to SINGLE SWEEP, the SCREEN SELECTOR to FULL and the AUTO ERASE lever to AFTER SWEEP. Depress both store buttons, and adjust the VIEWING TIME control to repeat the display approximately once per second. Increase the sweep speed by advancing the A TIME/CM switch. At approximately 5 μs/cm the display will be only partially stored as evidence by some gaps in the stored trace. Increase the INTENSITY control slightly and readjust FOCUS and ASTIGMATISM if required. The display should again be stored completely. Increase the sweep rate until gaps again appear in the stored trace. What is the writing rate?

Turn the ENHANCE LEVEL knob fully counterclockwise and turn the ENHANCE MODE switch to FULL. Slowly turn the ENHANCE LEVEL knob clockwise. Increased writing speed will be apparent. Increase the sweep rate and continue adjusting the ENHANCE LEVEL control until light spots begin to splatter the background. What is the writing rate under these conditions?

Return the AUTO ERASE lever, SCREEN STORAGE buttons, and ENHANCE MODE knob to OFF, and the HORIZONTAL DISPLAY lever to NORM. Turn off the oscilloscope.

12.3 sampling oscilloscopes

The response of a conventional oscilloscope to fast-rise pulses or high frequency sinusoids is limited by the bandwidth of its vertical amplifier. This bandwidth in turn is limited by the high frequency capability of the active devices used in the amplifier. Current technology permits the design of conventional oscilloscope amplifiers which provide bandwidths from DC to approximately 150 MHz, and pulse rise times of approximately 2.5 ns. While it is certain that these figures will improve in the future, the improvements will derive from advances in semiconductor device and cathode ray tube technology. *Sampling*, on the other hand, offers a circuit technique which permits the display of much faster signals provided the observed signals are repetitive. (Theoretically, sampling techniques can be used to display nonrepetitive events, but in practice the amount of equipment required is prohibitive.)

The sampling oscilloscope is effectively the electrical equivalent of the stroboscope, the instrument which permits visual observation of rapidly rotating equipment by momentarily illuminating it at slightly advanced positions on successive revolutions. The basic technique is illustrated in Figure 12.6. The original signal is shown as a periodic triangular pulse train. Sample instant t_A is located near the start of the first pulse of the original waveform. The value of the signal at instant t_A is determined and is displayed until sample instant t_B. The second sample occurs at a relatively

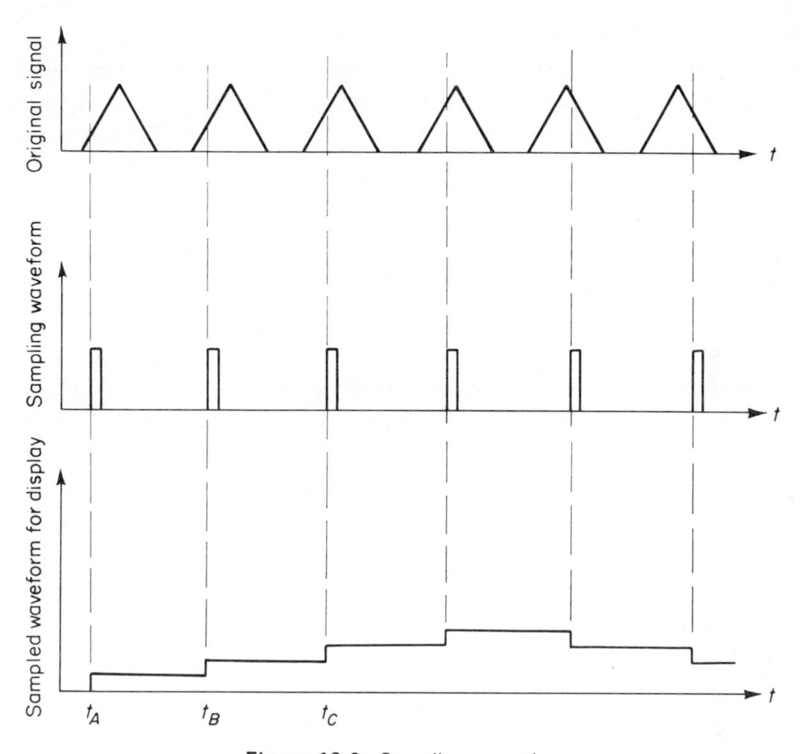

Figure 12.6. Sampling operation.

later portion of the second triangular pulse. The signal value at instant t_B is sampled and displayed during the time interval from t_B to t_C. This process of sampling successive values of the input waveform and displaying a sampled value until a new value is obtained produces a stretched out replica of the original signal. The amount by which the displayed signal is slowed relative to the original signal depends on the relative timing of adjacent sampling pulses.

The success of the sampling operation depends on designing a circuit which samples a signal in a very short period of time, since information is lost if the sampled signal changes by an appreciable amount during the sampling interval. A very simple sampling circuit is shown in Figure 12.7.

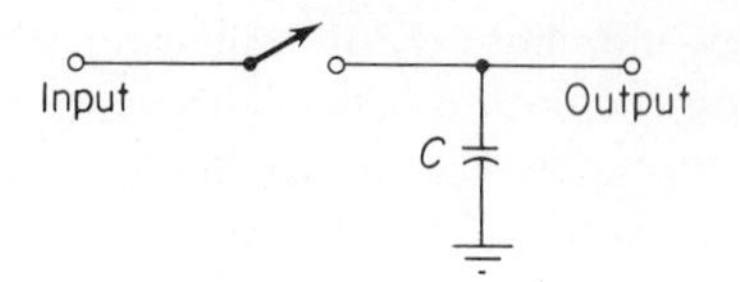

Figure 12.7. Sampling circuit.

When the switch is closed, the capacitor is charged to the value of the input voltage. When the switch is open, the voltage on the capacitor remains constant. In practice, the switching is accomplished with high-speed diodes. The capacitor value must be small so that it can be charged quickly, and a short piece of coaxial cable is often used for this capacitor. Sampling times on the order of 100 ps are possible with a circuit of this type. However, the stored voltage decays rapidly because the effective storage capacitor is small. For this reason a second sampling operation is performed with an active circuit which provides power gain and which stores the signal on a large capacitor.

Although the sampling gate and associated storage circuitry are critical for fast response, much of the circuitry in a sampling oscilloscope is used for "bookkeeping," to determine when samples should be taken. Figure 12.8 illustrates one technique used to determine sample times. A fast ramp is generated, and its start is synchronized with the trigger pulses. The trigger pulses are applied to the sampling oscilloscope from an external source, or in some cases derived from the vertical input signal. A staircase waveform

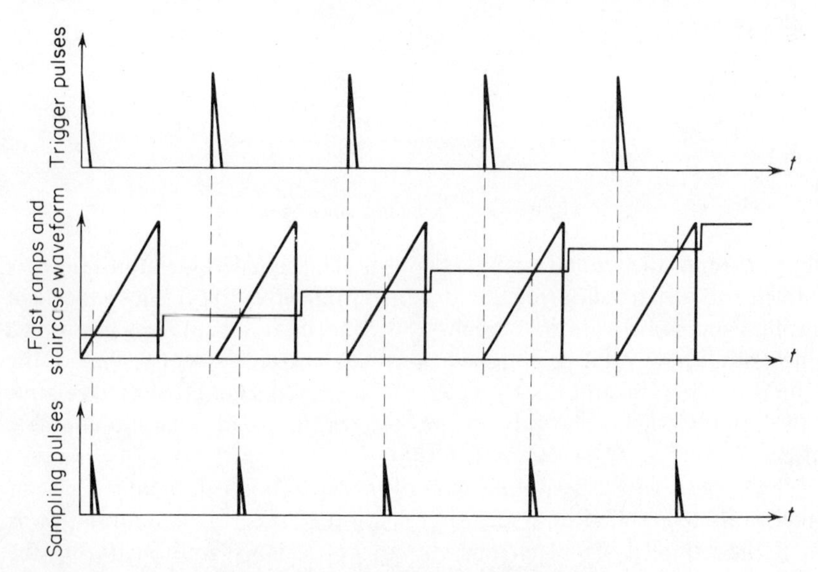

Figure 12.8. Timing signals for sampling.

is also generated, and the magnitude of this signal is incremented one step at the end of each ramp. Samples are taken when a positive-going ramp intersects the staircase waveform. In this way, successive samples occur at relatively later times on each cycle of the input signal. The staircase signal is also used as the horizontal deflection signal, since its value is proportional to the time of occurrence of the sampled portion of the waveform. If the time increment corresponding to one step is kept small, the display appears continuous in much the same way as in a dual-trace chopped display.

Many special features are available from various manufacturers, and some of the more important of these features are indicated below. The special circuitry required in a sampling oscilloscope consists of the sample and hold and timing circuits. The cathode ray tube and vertical and horizontal amplifier requirements are identical with those of conventional oscilloscopes. Sampling units are available as plug-in vertical amplifiers for certain oscilloscopes which accept plug-in preamplifiers. These sampling units also provide a horizontal signal such as the above-mentioned staircase waveform which is connected to the external horizontal input of the oscilloscope. This approach reduces the cost of a sampling system when a plug-in type oscilloscope is available. Figure 12.9 shows one sampling plug-in unit.

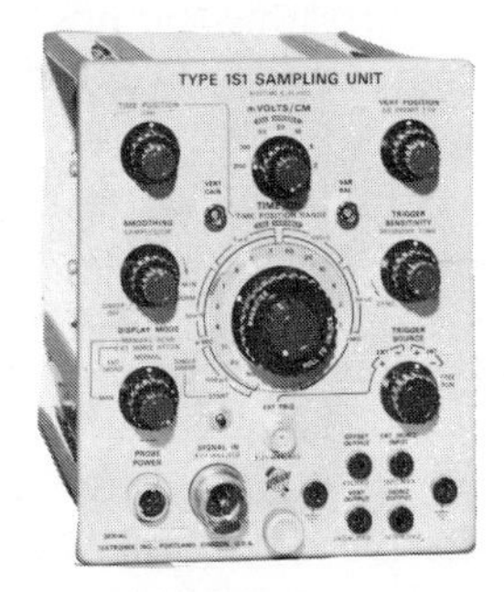

Figure 12.9. Sampling plug-in vertical amplifier.

Most sampling units have a 50-ohm input impedance, and are normally used to display pulses supplied from 50-ohm sources. Advantages of a sampling system of this type include the ability to derive an internal trigger signal and observe the leading edge of the triggering waveform. However, many signals are not available at low-impedance points, and therefore are altered considerably by a 50-ohm load of the probe. Special probes which use either tubes or transistors in the probe itself can be used to isolate the low input impedance of the sampling oscilloscope from the circuit under test. Normally such probes lower the bandwidth of the sampling system. Furthermore, the tube versions are quite large and fragile. An alternative is the use of a *direct sampling system.* The sampling diodes are located in the probe in this type of system, and the input impedance at the probe tip is typically 100 kΩ shunted by several pF. This type of a sampling oscilloscope generally uses an externally supplied trigger signal.

Sampling techniques are used to implement *time domain reflectometry* (TDR) systems. This is a special measurement procedure which can be used to evaluate high-bandwidth signal transmission media. Section 14.3.3 illustrates how reflections are caused when pulse signals encounter discontinuities in a transmission line. The generation of reflections is not limited to coaxial transmission lines. An antenna improperly matched to its feeder line causes reflections, as does a change in the width of a printed-circuit conductor. A TDR unit applies a fast-rise step from a known source impedance to the input of a transmission system and displays the signal amplitude as a function of time with a sampling oscilloscope. Alterations of pulse amplitude indicate discontinuities in the transmission system. The amplitude of the pulse disturbance is related to the magnitude of the discontinuity, and the time of occurrence relative to the start of the step indicates the location of the discontinuity.

12.3.1 sampling controls

Like storage oscilloscopes, operating features (and therefore the available controls) vary from manufacturer to manufacturer, and among the different models available from one manufacturer. This section describes the function of the controls common to most sampling units.

Vertical gain is selected in the same way as it is on a conventional oscilloscope by a combination of a step attenuator and a vernier. However, the minimum sensitivity of sampling oscilloscopes is normally considerably higher (several hundred mV/cm) than that of conventional oscilloscopes (tens of volts/cm). Moreover, the manufacturer specifies a maximum input voltage which is generally on the order of 1 to 2 volts. *This limit must be observed or the sampling diodes may be destroyed.* If larger signal voltages are to be measured, they must be attenuated before they are applied to the input of the sampling oscilloscope. Attenuators which attach to the probe tip are available for this purpose.

It is difficult to provide AC coupling with a 50-ohm input, since a large blocking capacitor is required and such a capacitor tends to deteriorate rise time. For this reason, a control which adds a DC offset or bias to the input signal is included. Its purpose is to permit observation of a small transient signal added to a relatively large DC level by canceling this DC level rather than by AC coupling. A conventional vertical position control is also available. In contrast to the position control which provides one to two screen diameters of adjustment independent of gain, the effects of the offset control are greatest at high gain settings.

The controls for trigger source, trigger slope, and trigger level or sensitivity are similar in function to those of conventional oscilloscopes. A control which adjusts the recovery time of circuitry following one trigger pulse is often included, and this control facilitates observation of sinusoidal signals.

Effective sweep time per cm is adjusted by means of a step switch and a vernier. The sweep speed is varied by changing the slope of the ramps used for timing (see Figure 12.8). Provision for horizontal magnification is also normally included. The magnification control may be separate from the sweep speed control, or it may be located so as to operate in conjunction with this control.

A control is also provided which adjusts the delay between the trigger signal and the start of the timing ramp. This control provides a type of delayed sweep capability and adjusts the location of the "time window" which is displayed on the screen. The maximum delay is a function of the sweep time and magnification values.

Some sampling units also can accept an externally supplied horizontal signal. The externally supplied signal is used in place of the staircase waveform (Figure 12.8), and therefore the magnitude of the external sweep signal controls sample times.

Two controls unique to the sampling process are usually provided. The number of samples per centimeter of display can be varied by changing the difference in amplitude between successive steps of the staircase waveform. At high input-signal repetition rates a better display is usually obtained with this control set for a large number of samples per centimeter, whereas if the input signal repeats relatively slowly, the use of fewer samples will speed up the presented display.

Provision is also included for smoothing or filtering the display. This is necessary since the sampling process results in a relatively high noise level at high gains. The use of smoothing, however, deteriorates rise time in a somewhat peculiar way. The number of samples required to complete the rise time increases as smoothing is increased. Thus, if a large number of samples per centimeter are used, smoothing does not alter the rise time of the unit, but if few samples are used, the displayed rise time can be slowed significantly.

Jacks are also usually provided to monitor certain internal quantities such as horizontal and vertical signals. These signals can be used as inputs to external display equipment. A jack which supplies power for the active probes often used with sampling units may also be provided.

12.3.2 sampling oscilloscope operation

For the exercises in this section it is assumed that the Tektronix Type 1S1 Sampling Unit will be used in conjunction with a Type 547 or equivalent oscilloscope. It may be necessary to modify the exercises slightly if another sampling oscilloscope is used.

Exercise 12.5

Obtain a Type 1S1 plug-in and an oscilloscope which accepts this plug-in. Since the 1S1 supplies an external horizontal signal to the display oscilloscope, it is necessary to first calibrate the oscilloscope horizontal gain. Any plug-in may be used for this purpose. Set the oscilloscope controls to disable the trigger, and select an external × 1 horizontal display. Apply a 5-volt horizontal signal from the calibrator. The display will appear as two dots on the screen spaced some distance apart horizontally. Use the horizontal position and horizontal gain controls so that these two dots fall on the 1-cm and 6-cm lines of the screen.

With the power off, insert the Type 1S1 plug-in into the oscilloscope. Connect the HORIZ OUTPUT of the 1S1 to the external horizontal input of the oscilloscope. Turn on power and adjust the Type 1S1 vertical sensitivity controls for a calibrated deflection of 200 mV/cm. Center the VERT POSITION and DC OFFSET controls. (Note that DC OFFSET has a 10-turn range.) Select the normal display mode and free run the sweep by selecting a positive internal trigger source and rotating the TRIGGER SENSITIVITY control on the 1S1 fully clockwise. Center the RECOVERY TIME control.

A display should be present on the screen. Adjust the oscilloscope horizontal position control so that the display starts at the left edge of the screen, and center the display vertically by adjusting DC OFFSET. Rotate the SAMPLES/CM control until individual dots are visible, and adjust the CRT controls for best spot shape. These represent the initial adjustments which should be made whenever the Type 1S1 is used.

Almost any pulse generator which supplies signals into a 50-ohm line can be used as a signal source. However, the capabilities of the sampling unit are best demonstrated if narrow pulses (20 ns or less duration) are available with rise time of 1 ns or less. *Caution:* It is imperative that the signals applied to the 1S1 input be less than ± 2 volts or the sampling diodes may be destroyed. If the generator used for the following exercise can supply larger signals, an attenuator must be used.

Exercise 12.6

Apply a pulse signal from an appropriate generator to the 1S1. Turn the TIME POSITION RANGE control fully clockwise and set the TIME/CM control to 5 ns/cm unmagnified. Magnification is controlled with the TIME/CM knob on the Type 1S1. Rotating this knob fully counterclockwise locks it in unmagnified operation. Magnification of up to 100 times is available at most sweep speeds by unlocking the control and rotating it clockwise. The actual sweep speed, independent of degree of magnification, is always indicated by the white dot on the TIME/CM knob. Rotate the TRIGGER SENSITIVITY control until a stable display is obtained. Note that the trigger is held off when the control is fully counterclockwise, that it is triggered with the control in midrange, and that it free runs with the control fully clockwise. Adjust the vertical gain for a convenient amplitude display, and center the display with the DC OFFSET control.

Select a normal, unsmoothed display. Also select an unmagnified sweep speed between 1 and 5 ns/cm so that the entire pulse is displayed. Use the TIME POSITION control to move the leading edge of the pulse near the left side of the screen. The total range of the TIME POSITION control is dependent on sweep speed and magnification. This range is indicated by the blue mark on the TIME/CM control. Rotate the SAMPLES/CM control throughout its range. Note that the display presentation is slowed at high sample density. Select a setting which provides a dense trace with minimum display flicker.

Measure the pulse width between 50% amplitude points. Magnify the trace by unlocking the magnifier knob and rotating it clockwise. Note that magnification takes place about the left edge of the screen. Measure the 10 to 90% rise time of the signal. (See Figure 14.5.) Use the TIME POSITION control to position the falling edge of the signal near the center of the screen, and measure fall time.

Return to an unmagnified display which shows the entire pulse. Place attenuators in the line between the generator and the 1S1 so that a vertical sensitivity of 5 mV/cm can be used. Use the DC OFFSET control to center the trace. Note the greater relative effect of the DC OFFSET control at high vertical gains. Observe the effects of the SMOOTHING control on noise. Note that smoothing does not alter rise time when a large number of samples per cm are used, but does deteriorate rise time when a low sample density is used.

The basic sampling technique can be used to display high-speed phenomena on display systems of limited bandwidth. For example, it is possible to sample signals from the deflection plates of a conventional laboratory oscilloscope and process them so that they can be displayed on a large-screen demonstration oscilloscope or X-Y recorder. (The magnetic deflection used with most large-screen CRT displays limits bandwidth.) The 1S1 can also be used to supply sampled signals for driving external equipment, such as an X-Y recorder.

Exercise 12.7

Remove the attenuators added in the last exercise. Readjust controls for a convenient display of the entire waveform. Switch the DISPLAY MODE control to MAN. Slowly rotate the MANUAL SCAN control and note that the display is traced with an instantaneous horizontal position proportional to MANUAL SCAN setting. Obtain an X-Y recorder which uses $8\frac{1}{2} \times 11$ in. paper, and load it with paper with major divisions at 1-in. increments. Adjust the vertical recorder gain to 0.2 volt/in. and the horizontal gain to 1.0 volt/in. (Refer to Exercise 13.10 if necessary to make these adjustments.) Connect vertical and horizontal inputs of the recorder to the VERT OUTPUT and the HORIZ OUTPUT, respectively, of the Type 1S1. (Also be sure to connect a ground wire between the two units.) Record the sampled display by slowly and smoothly rotating the MANUAL SCAN control.

12.4 references

1. *Storage Cathode-Ray Tubes and Circuits*, Tektronix, Inc.

2. *Type 549 Storage Oscilloscope Manual*, Tektronix, Inc.

3. *Type 1S1 Sampling Unit Manual*, Tektronix, Inc.

13

advanced voltage and current measurements

13.1 introduction

In Chapter 5, electromechanical devices for the measurement of voltage and current were discussed. Although these devices are useful for most routine laboratory work, their limited accuracy and sensitivity demand that alternative techniques be employed when precision measurements are required. In this chapter the potentiometric technique will be studied as a high-accuracy measurement scheme. The use of digital meters for high-precision measurements will also be discussed. Finally, some special meters will be introduced which extend the useful sensitivity range of the d'Arsonval movement.

13.2 potentiometric measurements

In the d'Arsonval meter movement, the electric force produced by the current to be measured is balanced against the mechanical force of the restoring spring. The accuracy of this arrangement is limited by such factors as the linearity of the restoring spring, friction in the pointer bearings, the actual width of the pointer, and viewing parallax. Under ideal conditions the accuracy is no better than 0.1%. To improve accuracy a system is required which eliminates the mechanical aspects of the measurement. One such system is the *potentiometric* or *null measurement* technique, shown in Figure 13.1.

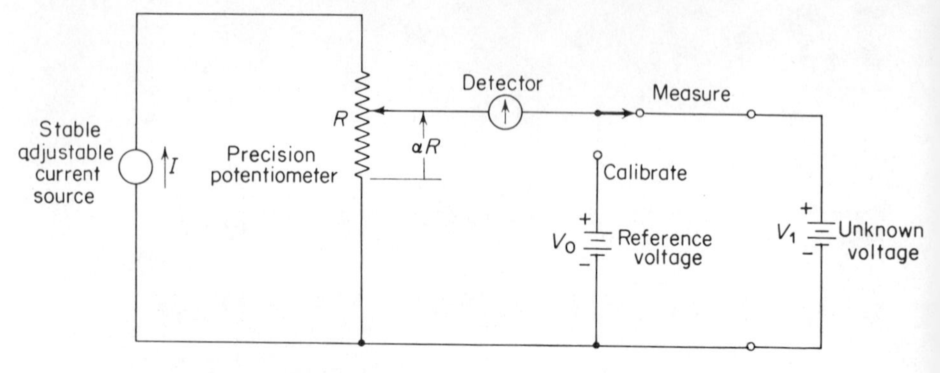

Figure 13.1. Basic potentiometer measurement scheme.

The potentiometric technique requires four basic elements: a precision potentiometer, a stable current source, a sensitive detector and a reference voltage source. The precision potentiometer is constructed so that the mechanical position of the sliding tap is directly proportional to the fractional resistance from the tap to the end terminals of the potentiometer. In this way the fractional resistance αR may be determined with high accuracy from the mechanical position of the sliding tap, which is indicated by an appropriate dial or scale arrangement. The degree to which this proportionality relationship holds is termed the *linearity* of the potentiometer.

The basic potentiometric measurement of an unknown voltage is carried out as follows. With the sliding tap at a position α_0, the switch is placed in the CALIBRATE position and the current source adjusted for zero deflection or null of the detector. Since no current is flowing through the detector at null, we find

$$\alpha_0 RI = V_0 \tag{13.1}$$

Next the switch is placed in the MEASURE position, connecting the detector to the unknown voltage V_1. The tap is then moved to a new position α_1, which produces a new null indication of the detector. Again, since there is zero current flowing in the detector at null and since the current I has not changed, we have

$$\alpha_1 RI = V_1 \tag{13.2}$$

Dividing (13.2) by (13.1) we find

$$\frac{V_1}{V_0} = \frac{\alpha_1}{\alpha_0} \tag{13.3}$$

which shows that the ratio of the unknown voltage to the reference voltage depends only upon the *ratio* of the tap positions. Therefore, the unknown voltage is given by

$$V_1 = \left(\frac{\alpha_1}{\alpha_0}\right) V_0 \tag{13.4}$$

and the accuracy of the measurement therefore depends directly on the linearity of the potentiometer (0.05% or better) and the accuracy of the reference voltage. Note that the absolute value of the potentiometer resistance R and the current I are not directly involved in the measurement. Also, the detector need not be calibrated, rather it need have only sufficient sensitivity to indicate the null with precision. Finally, since at null no current flows in the detector, the measurement scheme approaches the ideal voltmeter in that no current is drawn from the circuit under test. By the employment of a scheme in which two electrical forces are balanced against each other, the undesirable mechanical properties of the meter movement are eliminated from the measurement.

We have seen from (13.4) that the measurement accuracy depends directly on the linearity of the potentiometer and the accuracy of the reference voltage source. Since the reference voltage source is used to calibrate the system by adjustment of the current I before each measurement, this reference voltage must have excellent long-term stability. However, since the current is adjusted before each measurement, the current source need only be stable for the short time duration of the measurement, normally on the order of minutes. The only influence the detector will have on the accuracy of the measurement is through its ability to indicate the null precisely. If a percentage change in tap position away from null by an amount equal to the percentage linearity of the potentiometer produces a positive indication of the detector, its sensitivity is sufficient to contribute negligible error.

13.2.1 voltage references

We have seen that one essential component of a high-accuracy potentiometric measurement is a voltage reference of high accuracy and long-term stability. One common voltage reference meeting these requirements is the Weston *standard cell*, Figure 13.2, which produces a nominal voltage of 1.018636 volts. Individual standard cells are usually within 1 mV of the nominal value. The voltage of an individual cell will drift less than 30 μV per year and has a temperature coefficient of less than −10 μV/°C provided the cell is properly handled; thus the absolute accuracy of an individual cell is better than 1 mV once the actual cell voltage has been carefully measured.

For maximum accuracy there are several precautions that must be observed when using a standard cell:

1. The MAXIMUM current should not exceed 100 μA.

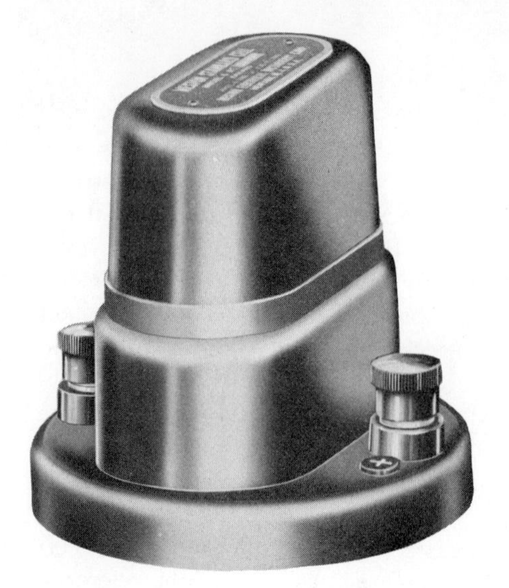

Figure 13.2. Laboratory standard cell. Courtesy Weston Instruments, Newark, N.J.

2. Avoid severe shock and vibration. Do not shake or invert the cell. Allow it to rest in place for a few hours for maximum accuracy.
3. Cell temperature should be kept between 4° and 40°C.
4. Standard cells should only be used in null circuits where the current in the cell is reduced to zero during the measurement.

References 2 and 3 contain additional information on standard cells.

A second voltage reference device that is widely used is the *zener diode*. A special type of silicon semiconductor diode, it may be fabricated to provide a constant reference voltage in the range from a few volts to several hundred volts. The voltage reference properties of a zener diode are derived from the region of the diode I-V characteristics where reverse-bias breakdown occurs, as shown in Figure 13.3. The reverse voltage at which breakdown occurs is called the zener voltage V_Z and the reverse current that flows in breakdown is termed the zener current I_Z. Once conduction in the zener region begins, I_Z increases very rapidly whereas V_Z remains essentially constant. A minimum value of I_Z is required to insure that the breakdown mechanism, and hence V_Z, is well established. The maximum value of I_Z is limited by the power dissipation in the device. In the forward conducting region, the zener diode is similar to any other silicon semiconductor diode.

For use as voltage reference devices, zener diodes are designed for minimum temperature coefficient and maximum long-term stability. By compensation techniques, a stability of 5 ppm/°C can be achieved over the

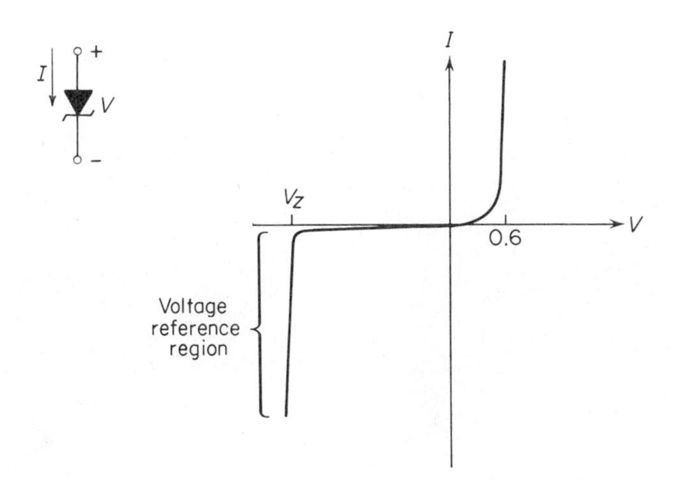

Figure 13.3. Circuit symbol and I–V characteristic of a Zener diode.

temperature range of $-55°$ to $+100°$C. Near room temperature, diodes with V_Z in the range of 5 volts can have zero temperature coefficient. By careful selection of devices, long-term drift can be held to 10 parts per million over a 1000-hour period. Thus, temperature-compensated zener diodes rival the performance of the standard cell in voltage reference applications. It should be pointed out, however, that a stable zener voltage is obtained only when the zener current is held constant by auxiliary circuitry. If I_Z is allowed to vary, then the reference voltage may vary over a much greater range than that implied by the temperature coefficient and drift specifications. For more detailed information on the properties and applications of zener diodes, see Reference 4.

Exercise 13.1

Using the transistor curve tracer, study the I-V characteristic of a 7.5-volt, 400-mW zener diode (IN755 or equivalent). What is the change in the zener voltage as the zener current changes from 2 to 10 mA? Estimate the accuracy of this measurement.

13.2.2 detectors

The basic requirement of the detector element is to provide a positive null indication within the resolution limit of the sliding tap position. The most common type of detector for potentiometric measurements is a very sensitive d'Arsonval movement, capable of full-scale deflection for currents of a few microamperes or less. These movements, called *galvanometers*, are constructed with pointer indicators in the less sensitive models and reflected light indicators for the more sensitive varieties. Figure 13.4 illustrates both

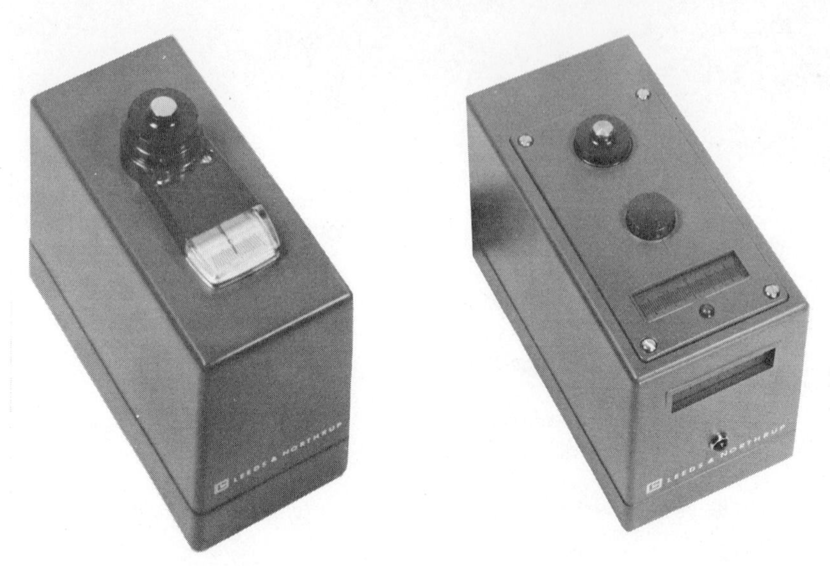

Figure 13.4. Typical null-detecting DC galvanometers. Courtesy Leeds & Northrup Company, North Wales, Pa.

types. Galvanometers are provided with a mechanical adjustment to align the zero-current indicator with the scale zero, and a mechanical clamp to prevent damage from shock and vibration whenever the unit is physically moved. *Be sure to clamp the movement when the galvanometer is not in use.*

Because of the high sensitivity of the galvanometer movement, it may be damaged by excessive current flow when the sliding tap is far from the null position. To reduce the sensitivity, a shunt is frequently used in parallel with the galvanometer as shown in Figure 13.5. At the initial balance, the shunt is set for zero resistance and then increased until a deflection is observed. At that point the circuit is adjusted for a null. By successive increases in the shunt resistance followed by readjustments for a null, the galvanometer is never driven off scale. For the final balance the shunt may be removed for maximum sensitivity.

Newer types of detectors are becoming available which combine solid-state amplifying circuits with a less sensitive d'Arsonval movement to obtain the required sensitivity in a more rugged electrical and mechanical package. Other choices for detectors include the use of an oscilloscope with a high-

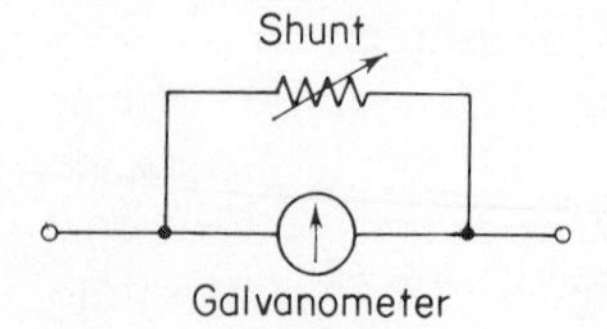

Figure 13.5. Galvanometer with shunt.

gain vertical amplifier. If high voltages are being measured, a conventional multitester may have sufficient sensitivity for use as a detector.

Exercise 13.2

Construct the potentiometric voltmeter as shown in Figure 13.6. The 22.5-volt battery along with the 100-kΩ resistor and the 33-kΩ and 1-kΩ potentiometers establish the required potentiometer current. The 100-kΩ, ten-turn potentiometer is the precision potentiometer, and should be equipped with a dial graduated in at least 100 divisions per revolution. The 0 to 10-kΩ decade resistor is used as the protective shunt for the galvanometer, and should be set at 0 ohms before the start of each measurement. The tap key provides a method of momentarily connecting the meter to the unknown voltage for measurement. This provides the maximum ability to detect a null through comparison of the pointer position at null with the zero-current position at open circuit. The tap key also disconnects the meter when a measurement is completed, thereby providing additional galvanometer protection.

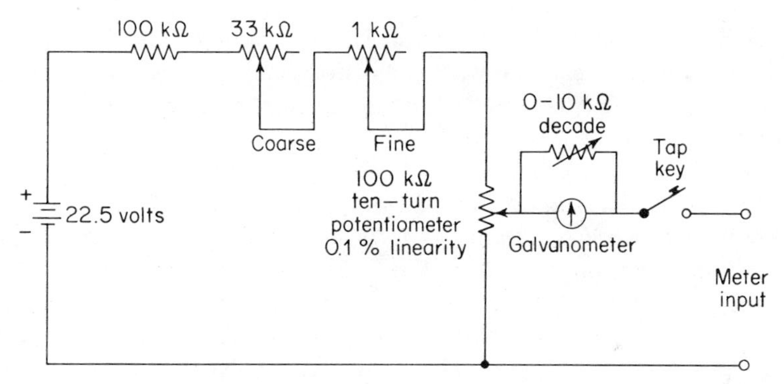

Figure 13.6. Potentiometer voltmeter.

Calibrate the potentiometric voltmeter using a standard cell by setting the precision potentiometer dial to correspond to the standard cell voltage, connecting the standard cell to the METER INPUT, and adjusting the COARSE and FINE current controls for a null. Be sure to set the shunt to 0 ohms before closing the tap key for the initial balance. Note that by choosing the dial reading α_0 equal to the standardizing voltage V_0, the dial is calibrated to read the unknown voltage directly within a decimal point. This corresponds to making the ratio V_0/α_0 in Equation (13.4) an integer power of 10, so that the digits of α_1 correspond to the digits of V_1.

Exercise 13.3

Use the potentiometric voltmeter constructed in the previous exercise to measure the zener voltage of a 7.5-volt, 400-mW zener diode (IN755 or equivalent) over the

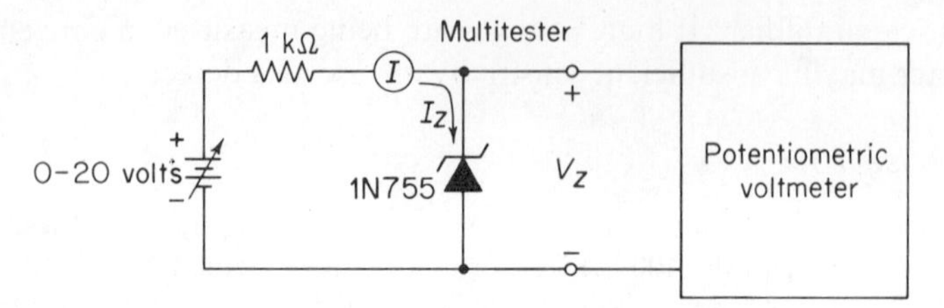

Figure 13.7. Measurement of zener diode.

range of zener current of 0.1 mA to 10 mA using the circuit of Figure 13.7. Plot the data on semilog coordinates. Compare the accuracy of the zener voltage measurement at 2 mA and 10 mA with the results obtained on the curve tracer.

13.2.3 commercial potentiometers

Potentiometric voltmeters as complete measurement systems are available in voltage ranges from 0.01 μV to 1.5 kV. However, the most common units have 150-mV and 1.5-volt ranges. The basic accuracy is usually 0.05% of *reading*, not full scale as with d'Arsonval movements. The complete instrument has come to be called a "potentiometer" and we will refer to them as such henceforth.

An example of a commercial potentiometer is shown in Figure 13.8. The operation is identical with the circuit of Figure 13.6, with the exceptions that the standard cell is internal and selected by a switch or separate tap key, and the precision potentiometer consists of a stepped switch plus a continuous slidewire. Thus the measured voltage is the sum of the step switch and slidewire indications. On some potentiometers a control marked REFERENCE JUNCTION is found. This is for use in thermocouple measurements of temperature (see Section 20.2.3) and should be set to zero for normal voltage measurements.

Exercise 13.4

The setup in Figure 13.9a may be used to determine the resistance of a wire and the resistance of the wire-binding-post contact. Construct the circuit as shown, and adjust the current in the wire to 1.0 amp. Calibrate a commercial potentiometer and use it to measure the voltage V as a function of position along the wire. Be sure the negative potentiometer terminal is connected to the binding-post terminal and not to the wire under test. Plot the data as shown in Figure 13.9b and determine the resistance per centimeter of the wire. (The handbook value for No. 22 copper wire at 25°C is 0.57 mΩ/cm.) From the voltage offset at $x = 0$ find the contact resistance between the wire and the binding post. Note that since the current is 1.0 amp, voltage readings translate directly into resistance.

Figure 13.8. Commercial potentiometer. Courtesy Leeds & Northrup Company, North Wales, Pa.

13.3 digital voltmeters

Digital voltmeters (DVM's) display measurements as discrete numerals, rather than as a pointer deflection on the continuous scale commonly used in analog devices. Several advantages in DVM characteristics lead to selection of a DVM in preference to some analog measurement methods. Direct numerical readout in DVM's reduces human error and tedium, eliminates parallax error and increases reading speed. Automatic polarity and range-changing features reduce measurement error and possible instrument damage through overload or reversed polarity. Measurement capabilities of AC voltages, DC currents, and resistance are available. Permanent records of measurements are available with printers, card and tape punches, and by magnetic tape equipment. With data in digital form, it may be processed with no loss of accuracy. Some examples of digital voltmeters are shown in Figure 13.10.

The heart of a digital voltmeter is the circuitry which converts analog voltage to a digital form. Of the several techniques employed, one of the simplest

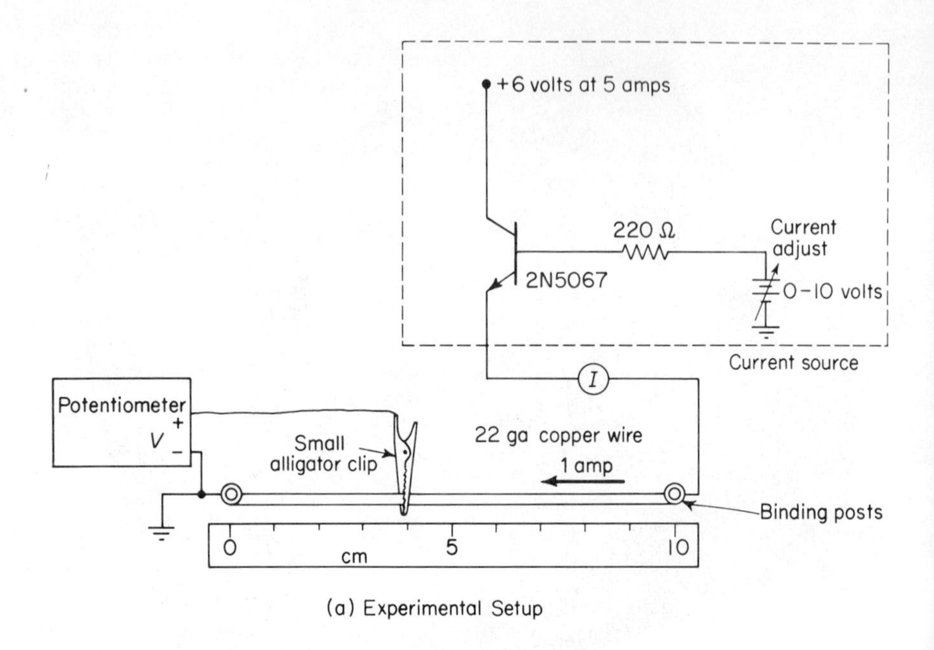

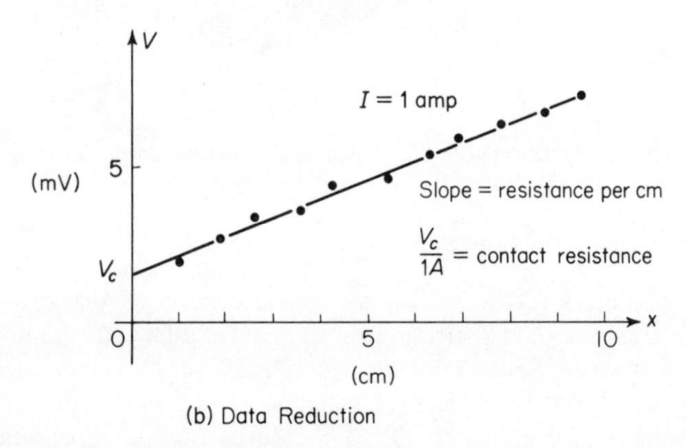

Figure 13.9. Measurement of wire and contact resistance. (a) Experimental setup. (b) Data reduction.

is the *ramp* or voltage-to-time conversion method, which we will investigate in detail as an example of DVM operation. For a detailed description of other digitizing techniques, consult Reference 5.

13.3.1 ramp DVM

The operating principle of the ramp digital voltmeter is to measure the length of time it takes for a linear ramp of voltage to become equal to the unknown input voltage after starting from a known level. This time period is measured with an electronic time-interval counter and displayed on indicating tubes.

Conversion of a voltage to a time interval is illustrated by the timing diagram of Figure 13.11. At the start of a measurement cycle, a ramp voltage is initiated. The ramp is compared continuously with the voltage being measured; at the instant they become equal, a coincidence circuit generates a pulse which opens a gate. The ramp continues until a second comparator circuit senses that the ramp has reached zero volts. The output pulse of this comparator closes the gate.

It is readily seen that the time duration of the gate opening is proportional to the input voltage. The gate allows clock pulses to pass to totalizing circuits and the number of pulses counted during the gating interval is a measure of the voltage. Choice of ramp slope and clock rate enables the totalizing circuit readout to read directly in millivolts (e.g., a slope of 400 volts/sec and clock rate of 400 kHz).

If the input were a negative voltage, coincidence with it would occur after zero voltage coincidence. Circuitry senses which coincidence occurs first and switches the polarity indicator accordingly.

The virtue of the voltage-to-time conversion as a digitizing technique lies in its simplicity. Furthermore, slowly varying input voltages do not disturb the operation of a voltmeter, as often happens with null-seeking DVM's which may continually hunt for, but never achieve, a balance.

13.3.2 DVM accuracy and resolution

The accuracy of digital voltmeters is usually expressed as a percentage of the reading and ranges from 0.1 to 0.001%, depending on the quality of the instrument. In any case, the last digit is always in doubt by ± 1 count. Some instruments specify accuracy on the basis of full-scale reading and thus may appear to be superior. If accuracy is essential, careful reading of the manufacturer's specifications is the best advice.

The *resolution* of a DVM is related to the number of digits in the display. For example, a 5-digit instrument is capable of resolving 1 part in 10^5. Generally speaking, the resolving power exceeds the absolute accuracy, which is dependent upon the stability of the reference voltage, temperature, etc., and one must be careful not to assume that a 6-digit DVM is necessarily more accurate than a 5-digit model.

Another factor which is important in high-accuracy DVM measurements is the loading effect of the DVM on the circuit under test. Input impedances of DVM's range from 10 MΩ to 10 GΩ. To have negligible effect on a measurement the input impedance of the meter must exceed the measured source impedance by at least a factor of 10^n, where n is the number of digits

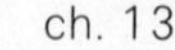

H-P Model 3430A

Hickok Model DP 140

Figure 13.10. Examples of digital voltmeters.

H-P Model 3460A

NLS Series X-1

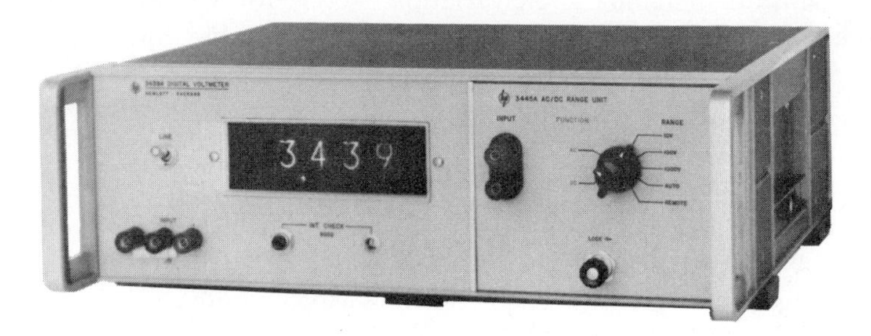

H-P Model 3439A

Figure 13.10. (*cont.*).

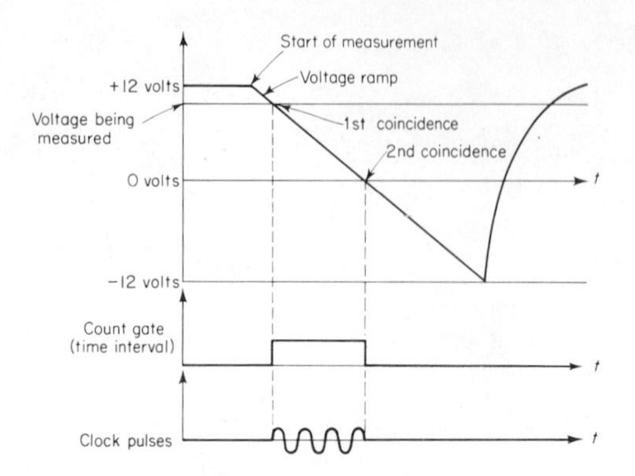

Figure 13.11. Voltage-to-time conversion in ramp DVM.

in the display. Otherwise, a correction factor must be calculated to obtain the true voltage from the DVM reading.

Other sources of errors in high-accuracy DVM measurements include drift of the reference voltage since last calibration, thermal voltages generated at dissimilar metal contacts (thermocouple effect) and responses of the DVM to noise and common-mode signals, i.e., voltages that are common to both input terminals and produce no voltage difference at the input. Common-mode signals are encountered only when the DVM is operated in a floating mode.

13.3.3 transistor emitter-base voltage

In Exercise 6.4 the output characteristics and current gain of a transistor were measured. Another transistor property, which is especially important in temperature matching, is the variation of collector or emitter current with the base-to-emitter voltage. The theoretical relationship, for the case where base and collector are shorted together, is

$$I_E = I_{ES}(e^{qV_{BES}/AkT} - 1) \tag{13.5}$$

where k is Boltzmann's constant, q is the electronic charge, T is the temperature, A is a parameter of value between 1 and 2, and I_{ES} is a characteristic current which depends on the device structure and temperature. The third subscript s on V_{BE} denotes the fact that the collector and base terminals are shorted. At room temperature the value of kT/q is 26 mV. Because I_E depends exponentially on V_{BES}, small changes in V_{BES} will make large changes in I_E. Therefore an accurate determination of the parameters I_{ES} and A require high resolution in the measurement of V_{BES}.

Exercise 13.5

The circuit of Figure 13.12 may be used to measure the variation of V_{BES} with I_E. Plot I_E vs. V_{BES} over the current range of 0.1 to 10 mA on semilog coordinates. Fit a straight line to the data points and determine A and I_{ES} for the transistor. (See Section 3.3.1 for the interpretation of data on semilog coordinates.)

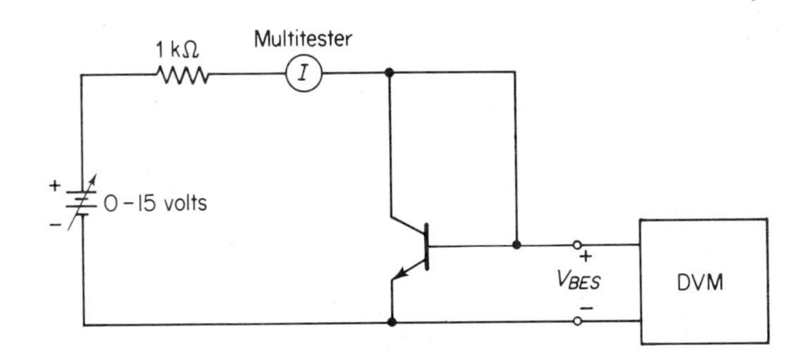

Figure 13.12. Circuit for measurement of V_{BES}.

13.3.4 multifunction DVM's

The basic digital meter is designed to measure DC voltages. However, by incorporating additional circuitry, they can be made to indicate DC current, resistance, AC voltages, and voltage ratios with accuracies approaching their basic accuracy as a DC meter. The selection of these additional measurement functions is usually accomplished by a function switch on the meter panel or by use of an appropriate plug-in unit.

The measurement of DC current is the most straightforward function conversion from DC voltage. A high-accuracy resistor is simply connected in parallel with the DVM input terminals. This resistor is chosen such that desired full-scale current will produce a full-scale voltage indication on the most sensitive DC voltage range. For example, to provide a 10-mA full-scale reading on a DVM with a maximum full-scale sensitivity of 100 mV, a high-accuracy shunt resistance of 10 ohms is required. This arrangement is shown in Figure 13.13a. In order to prevent resistance variation in the high-accuracy shunt through heating, it is most important that the maximum current rating never be exceeded. Always start on the least sensitive current range to prevent overheating or burn-out of the shunt resistor.

Figure 13.13b illustrates the method of converting a DVM to an ohmmeter. A high-accuracy current source provides a constant test current at the terminals which flows through the unknown resistor. The DVM simply measures the voltage across the resistor to determine its value. The voltage

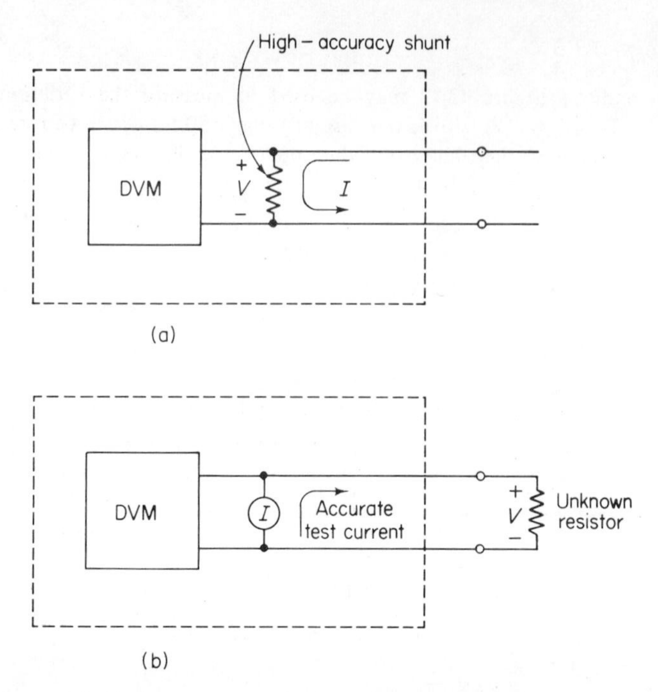

Figure 13.13. DC current and resistance measurements with a *DVM*. (a) *DVM* as a DC ammeter. (b) *DVM* as an ohmmeter.

range of the DVM during a resistance measurement is given by the product of the test current and the full-scale resistance range. For example, a digital ohmmeter using a 1-mA test current on the 1-kΩ range is measuring 1 volt full-scale on the DVM. On any given range, the current source is accurate only for voltages below the full-scale DVM voltage. If the voltage exceeds this value through too large a resistance or an open circuit, the test current decreases from its nominal value.

Exercise 13.6

Using a multitester set as a 10-mA ammeter for the unknown resistance, measure the magnitude and polarity of the test current on each resistance range of a digital ohmmeter. Increase the current sensitivity of the multitester as required to obtain a significant deflection. From the resistance ranges and test current magnitudes, determine the full-scale voltages of the DVM on each resistance range.

As long as the terminal voltage of the digital ohmmeter does not exceed the full-scale DVM voltage for the particular resistance range, the digital ohmmeter can be used as an accurate constant current source. One use of this accurate current is checking and calibrating d'Arsonval milliammeter movements. A second use is the testing and matching of some semiconductor device parameters.

In Figure 13.12, a circuit for the measurement of V_{BES} of a transistor was given. A digital ohmmeter with a full-scale voltage of 1 volt or more is all that is required to make the same measurement at emitter currents equal to the test currents of the ohmmeter. Simply select the ohmmeter range which provides the desired test current and connect the emitter and collector-base terminals of the transistor to the ohmmeter such that the test current flows in the proper direction in the transistor (indicated by the arrow on the emitter symbol). The digital display, *read as a voltage*, yields V_{BES} at the test current directly. If the full-scale DVM voltage is 10 volts or more, the digital ohmmeter may be used to check zener diodes for the value of V_Z at I_Z equal to the ohmmeter test currents. Of course, the maximum value of V_Z that can be measured is limited by the full-scale DVM voltage in the ohmmeter mode.

Exercise 13.7

Using a digital ohmmeter, remeasure V_{BES} of the transistor used in Exercise 13.5 and (if possible) V_Z of the zener diode used in Exercise 13.3 at test currents between 0.1 and 10 mA. Compare the results with the previous measurements.

The use of a DVM for AC voltages requires a waveform converter circuit, similar to those discussed in Section 5.5, to provide a DC signal which is proportional to some feature of the AC waveform. The overall accuracy of the DVM is then governed by the combined accuracies of the converter circuit and the DC DVM. Typical accuracies are on the order of 0.1% over the frequency range of 50 Hz to 100 kHz.

Standard analog AC voltmeters can also be used as AC-to-DC converters for a DC DVM. This only requires that the analog meter have a DC output available, normally used to drive an X-Y or strip-chart recorder. The accuracy of the overall system is usually better than the stated accuracy of the analog meter, since the mechanical movement is eliminated from the reading. By using analog meters in this fashion, a digital display can be obtained over extended frequency ranges and waveform responses.

Many DVM have provision for an external reference voltage input. Since the instrument basically performs a comparison between the input voltage and the reference voltage by forming the ratio of the input voltage to the reference voltage, it may be used to measure the ratio of any two voltages directly by applying the second signal to the external reference input. Before attempting this type of use, however, check the manufacturer's instructions concerning the allowed voltage range at the reference input. The impedance level at the reference input should also be considered since

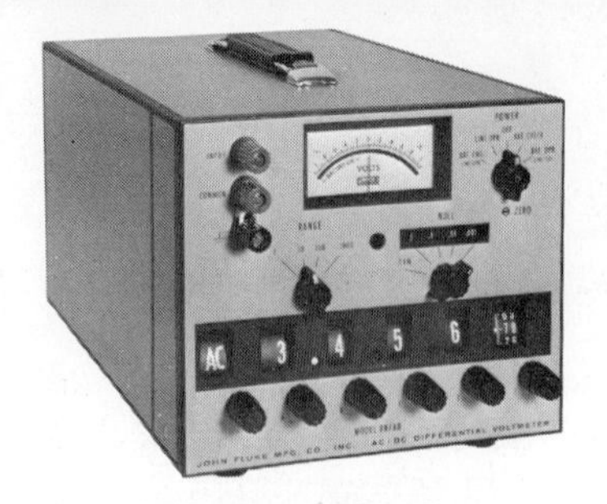

Figure 13.14. Differential voltmeters.

Fluke Model 887AB

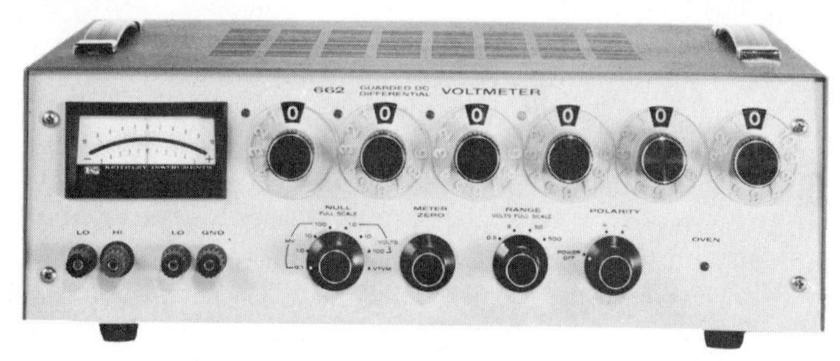

Keithley Model 662

it is often much lower than at the normal input, and may therefore provide an unwanted loading effect. The use of the DVM as a ratio meter is essentially the same as the use of a digital frequency counter to measure the ratio of two frequencies, as described in Section 15.3.1. Consult that section for the details of determining the scale factor in ratio measurements.

13.4 special meters and range extension

We have considered the potentiometer and digital voltmeter as two examples of high-accuracy meters for voltage measurements in the millivolt to volt ranges. We will now discuss some special meters which extend these voltage and current measurement capabilities.

13.4.1 differential voltmeter

The differential voltmeter is basically a potentiometer with a sensitive, calibrated VTVM used as the null detector. Some examples of these instruments are shown in Figure 13.14. In the differential mode, the voltage dials are adjusted to provide a null indication, whereupon the dial indications present the measured voltage. Capable of 6- to 7-digit resolution with accuracies of 0.01% or better, they are often used in research laboratory applications.

13.4.2 microvoltmeters and nanovoltmeters

For measurement of minute DC voltages, the microvoltmeter and nanovoltmeter are employed. These are basically a d'Arsonval movement coupled to a stable, high-gain amplifier. Usually this amplification is achieved by converting the DC signal to AC, amplifying the AC, and converting it back to DC to drive the meter. This technique, sometimes called carrier amplification, eliminates the DC drift errors inherent in high-gain DC amplifiers. Although sensitivities of 10 nV full scale are possible, the usual accuracy is on the order of $\pm 2\%$, and typical input impedances range from 1 kΩ to 1 MΩ, depending on voltage scale. Some types provide a differential mode of operation that increases the accuracy by a factor of 10. Figure 13.15 illustrates some typical instruments. Meters of this type make excellent detectors for potentiometers.

13.4.3 nanoammeters and picoammeters

By combining an internal shunt resistor with a DC microvoltmeter, as illustrated in Figure 13.13a with a DVM, an ammeter with sensitivities to 10 nA and $\pm 2\%$ accuracy is obtained. Further improvement in sensitivity to 0.3 pA full scale is available in the picoammeter shown in Figure 13.16. Again the accuracy is in the range of ± 2 to 4% full scale, depending on current range.

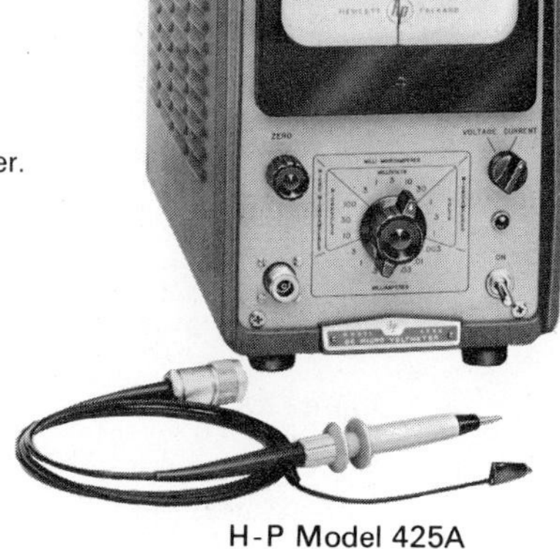

H-P Model 425A

Keithley Model 148

Figure 13.15. Microvoltmeter and nanovoltmeter.

Figure 13.16. Keithley Model 410A picoammeter.

Exercise 13.8

Using a nanoammeter or a microvoltmeter with current-measuring capability, measure the reverse leakage current of a 7.5-volts, 400-mW zener diode (1N755 or equivalent) at a reverse voltage of 1.5 volts. A suggested circuit is shown in Figure 13.17. The 1-kΩ resistor provides some protection for the nanoammeter with negligible voltage drop at the currents involved.

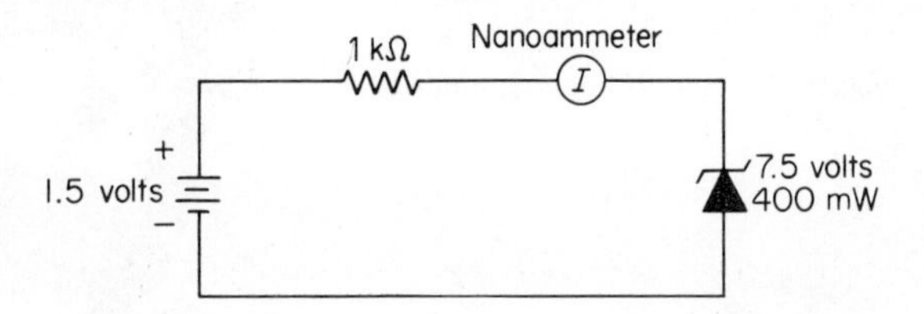

Figure 13.17. Measurement of diode reverse leakage current.

13.4.4 electrometers

Electrometers are voltmeters which feature input impedances as high as 10^{16} ohms, and are therefore useful in making voltage measurements from very high impedance sources. An example of an electrometer is shown in Figure 13.18. Although they are simple to use at conventional impedance levels, several problems arise when their use is attempted at extremely high impedance levels—10^8 ohms and higher. These difficulties include electrostatic pickup, physical motion, and, especially, leakage currents. Conventional rubber insulation may have a resistance of only 10^9 ohms, causing

Figure 13.18. Keithley Model 610C electrometer.

substantial errors in circuits with impedance levels of 10^8 ohms or greater. Teflon insulation is far superior to rubber, and is normally satisfactory for work at 10^{14} ohms or lower. Care must also be taken to avoid leakage paths due to dirt, moisture, oil films, and solder flux. The best insulation is of no value if it is contaminated.

Exercise 13.9

Connect a 10-pF mica capacitor *directly* to the input terminals of an electrometer, as shown in Figure 13.19. Be sure to make a connection to the laboratory ground. With the electrometer set for about 10 volts full scale, charge the capacitor to 6 volts by momentarily connecting the positive lead from the battery to the positive

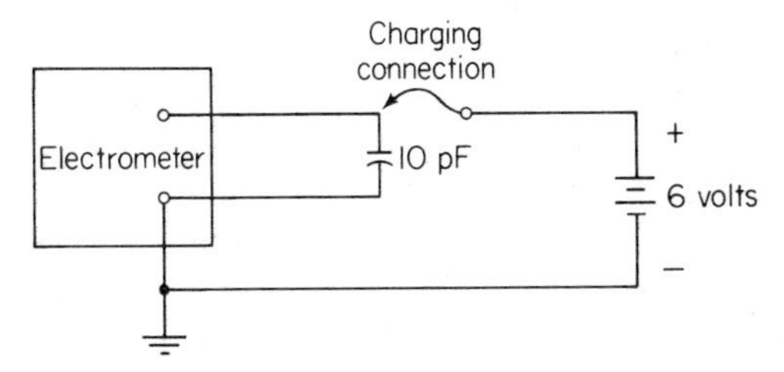

Figure 13.18. Leakage measurements with an electrometer.

terminal of the electrometer. Observe the electrometer reading as a function of time after removing the charging connection. The subsequent exponential decay of the voltage is due to the leakage resistance of the capacitor in parallel with the input impedance of the electrometer. A good capacitor should have a leakage resistance that is sufficiently large so that this decay should be very slow. If this is not the case, reduce the surface leakage by cleaning the capacitor with alcohol, or try another capacitor. Estimate the order of magnitude of the capacitor leakage resistance on the basis of the decay-time constant.

Explore electrostatic pickup by bringing your finger close to, but not touching, the capacitor body. What is the effect on this pickup of grounding your body with your other hand? Momentarily touch the positive electrometer terminal with your finger. From the change in voltage, estimate the charge transfer from your body to the capacitor ($\Delta Q = C\Delta V$). Again, investigate the effect of grounding your body.

13.4.5 high-current shunts

Up to now we have been concerned with extension of current measurement to high sensitivities. However, as currents exceed the order of 1 ampere, it is not feasible to construct meter movements to carry the entire current to be measured. The basic method of extending the range of an ammeter is to employ a *shunt*, as has been already described in Section 5.2. In this section, however, we will consider the use of precision shunts specifically designed to extend current measurements to the range of several thousand amperes.

The precision shunt is simply a precision resistance which has been designed to handle the current to be measured with negligible heating and nonlinearity. The value of the precision resistance is determined such that the rated current of the shunt will produce a small voltage drop, normally 50 or 100 mV. Thus a shunt rated at 200 amp and 50 mV produces a 50-mV voltage between its terminals when the current through the shunt is 200 amp. Since the shunts are designed to provide a fixed voltage for their rated current and since their resistances range from 1 to 10^{-4} ohm, the associated current-indicating meter is normally a voltmeter capable of a full-scale deflection of 50 or 100 mV. Of course, any instrument capable of indicating voltages in the 50- or 100-mV range, including a potentiometer, oscilloscope, or X-Y recorder, may be used to indicate the current through the shunt.

Figure 13.20 illustrates two types of precision shunts. In each case the shunt is provided with two sets of terminals: one heavy set to carry the current to be measured and one small set to provide connections to the indicating voltmeter. This arrangement eliminates the error introduced by the unknown contact resistance of the current-carrying contacts. Precision shunts are available in current ranges from 0.1 to 2000 amp, and their accuracy is normally 0.1%.

Figure 13.20. Precision shunts for high-current measurements. Courtesy Weston Instruments, Inc., Newark, N.J.

Exercise 13.10

The volt-ampere characteristics of a 1-amp fuse, commonly found in electronic instruments, may be measured with the circuit of Figure 13.21.

In order to calibrate the X-Y recorder, a commercial potentiometer can be used as a precision voltage source. Calibrate the potentiometer with its standard cell. Connect the EMF terminals of the potentiometer to the recorder input and

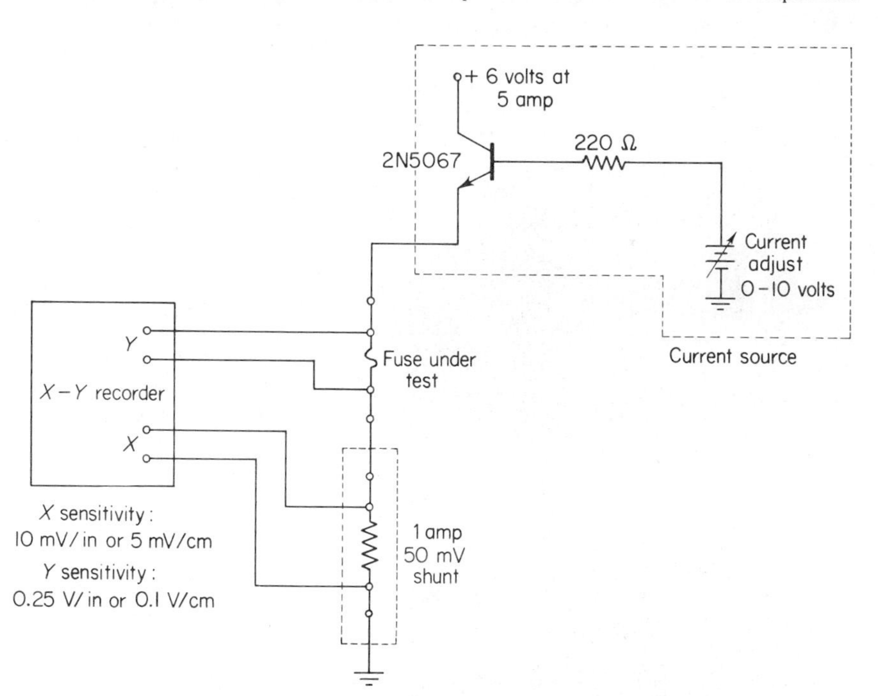

Figure 13.21. Measurement of V–I characteristic of a fuse.

set the potentiometer for the desired voltage. (The recorder sensitivity in inches or centimeters is selected to agree with the graph paper to be used.) Press the tap key to apply the voltage to the recorder input and adjust the recorder gain for the desired pen deflection. There should be no noticeable galvanometer deflection, indicating that the recorder input impedance is high enough not to load the potentiometer.

Provide two connections at each terminal of the fuse; one pair to apply current, the other to measure voltage. This *four-terminal* measurement eliminates the effect of contact resistance in the fuse holder in the same manner as just described for the precision shunt. Connect the recorder to the voltage terminals of the fuse and the shunt. Slowly increase the current until the fuse burns out and record the *V-I* characteristic.

A class of fuses, called *slow-blow*, are designed to withstand momentary overloads. Repeat the measurement with a 1-amp slow-blow fuse and compare the *V-I* characteristics of the two types.

13.4.6 high current transformers

The precision shunts described in Section 13.4.5 may be used for AC current measurements, although this is not normally the case. Such use would naturally require an AC voltmeter of 50- or 100-mV sensitivity. Rather, current transformers are employed which provide complete isolation between the measured current and the metering circuit. A typical current transformer, shown in Figure 13.22, provides a 5-amp secondary current for

Figure 13.22. High current transformer. Courtesy Weston Instruments, Inc., Newark, N.J.

up to 800 amp of primary current. Thus a 5-amp full-scale ammeter may be used to indicate up to 800 amp. For the current ranges of 100 amp and below, a series of primary current terminals are provided. Above 100 amp, the current-carrying wire is simply passed through the hole in the center of the transformer one or more times. Current transformers of the type illustrated are accurate to 0.1% over the frequency range of 60 Hz to 2.5 kHz.

13.5 references

A comprehensive reference of precision electrical measurements is

1. Precision Measurement and Calibration, *National Bureau of Standards Handbook 77*, Vol. 1, U.S. Government Printing Office, Washington, 1961.

Other references on this topic include

2. R. B. Marshall, *Measurements in Electrical Engineering*, John Swift Co., Cincinnati, 1948.

3. M. H. Aronson, *Handbook of Electrical Measurements*, Instruments Publishing Co., Pittsburgh, 1961.

4. *Zener Diode Handbook*, Motorola Semiconductor Products, Inc., 1967.

5. *Electronic Instrument Digest*, "Digital Voltmeters for the Laboratory", pp. 42–48, October, 1967.

14

signal and pulse generators

14.1 introduction

This chapter introduces several types of signal sources and discusses some of their more important characteristics, specifications, and limitations. Signal sources are normally used to generate AC waveforms with controlled characteristics which can be applied as inputs to other equipment. A wide variety of signal sources are currently available, and new types are constantly being developed as requirements and technology advance. Examples range from the simple sine wave oscillator introduced in Chapter 4 to radar modulators generating pulses that supply megawatts of power.

We will subdivide the generators considered in this chapter into three broad (and somewhat overlapping) categories:

1. Oscillators which provide a sine wave output.
2. Pulse generators, which produce repetitive patterns of rectangular pulses.
3. Function generators, which provide outputs such as sine waves, square waves, triangle waves, and sawtooth waves.

14.2 oscillators

While the nomenclature is by no means standard, we will categorize oscillators as instruments which provide purely sinusoidal output signals. The period of a sinusoidal signal is defined as the length of time required for the signal to repeat itself. The frequency in hertz (or cycles per second) is the reciprocal of the period. The frequency range which can be covered with available instruments extends from fractions of a hertz to more than 10 GHz. It should be noted that no one instrument covers this entire frequency range but rather some smaller range of frequencies with a ratio between the maximum and minimum frequency typically 10^4 to 10^6 for any single generator. Although oscillators which can supply kilowatts of power are available, the maximum output power of a typical laboratory oscillator is on the order of a watt. A typical maximum output voltage is 20 volts rms. Figure 14.1 illustrates several commercially available oscillators.

14.2.1 oscillator performance

The output frequency of most oscillators is adjusted in steps with a *range switch*, with continuous adjustment between steps accomplished with

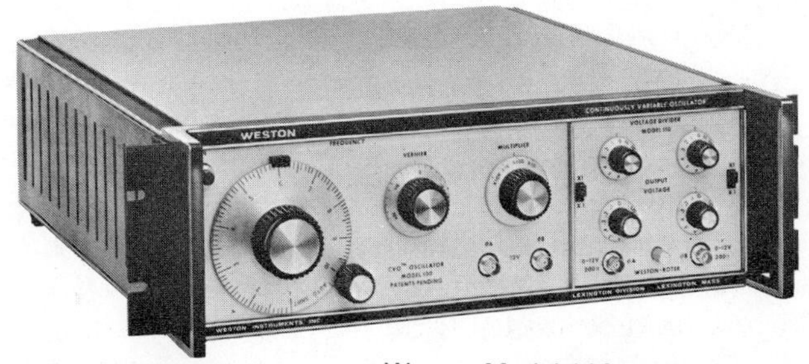

Weston Model 110

H-P Model 204C

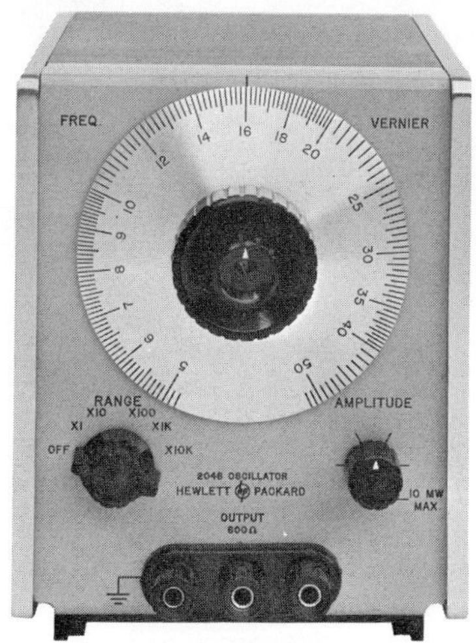

H-P Model 204B

Figure 14.1. Oscillators. Courtesy Hewlett-Packard Co., Palo Alto, Calif., and "Weston-Lexington" Weston Instruments Division, Lexington, Mass.

a calibrated dial. *Dial accuracy* (normally expressed as a percentage of actual dial reading) indicates the degree of correspondence between the actual frequency of the oscillator and that indicated on its dial.

Exercise 14.1

Determine the dial accuracy of an oscillator at 60 Hz by applying the oscillator signal to one axis of an oscilloscope and a 60-Hz signal derived from the power line to the second axis. Adjust the oscillator output frequency until a stable Lissajous figure is obtained. (See Section 4.3.3.) Compare the dial reading with 60 Hz. Use the same technique to determine dial accuracy at some higher frequency by comparing the output of the variable oscillator with that of a precision oscillator having an accurately known output frequency.

The output of an oscillator can be modeled as a sinusoidal generator in series with a Thevenin equivalent resistance R_G, as shown in Figure 14.2a. The output may be supplied directly to the output terminals of the instrument, and this type of output circuit is called DC coupled. A disadvantage of this connection is that drift associated with an output stage of this type can result in an unwanted DC component of output voltage. Another possibility is to include a capacitor in series with the output as shown in Figure 14.2b. Values are chosen so that the reactance of the capacitor is small compared to R_G at all frequencies of interest. If a DC current path is required by a circuit connected to an AC-coupled oscillator, it is necessary to provide this path external to the oscillator as shown in Figure 14.2b.

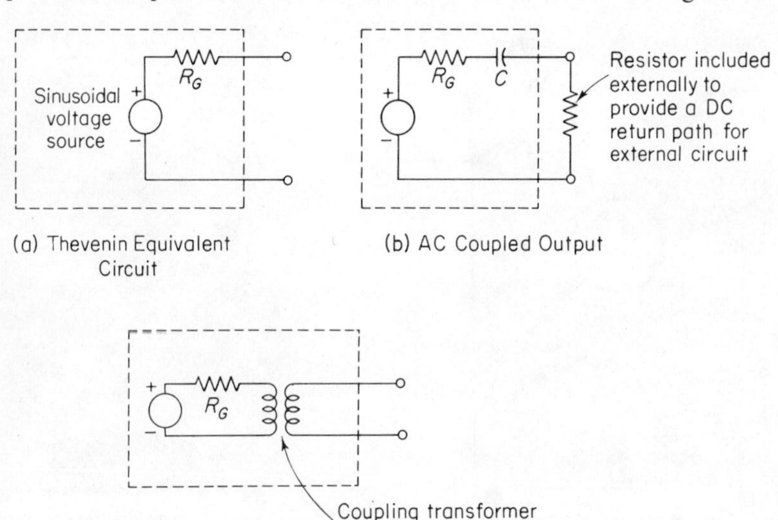

(a) Thevenin Equivalent Circuit

(b) AC Coupled Output

(c) Transformer Coupled Output

Figure 14.2. Types of generator output stages.

A third alternative, which combines a DC return path with zero DC output voltage component is the use of a coupling transformer as shown in Figure 14.2c. A disadvantage of this method is the relatively limited frequency response possible with a transformer.

The manufacturer normally specifies a recommended value for the load resistance applied to the oscillator. Common values are 600 and 50 ohms. Significant variation from rated load can cause *distortion* of the output signal.

Exercise 14.2

Determine the output resistance of an oscillator at 1 kHz. In order to measure this quantity, adjust the unloaded oscillator output to 10 volts peak-to-peak. Load the oscillator with a 1-kΩ resistor. From the decrease in amplitude with load, determine the output resistance R_G of the oscillator assuming the model of Figure 14.2a applies. If the change in output produced with the 1-kΩ load is too small to measure accurately, use a smaller load resistor.

Load the oscillator with progressively smaller resistors until the output signal becomes distorted. What is the peak load current under these conditions?

The second part of the preceding exercise shows how overloading the oscillator can cause a grossly distorted output waveform. Actually, oscillator outputs are always distorted to some degree, and this distortion can be measured even when it is too small to be apparent on an oscilloscope. Techniques for a detailed measurement of distortion are discussed in Section 15.4.2.

Frequency stability indicates the expected maximum change in frequency as a function of time, temperature, or input voltage and is usually expressed as a percentage of actual frequency. Stability with time may be further specified as long or short term, and the interval of "short" or "long" is defined for a particular instrument.

Exercise 14.3

Connect two oscillators to the horizontal and vertical inputs of an oscilloscope. Adjust both oscillators for a nominal frequency of 1 kHz. Adjust the dial setting of one of the generators so that a stable Lissajous figure is obtained. Wait 5 minutes and then measure the rate at which the pattern is rotating. The number of rotations per second is equal to the relative frequency change of the two oscillators over the 5-minute interval.

Amplitude stability indicates the change in amplitude which can be expected as a function of disturbing influences, such as temperature or changes in line voltage. One important component of amplitude change is that which occurs as the frequency is varied. Some oscillators include a meter which reads a scaled value of output voltage so that any change in amplitude is easily detected.

Exercise 14.4

Connect the output of an oscillator to an oscilloscope. Adjust the sweep to free run at some slow sweep time ($\sim$0.1 sec/division). If the oscillator frequency exceeds several hundred hertz, the trace becomes a solid band with a height proportional to oscillator-output amplitude. Adjust the oscillator amplitude so that this height is two divisions less than full scale. Slowly vary the oscillator frequency through its full range. Determine the maximum percentage change in output amplitude as a function of frequency. Switch between adjacent ranges and measure the instantaneous amplitude change while the oscillator waveform recovers from the transient. *Note:* Make sure the bandwidth of the oscilloscope is greater than the maximum frequency used.

14.3 pulse generators

The most common type of pulse generator produces a periodic train of equal amplitude pulses. Most general-purpose generators include provisions to vary at least three pulse parameters:

1. The *frequency*, or rate at which pulses occur. Some generators are calibrated in *period*, the time between similar portions of two adjacent pulses, and, as we have already seen, this quantity is inversely related to frequency.
2. The *width*, or time duration of the pulse.
3. The *amplitude*, or peak voltage of the pulse and its polarity.

By appropriately adjusting these three parameters, it is possible to generate test signals of particular interest. For example, by shortening the duration of a pulse it can be made to approximate an impulse, but if both the period and width are increased, an approximation to a step results.

Although our study in this assignment is limited to simple pulse generators as described above, generators with provisions for generating more complex pulse patterns exist. One possibility is the generation of two or more identical pulses in rapid succession, followed by a relatively long period before the pulse packet is repeated. Generators of this type are frequently used to test the resolving power of nuclear instrumentation, where the ability to discriminate between several closely spaced pulses is required. Since average frequency is often quite low in such applications, excitation with a high frequency, uniform pulse train is an unrealistic test.

Digital word generators can provide any pattern of identical pulses, within some maximum length, before an identical pulse pattern is repeated. For example, such a generator might provide a pulse train as shown in Figure 14.3. Waveforms of this type are useful in testing digital systems,

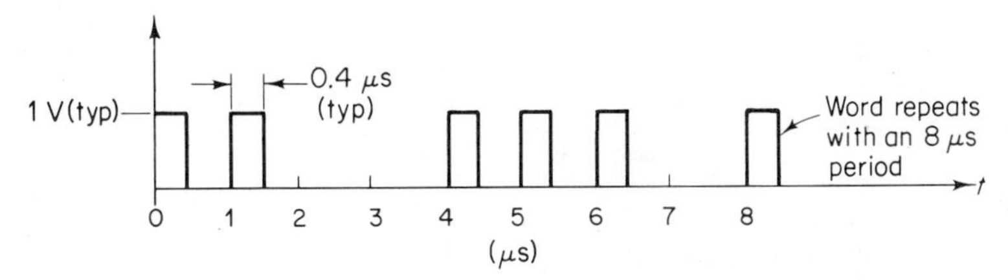

Figure 14.3. Output of a word generator.

such as those discussed in Chapter 17. This type of flexibility can be increased for a price, and more sophisticated generators permit the amplitude, width, and location of individual pulses in a word to be independently adjusted.

A related feature is the external trigger. If a pulse generator is operated in this mode, it produces a pulse wherever it is triggered by an external pulse. This feature greatly increases the number of patterns which can be generated. For example, trigger pulses can be generated with a computer and therefore can have very complex patterns. The minimum time between consecutive trigger pulses for reliable pulse generation is an important specification for generators with this feature.

Many generators provide two output pulses, the main pulse and an auxiliary pulse which is intended to trigger other equipment such as an oscilloscope in synchronism with the main pulse. Since the main pulse and the trigger pulse are generated by independent circuits, loading by the circuit to be triggered does not affect the fidelity of the main pulse. Further, in some generators it is possible to change the relative timing of the trigger and main pulses. For example, the trigger pulse may occur simultaneously with the leading or falling edge of the main pulse, or it may be possible to have the trigger pulse occur some adjustable time before the main pulse. The control which adjusts the relative timing of these two pulses is often labeled *delay*, signifying the time delay of the main pulse following the trigger pulse. This flexibility may permit the observation of initial portions of transients which are difficult to observe by other means.

Another feature is the ability to increase the rise and/or fall time (see Section 14.3.1) from some minimum value. In some generators these two parameters are varied simultaneously, while in others they can be independently adjusted.

Figure 14.4 illustrates several pulse generators.

14.3.1 pulse specifications

If we observe the output of a pulse generator, we find that the idealized description of a single member of a pulse train in terms of width and amplitude is not sufficiently detailed for many applications. Numerous discrepancies

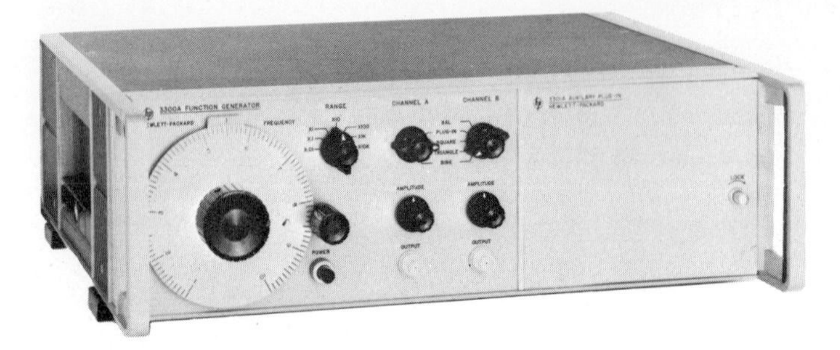

H-P Model 3300A

E-H Model 139B

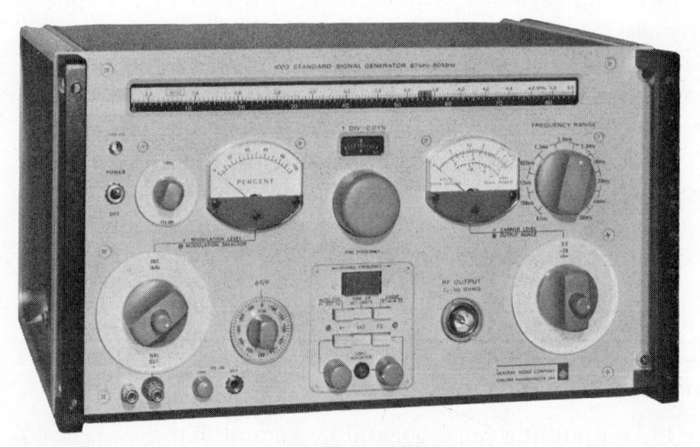

GR Model 1003

Figure 14.4. Pulse generators.

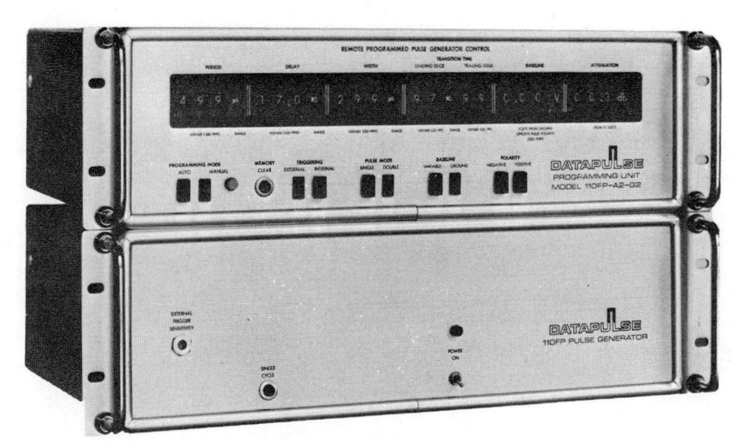

Datapulse Model 110FP/A2

Tektronix Model 114

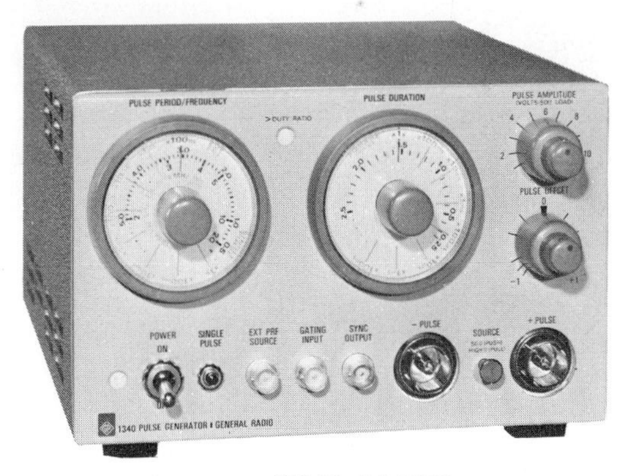

GR Model 1340

Figure 14.4. (*cont.*).

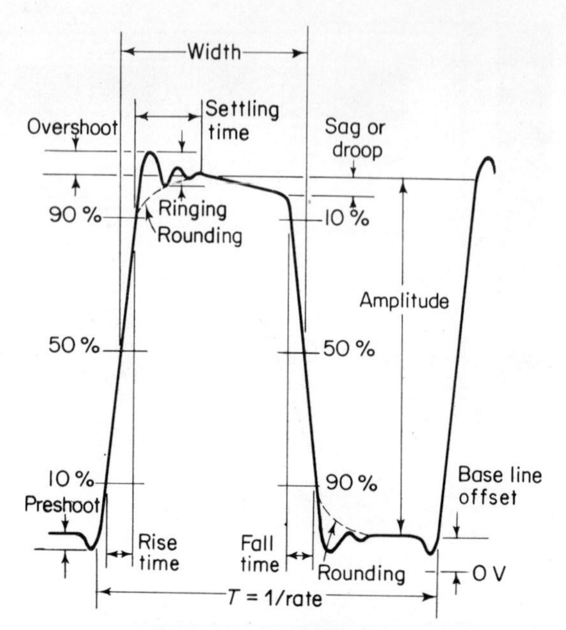

Figure 14.5. Pulse description.

between an actual pulse and an idealized rectangular pulse exist. For example, small oscillations or *ringing* is often present following rapid transitions, or the top of a pulse may not be flat. Figure 14.5 shows the terms suggested by Hewlett-Packard (Reference 1) for the detailed description of pulses.

Exercise 14.5

Obtain a pulse generator and set it for 100-μs wide, 1-volt amplitude, positive pulses with a repetition frequency of 1 kHz. Apply this signal to the network shown in Figure 14.6. Determine approximate values of amplitude, rise time, fall time, width, and droop for the pulse displayed on the oscilloscope. *Note:* Measurements are more easily made if the oscilloscope is triggered externally directly from the pulse generator.

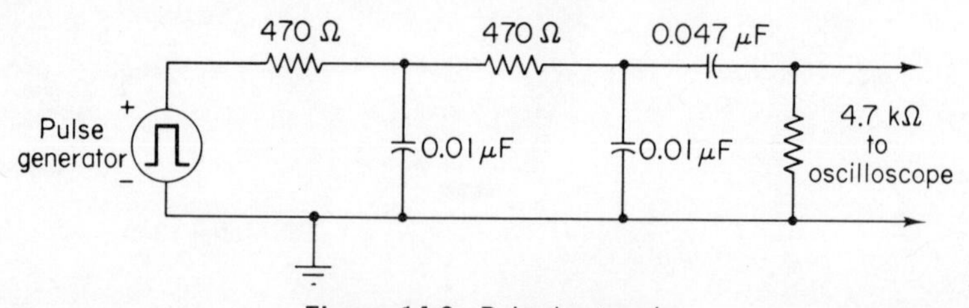

Figure 14.6. Pulsed network.

14.3.2 other pulse generator specifications

In addition to pulse characteristics as discussed in the preceding section, certain other parameters are often included in pulse generator specifications. Two of these, duty cycle and jitter, are discussed in this section.

Jitter is defined as the maximum variation in period from cycle to cycle. For example, suppose a pulse generator is operating at a nominal 1-kHz repetition frequency and it is found that the start of adjacent pulses are separated by times which vary from 999 to 1001 μs. In this case the uncertainty in the period is ±1 μs and the jitter is ±0.1% (or 0.2% peak-to-peak) of the period.

Generate a train of 1-μs wide pulses at a 1-kHz repetition frequency. Trigger an oscilloscope externally from one pulse and observe the start of the following pulse at a time scale of 1 μs/division or less using the delayed sweep mode. (Operation of the delayed sweep is discussed in Section 11.5.3.) Jitter will be displayed as a blurring or smearing of the leading edge of the delayed pulse. The maximum width of the smear divided by the nominal period is the peak-to-peak jitter expressed as a fraction of the period. Determine this quantity for your generator. *Note:* The jitter of a delayed oscilloscope sweep is on the order of 1 part in 20,000, or 0.005%. Unless the measured jitter is significantly larger than this, the result will include oscilloscope errors.

Duty cycle is the ratio of pulse width to pulse period. For example, the duty cycle of the 1-μs, 1-kHz wave train used in the preceding exercise is 0.1%. The maximum duty cycle of some pulse generators (particularly high-power units) is often significantly less than 100% because of circuit limitations. Some generators are equipped with a signal light to indicate when the maximum duty cycle is exceeded.

Exercise 14.7

Determine the maximum duty cycle of your pulse generator by slowly increasing the pulse width until it approaches the period. When maximum duty cycle is exceeded, the waveform becomes irregular or the width will no longer increase.

14.3.3 pulse transmission (reference 2)

Even if a pulse generator which can provide an adequate pulse shape at the generator is available, maintenance of pulse fidelity during transmission may be a significant problem. If the time required for the pulse to travel from the generator to the load is significant compared with the rise time of the pulse, it is not possible to assume instantaneous transmission.

In this case, discontinuities in the transmission medium cause pulse reflections, much as a solid wall can cause waves in water to be reflected. Before discussing this problem in detail, some of its effects will be illustrated in the following exercise.

Exercise 14.8

Adjust a generator with a rise time of less than 10 ns to produce 100-ns wide pulses. Measure the pulse shape directly at the generator output with a fast rise time oscilloscope. Use a short oscilloscope ground lead for this measurement, and make sure that nothing other than the oscilloscope probe is connected to the generator output. Record important features of the observed waveform. Add a 2-foot clip lead in series with the oscilloscope ground and note waveform changes. Next, record the waveform at the end of a 5-foot piece of RG-58/U cable, connected as shown in Figure 14.7. Finally, connect the cable to the oscilloscope with a 50-ohm through termination (see Figure 19.3). How does the observed pulse compare with that produced directly at the generator?

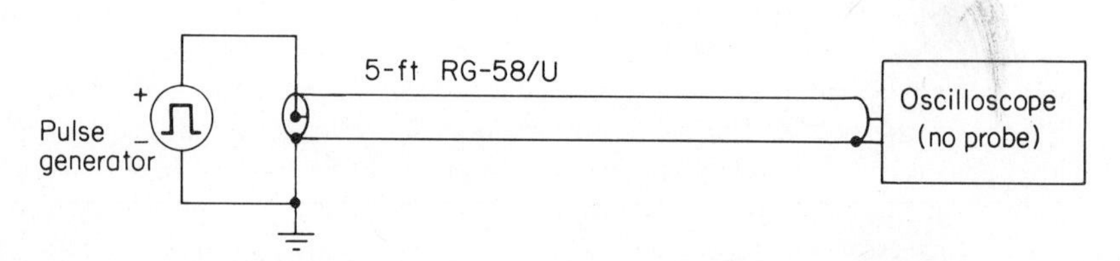

Figure 14.7. Transmission with unterminated line.

The previous exercise demonstrates the pulse deterioration which can occur with improper transmission techniques. One aspect of the general transmission line problem is considered in Chapter 19. The approach used in Chapter 19 is to determine the impedance at the input of a given length of line as a function of the characteristic impedance of the line, the length of line, the load or termination impedance, and the frequency of excitation. However, this type of analysis is somewhat cumbersome for determining the behavior of pulses transmitted on a coaxial line. An alternative approach is to consider how a pulse signal is changed or reflected when it encounters a load impedance at the end of a transmission line. The problem we will consider in detail is illustrated in Figure 14.8. The generator is represented as a voltage source with a Thevenin output resistance equal to R_G. The load end of the line is terminated in a resistance R_L. The coaxial line is characterized by three quantities, the physical length l, the phase velocity v_p (which for the coaxial line is also the pulse propagation velocity), and the characteristic impedance Z_0.

It is assumed that the line is *lossless*, which implies that the resistance of the center and outer conductors is zero, and that the conductance of the

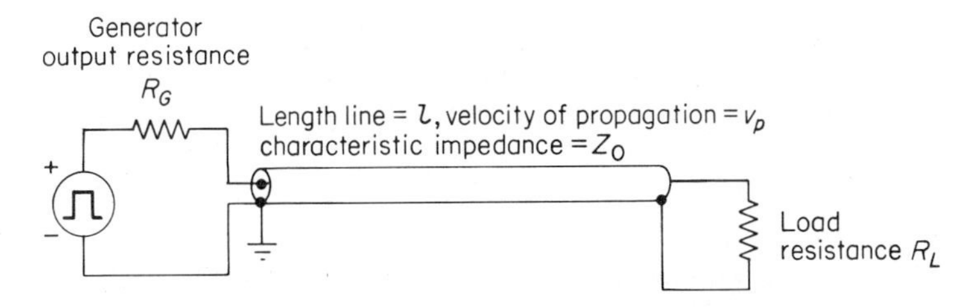

Figure 14.8. Line with source and load resistances.

dielectric which separates them is zero. It is further assumed that the line does not distort the signals on it. These assumptions are justified in many cases of interest.

Several questions are evident, including:

1. If the generator produces a given pulse with no load, how is this pulse altered when the line is connected to the generator?
2. How does this pulse travel down the line?
3. What happens when the pulse reaches a discontinuity (represented in this problem by a resistance) in the line?

One interpretation of *characteristic impedance* for a transmission line is the impedance which would be measured at the input of an infinitely long line. There is no way that the generator can tell that the line is not infinitely long until a pulse has traveled the length of the line, been reflected from the termination R_L, and returned the length of the line to the generator terminals, and this sequence requires a time equal to $2(l/v_p)$. Therefore it is possible to determine the signal initially applied to the generator end of the line from the model shown in Figure 14.9. From this model we see that the amplitude of the initial pulse applied to the line is equal to $Z_0/(R_G + Z_0)$ times the open-circuit amplitude of the generator.

If the line is lossless and uniform, the pulse applied to the line travels along the line without any change in amplitude or shape at a velocity v_p. At a time $t = l/v_p$ the leading edge of the pulse reaches the end of the line which is terminated in R_L. If $R_L = Z_0$, there is no electrical discontinuity,

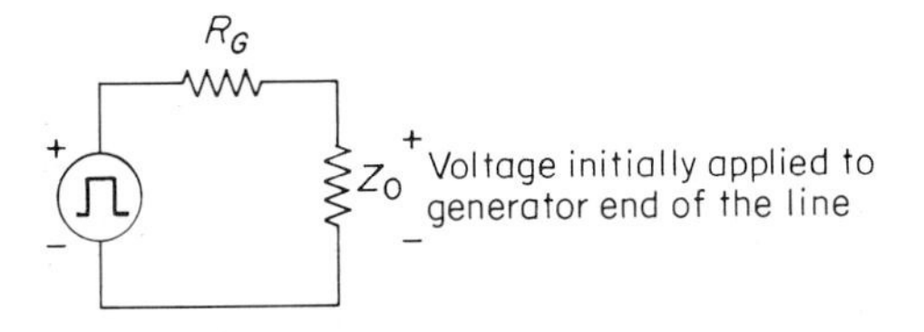

Figure 14.9. Determination of initial pulse on line.

since this load impedance is identical with that of an infinitely long line, and the pulse cannot distinguish the lumped resistor $R_L = Z_0$ from an additional length of line. Therefore, no reflection is generated by a load $R_L = Z_0$. If $R_L \neq Z_0$, a reflected pulse, which travels back along the line, is generated at the termination. It can be shown (see Reference 2, Chapter 4) that the amplitude of the reflected pulse is equal to the *reflection coefficient* Γ multiplied by the amplitude of the incident pulse, where

$$\Gamma_L = \frac{R_L - Z_0}{R_L + Z_0} \tag{14.1}$$

(The subscript L indicates the reflection coefficient at the load end of the line.)

Notice that $\Gamma_L = 0$ and no reflected pulse is generated if $R_L = Z_0$; the polarity of the reflected pulse is opposite that of the incoming pulse for $R_L < Z_0$; and that the polarity of the reflected pulse is equal to that of the incoming pulse for $R_L > Z_0$.

The reflected pulse travels back toward the generator with a velocity v_p. The generator resistance R_G is the termination for the line at the generator end. If this resistance is equal to Z_0, no reflection occurs when the first reflected pulse reaches the generator. If $R_G \neq Z_0$, a second reflected pulse is generated with an amplitude equal to Γ_G times the amplitude of the pulse incident at the generator end of the line, where

$$\Gamma_G = \frac{R_G - Z_0}{R_G + Z_0} \tag{14.2}$$

Note that $|\Gamma| < 1$ for any resistive termination, and therefore the amplitude of successive reflections decreases.

This process of reflection continues, and a new reflected pulse generated whenever a pulse encounters a discontinuity in the impedance of the line. Since the system consisting of the generator, line, and termination is linear, the voltages of all pulses present on the line at any particular time and location can be added to obtain the total voltage distribution along the line. If the generator output remains constant for a sufficiently long time, the reflections damp out, and the voltage becomes equal at every point on the line. This final value voltage can be determined by considering the cable as simply two pieces of wire.

Figure 14.10 illustrates the process with an example. A 50-ohm line (such as RG-58/U) is assumed connected to a generator with a 25-ohm output resistance. The line is terminated in a 100-ohm resistor. Figure 14.10b shows the output the generator would produce with no load connected to it. Use of the model of Figure 14.9 shows that when this generator is loaded with a 50-ohm line, the amplitude of the initial pulse is $\frac{2}{3}$ of the generator open-circuit voltage. Figure 14.10c shows the distribution of volt-

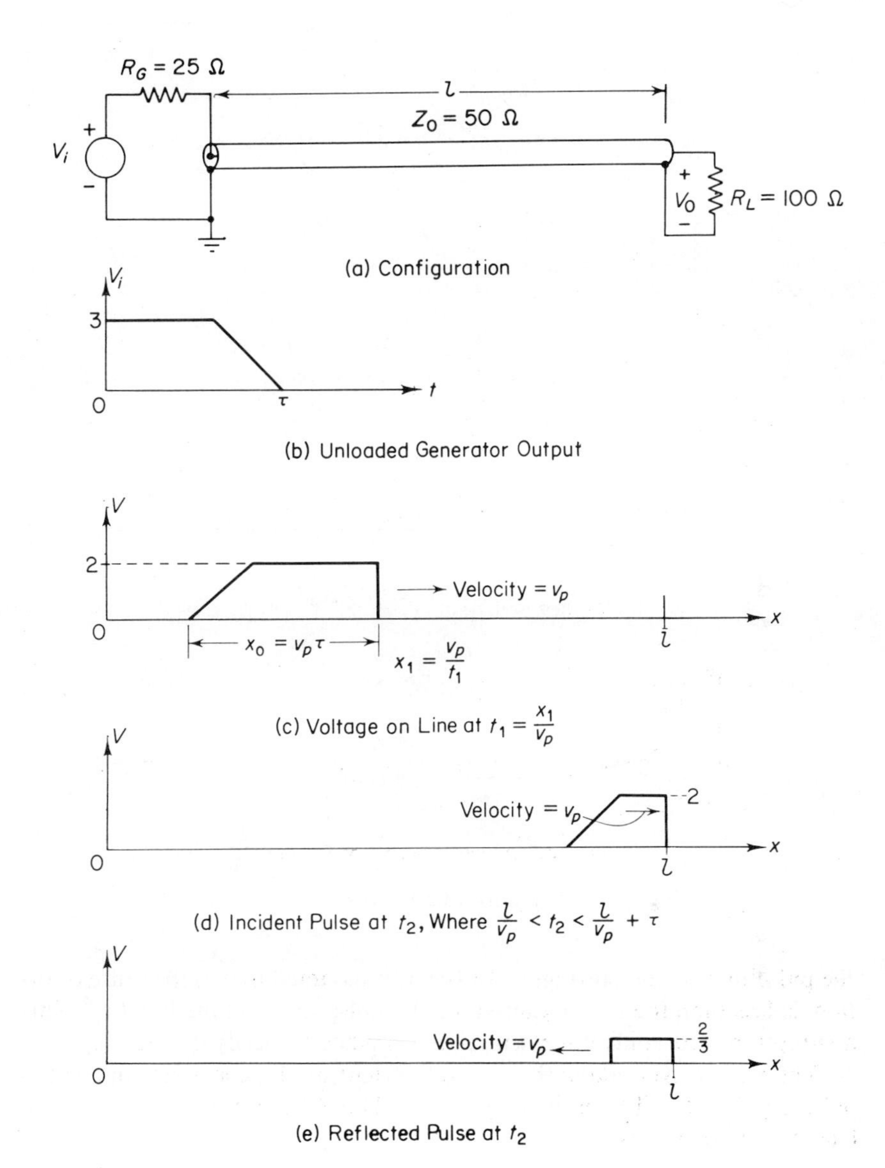

Figure 14.10. Example of Reflections.

age along the line *as a function of* x, the distance from the generator end of the line. Note in this figure we are looking at the voltage on the line at a *fixed* time $t_1 = x_1/v_p$. Also note that the leading edge of the pulse, which occurs at time $t = 0$ at the generator end of the line, is the leading portion of

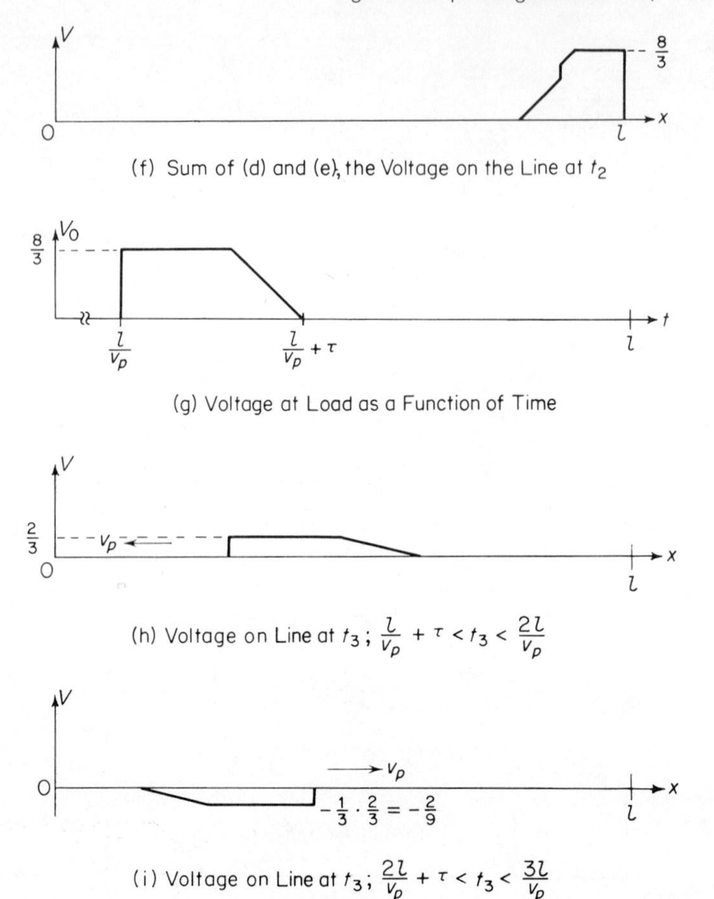

Figure 14.10. (*cont.*).

the pulse on the line. In Figure 14.10c, it is assumed that τ, the pulse duration, is less than the time required for the pulse to travel the line l/v_p. This assumption is certainly not essential, but is made to clarify the drawing.

The pulse travels along the line to the right until the leading edge of the pulse reaches the 100-ohm termination. The reflection coefficient at the load end of the line is

$$\Gamma_L = \frac{R_L - Z_0}{R_L + Z_0} = \frac{100 - 50}{100 + 50} = +\frac{1}{3} \tag{14.3}$$

Therefore, a reflected pulse traveling to the left with an amplitude equal to $+\frac{1}{3}$ times the amplitude of the original pulse is produced by reflection. Figures 14.10d, 14.10e, and 14.10f show, respectively: the voltage of the incident pulse, the voltage of the reflected pulse, and the voltage distribution along the line, which is the sum of the incident and reflected pulse at some time t_2, where $(l/v_p) < t_2 < (l/v_p) + \tau$. Figure 14.10g shows the voltage at the load end as a function of *time*. This is the sum of the voltages in the incident and reflected pulses at $x = l$.

In Figure 14.10h we see the voltage along the line at time t_3, where $(l/v_p) + \tau < t_3 < 2(l/v_p)$. The amplitude of this pulse, which travels to the left, is $\frac{1}{3}$ the amplitude of the original pulse since Γ_L is $\frac{1}{3}$. It encounters the discontinuity at the generator end of the line at time $2(l/v_p)$. This reflection coefficient is

$$\Gamma_G = \frac{R_G - Z_0}{R_G + Z_0} = \frac{25 - 50}{25 + 50} = -\frac{1}{3} \tag{14.4}$$

At some later time t_4, where $2(l/v_p) + \tau < t_4 < 3(l/v_p)$, the voltage distribution along the line is seen in Figure 14.10i. This figure shows the pulse generated by the second reflection traveling to the right.

This process of reflection continues indefinitely with decreasing amplitude if both the generator and load impedances differ from the characteristic impedance of the transmission line. It can be seen from Equations (14.1) and (14.2) that if the load impedance is equal to Z_0, there are no reflections, while if $R_L \neq Z_0$ but $R_G = Z_0$, there is only one reflection.

Exercise 14.9

Consider the connection shown in Figure 14.11. In this circuit, a pulse generator with an output resistance R_G is shown connected to a high frequency oscilloscope. Trigger the oscilloscope externally from the pulse generator's trigger pulse if one is available. The connection between the oscilloscope and generator should be as short as possible. A Tee connector is used to attach a 50-foot length of RG-58/U cable to the oscilloscope input terminal. The remote end of the cable is left open circuited. Adjust the generator for 100-ns wide pulses at a 100-kHz repetition frequency. This pulse width is less than $2(l/v_p)$, since the velocity of propagation for any transmission line is less than the velocity of light, and the velocity of light is approximately 1 foot/ns.

A pulse from the generator is displayed on the oscilloscope when it is generated and this pulse travels the length of the line and back following reflection at the

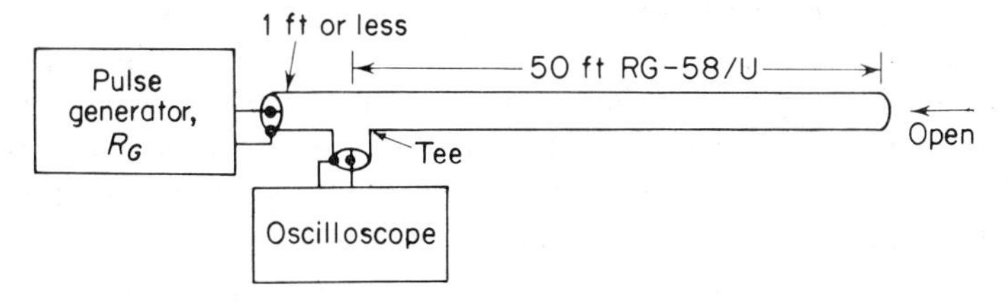

Figure 14.11. Connection for measuring reflections.

load end. The sum of the first and second reflected pulses is displayed on the oscilloscope, and the time separation from the first pulse is $2(l/v_p)$. If the amplitude of the first pulse is A_1 and the amplitude of the second displayed pulse is A_2, show that $A_2/A_1 = 1 + \Gamma_G$. Use this relationship to determine Γ_G and R_G for your generator. Also use the time spacing between the first two displayed pulses to determine v_p for RG-58/U cable.

If R_G is equal to 50 ohms, only two pulses, of approximately equal amplitude, are displayed. Notice that the value of R_G can be altered by adding a resistor in series or in parallel with the generator output. For example, if your measurements show that $R_G > 50$ ohms, shunting an appropriate resistor from the output of the generator to ground will lower the effective value of R_G to 50 ohms. Calculate the value of series on parallel resistor required to adjust the output resistance of your generator to 50 ohms. Verify your result by measuring Γ_G after this modification. Show that with $\Gamma_G = 0$, the ratio A_2/A_1 is Γ_L. Use this relationship to measure Γ_L for $R_L = 0$ ohms and $R_L = 50$ ohms.

14.4 function generators

Hewlett-Packard (Reference 1) defines a function generator as "a signal generator that delivers a choice of different waveforms with frequencies adjustable over a wide range." Available instruments cover a frequency range from less than 10^{-4} Hz to more than 100 kHz, and produce sine, square, triangle, and sawtooth waves. Oscillators as discussed in Section 14.2 are simplified function generators in that they provide only one type of output. Certain purely sinusoidal oscillators are available with some of the features described in this section.

Many generators provide two or more different types of waveforms. These multiple outputs may be available simultaneously, and can be used in a variety of applications. For example, if a sine wave and a triangle wave with identically timed zero crossings are generated, the triangle wave provides a voltage which is linearly related to the phase of the sine wave.

Square waves provided by function generators are particularly useful waveforms for testing electronic amplifers. A low-frequency square wave with fast rise time has frequency components which extend over a wide frequency range, and therefore can provide as much information about the performance of an amplifier as can be learned by testing with a number of different frequency sinusoids. The amount by which the rise time of a square wave is slowed up yields information about the high-frequency response of the amplifier, while the droop introduced by the amplifier is a measure of its low-frequency performance. Feedback amplifiers can add overshoot and ringing to square waves, and the amplitude of such characteristics is an indication of the stability of the amplifier.

A function generator with an electronically variable output frequency is called a *voltage-controlled oscillator*. The range of control of such units can

Beckman Model 9010

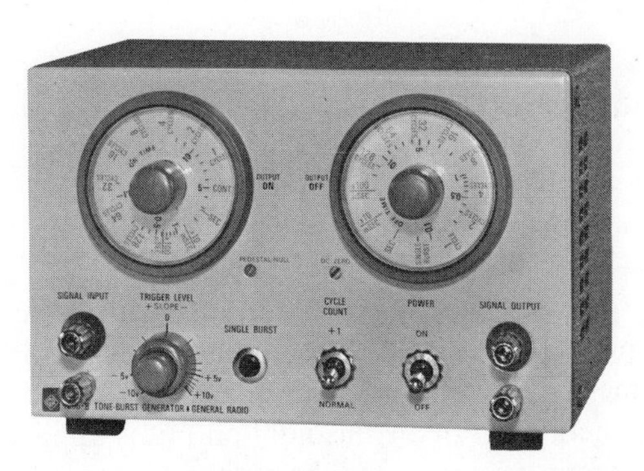

GR Model 1396B

Wavetek Model III

Figure 14.12. Function generators.

extend over several decades of frequency. This feature is valuable in automated test situations. Another voltage-controlled oscillator application is for voltage measurement. If the number of cycles produced by a voltage-controlled oscillator during a fixed time interval is counted, the total count is proportional to the average control voltage applied to the oscillator during the counting interval. This technique is used in several types of digital voltmeters (see Section 13.3), and such instruments have excellent noise rejection characteristics because of the averaging supplied by the counting operation.

Exercise 14.10

Obtain a multiple-output function generator which can be voltage controlled. Observe the shape of the available output waveforms with a fixed-output frequency. Set the output frequency for 20 kHz and apply a 1-kHz signal from a separate generator as a frequency-control signal. Trigger your oscilloscope from the 1-kHz signal. The displayed waveform is an example of a *frequency-modulated* signal.

Certain function generators provide a means for *gating* or turning the output signal on or off with an externally supplied signal. In some instruments the output signal always starts with the same phase, while in others no phase-lock provision is included. A related feature is the ability to *amplitude modulate* an output, or provide a signal with peak-to-peak magnitude linearly related to an applied control voltage.

Function generators which provide two or more outputs with an adjustable phase relationship are also available. One application of this feature is for accurate measurement of the phase shift produced by a circuit under test. One phase from a generator with a sinusoidal output is applied to the circuit, while the second generator output and the output of the circuit are used as inputs to the two axes of an oscilloscope. The relative phase of the two generator outputs is adjusted until a linear Lissajous figure is obtained. The phase shift produced by the circuit is then identical with that between the two generator outputs.

Figure 14.12 illustrates several function generators.

14.5 references

1. *1968 Instrumentation*, Hewlett-Packard Co.

2. Adler, Chu, and Fano, *Electromagnetic Energy Transmission and Radiation*, Chapter 4, John Wiley & Sons, Inc., New York, 1960.

15

time, frequency, and waveform analysis

15.1 introduction

Frequency is defined as the number of repetitive events occurring in a unit interval of time, or, less rigorously, as the number of cycles of a periodic waveform that occur in one second. A waveform $g(t)$ is periodic with period T seconds if T is the shortest interval of time for which $g(t) = g(t + kT)$ where $k = 1, 2, 3, \ldots$, for all values of t. The frequency f of a waveform with period T is then defined as $f = 1/T$. The unit of frequency is the hertz (Hz). Time and frequency are therefore intimately related. For example, if we are capable of measuring time, we may determine the frequency of a periodic waveform by measuring the duration of a number of cycles of the waveform and then dividing this number of cycles by the duration. Similarly, if we can measure frequency, we may determine a length of time by counting the number of cycles of the known frequency that occur during the time interval and dividing this number of cycles by the frequency.

In this chapter we will discuss the necessity of accurate and precise standards of time and/or frequency and explore digital and analog instruments for frequency measurements. Some elementary harmonic analysis (Fourier series) and frequency multiplication techniques will be introduced, and measurements of frequency content of waveforms discussed.

15.2 time and frequency standards

Originally, time standards were derived from astronomical observations. Today, exact time intervals are derived from atomic standards. International agreement has defined the second as 9,192,631,770 cycles of the resonance frequency of the cesium atom. This uniform time scale is called Atomic Time. The previous time scale (Ephemeris Time) was based on the orbital period of the earth about the sun in the tropical year 1900. For everyday living, Solar Time (Universal Time or Greenwich Mean Time), based on the mean rotational period of the earth, is used. Note that the conversion factor from Atomic Time to Ephemeris Time is a constant, while that from Atomic Time to Solar Time changes with the fluctuations of the mean rotational period of the earth.

A frequency standard is specified by its accuracy and its stability. The accuracy of the standard is the difference between the absolute frequency the standard is said to produce and the average frequency that it does produce. The stability or precision of the standard is the degree to which the standard retains a specific frequency within any specified time interval. It is common practice to specify both short- and long-term stability. "Short" implies the time interval required to make a measurement, on the order of seconds, and "long" implies time intervals on the order of days or weeks. The evolution and major uses of frequency standards are shown in Figure 15.1.

In all frequency standards, oscillation is produced by the insertion of the frequency-determining element into the feedback loop of an amplifier as shown in Figure 15.2. The most common frequency-determining element is the tuned *LC* circuit, which is accurate to 1 part in 10^4 under ideal conditions and is highly dependent on the Q of the coil and capacitor.

A mechanical tuning fork magnetically coupled to the feedback loop of an amplifier can be used as a standard frequency source accurate to 1 part in 10^7 under ideal conditions. Probably the most familiar example of this system is the Accutron watch. Electromagnets controlled by the output of the amplifier excite the tuning fork, and magnetic pickup coils send induced electric oscillations back to the amplifier input. The tuning fork vibrates at a frequency proportional to the material characteristics of the tines (density, elasticity, thickness, and length). A change in resonant frequency can result from a change in the temperature, which in turn changes the mechanical dimensions of the fork, or from a change in amplitude of vibration. For this reason, it is often necessary to enclose the tuning fork in a temperature-controlled environment and to maintain a small amplitude of oscillation if it is to be used as a frequency standard.

A quartz crystal is another example of an electromechanical frequency-determining element. It may be made to mechanically resonate by the application of an electric field of the proper frequency. Like the tuning fork, the resonant frequency of the crystal is determined by its physical dimensions and material parameters. In most crystal oscillators the temperature of the crystal is controlled to achieve greater frequency stability. Under ideal conditions accuracy of 1 part in 10^8 is possible.

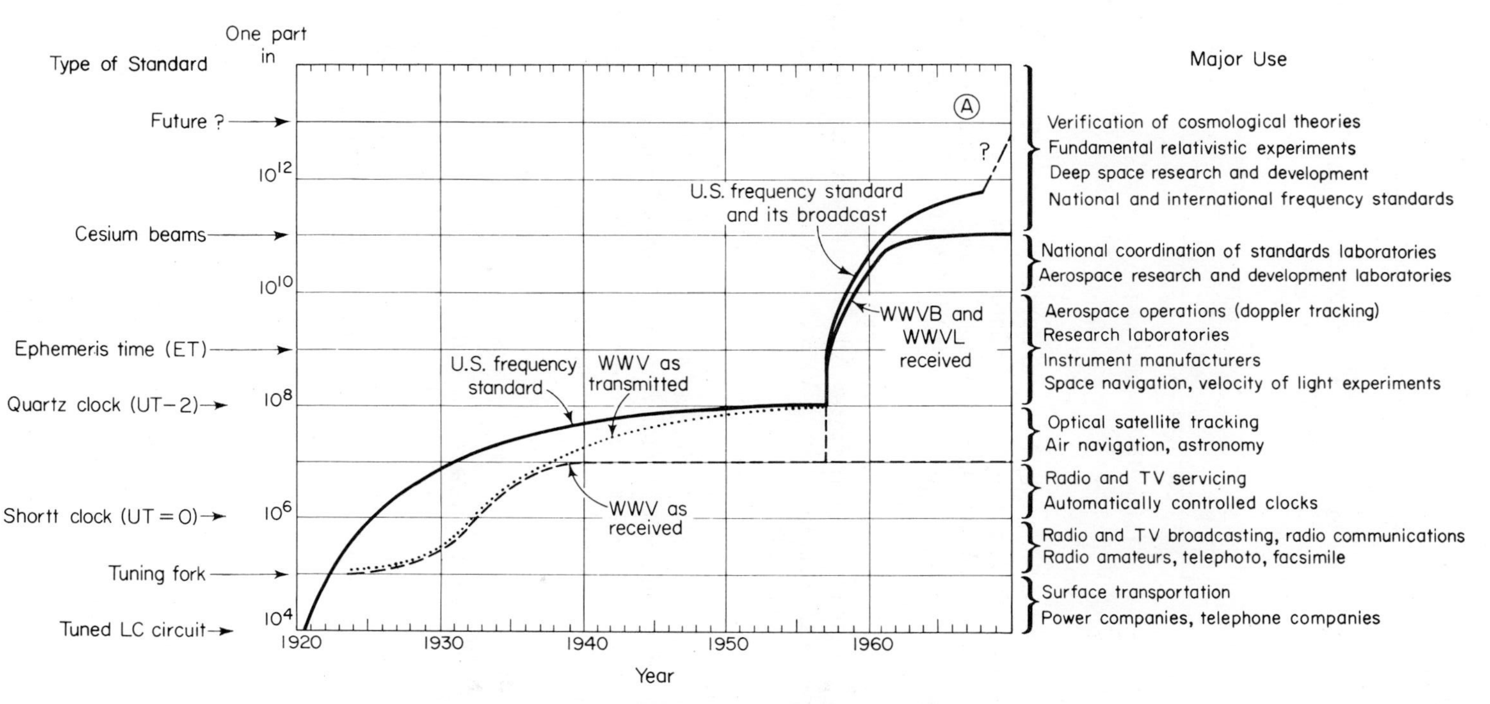

Figure 15.1. Evolution and major uses of precision frequency standards.

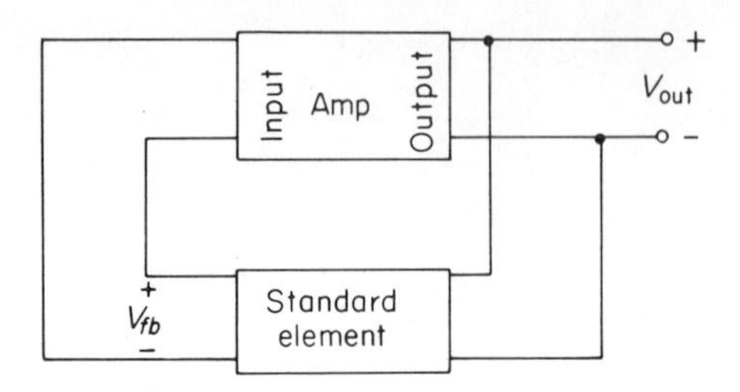

Figure 15.2. Basic frequency standard.

In the United States, the National Bureau of Standards time signals are continuously broadcast by the standard-frequency transmitters WWV (in Maryland) and WWVH (in Hawaii), on 5 and 10 MHz as well as other frequencies. In addition to WWV and WWVH, frequency broadcasts are made on WWVB (60 kHz) and WWVL (20 kHz). Owing to atmospheric properties, the reception of low-frequency broadcasts is more stable than that of high-frequency broadcasts. All standard-frequency broadcasts are based on Solar Time, except for WWVB which is based on Atomic Time, and are accurate to 5 parts in 10^{10}.

At present, the atomic clock is accepted as the most accurate frequency standard. It is accurate to approximately 1 part in 10^{10}, and with modifications the accuracy can be increased a hundredfold. This high degree of accuracy is accomplished by utilizing the resonant frequency of the cesium atom. In the near future the hydrogen laser will probably be used as a frequency standard with accuracies approaching 1 part in 10^{14}.

15.3 frequency measurement

In Chapter 4 and Chapter 11 it has been shown how the period of a waveform may be measured to an accuracy of a few percent. The accuracy of the oscilloscope for time measurements is limited to about 1% when the delayed sweep feature is employed. In this chapter we are concerned with much more precise time and frequency measurements.

Instruments for the measurement of frequency are divided into the categories of digital and analog. Generally speaking, the digital types provide higher accuracy and stability. Their simple visual readout is another advantage. However, they are more expensive than analog frequency meters, which may be adequate if cost is important and high accuracy is not important.

15.3.1 digital frequency meters

The *digital frequency meter* or *counter* is probably the most accurate, flexible, and convenient instrument for measuring frequency in general laboratory applications. These instruments are capable of measurements in the range of 0 to 40 GHz and their accuracy is limited only by the time-base oscillator stability of the counter. Although the absolute accuracy will depend on the frequency being measured, it is possible to approach the accuracy of the reference frequency standard or time base used with the counter. Some examples of digital frequency meters are shown in Figure 15.3.

The basic operation of a digital counter is shown in Figure 15.4. The input signal is converted to a series of pulses, one pulse per cycle, controlled by some feature of the input waveform. In Figure 15.4, the negative-going zero crossings are shown as the input waveform feature that produces the trigger pulses. A crystal-controlled internal time base, typically operating at frequency of 100 kHz or 1 MHz, feeds a frequency divider. This divider counts a predetermined number of cycles of the internal time base and produces an output pulse of length from 0.1 to 10 sec, the duration of which is the precise counting time or sampling time. A gate permits the trigger pulses from the input waveform to be counted only during the counting time, and thus the number of cycles occurring during the fixed time interval is indicated. In this way the frequency is measured. In Figure 15.4, four pulses will be counted during the counting time pulse. After a short period, the system is reset and a new measurement made. The frequency of this reset is called the *sampling rate* or *display time.* From this discussion, it is clear that the critical problem is the length of the counting time, which is governed by the stability of the internal time base. Typically, the internal time base is accurate to 1 part in 10^8. For higher accuracy, an external time base signal obtained from a frequency standard may be employed.

Exercise 15.1

Connect a signal generator to a digital frequency meter. Check the frequency meter by means of its internal calibration. Measure the frequency of the generator on three different ranges and note the stability of the frequency over a 1-minute period.

Vary the sampling time and display time controls to become familiar with their functions.

In addition to measuring frequency directly, a digital frequency meter may be made to perform other tasks such as direct display of the period of a waveform, showing the ratio of two frequencies, or counting the total of individual events, either random or periodic. To see how this is achieved,

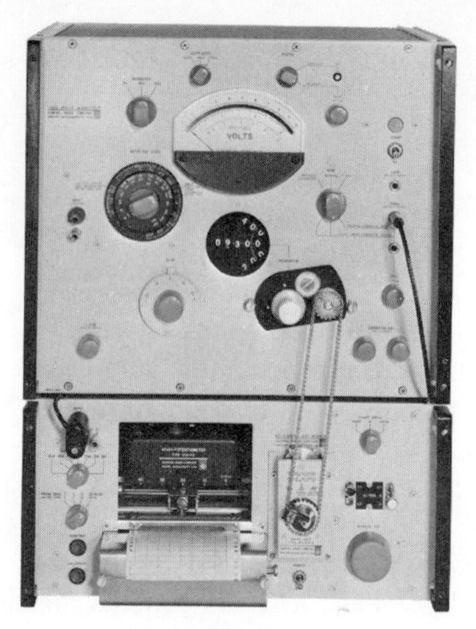

GR Model 1900

H-P Model 5221A

GR Model 1191

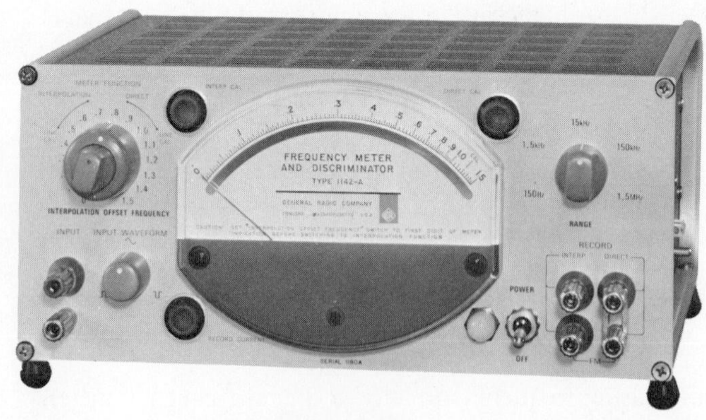

GR Model 1142-A

Figure 15.3. Examples of frequency meters and waveform analyzers.

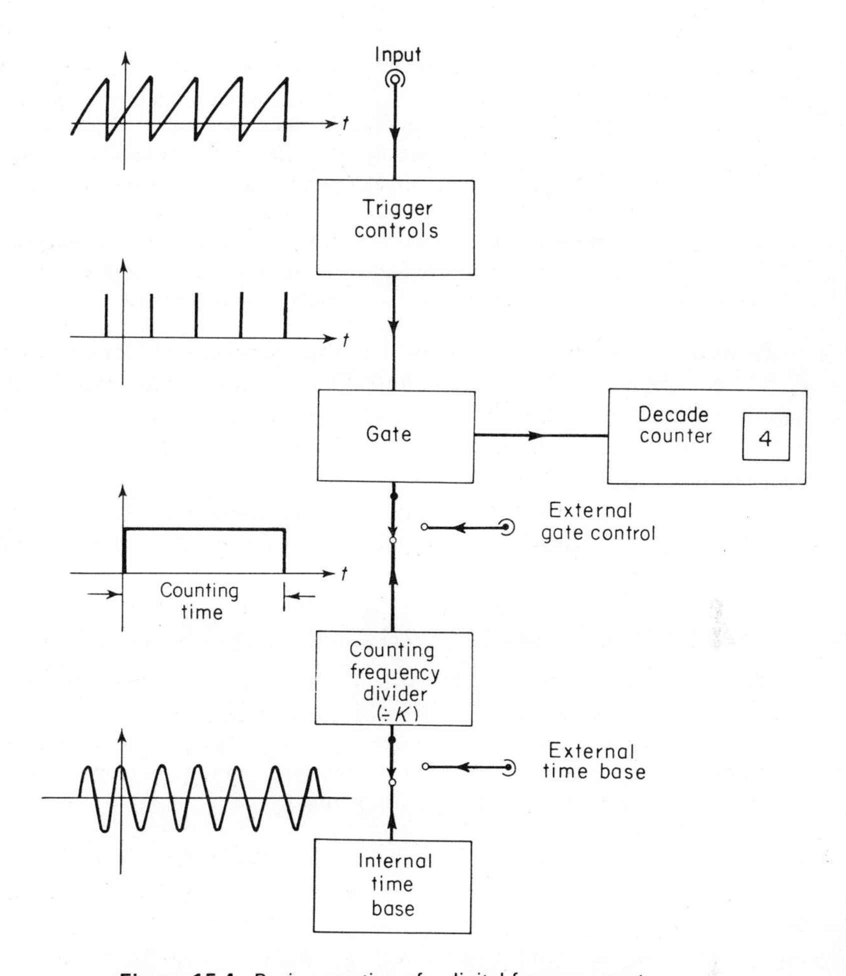

Figure 15.4. Basic operation of a digital frequency meter.

let us return to Figure 15.4. If we denote the frequency of the time base as f_T, then the counting time is K/f_T, where K is the division factor of the counting frequency divider. For an input signal of frequency f_I, the decade counter display indicates the quantity Kf_I/f_T. Since K is a constant, dependent on the setting of the sampling time control, the actual digital display is simply the ratio f_I/f_T multiplied by the scale factor K.

Exercise 15.2

To determine the scale factor K for a digital frequency meter, connect the output of a signal generator simultaneously to the normal input terminals and the external time-base input. Consult the instruction manual for the proper signal level at the time-base input. With the time-base selector switch on internal, set the signal generator frequency to that of the internal time base. Next set the time-base selector switch to the external position and note the digital display. Since $f_I = f_T$, the reading is the scale factor K. Vary the sampling time and note the value of K for each setting. If a reading is outside the scale for the display, calculate the value of K from the sampling time. Finally vary the generator frequency from one-half to twice the internal time base frequency. Does K remain constant? Do you observe any change in sampling rate or time? (*Note:* On some counters, varying the frequency at the external time-base input may require readjustment of the signal level to provide positive action of the counting frequency divider.)

Most digital frequency meters have a front panel control for calibration or internal frequency check. This control merely applies the internal time-base signal to both the signal input and the counting frequency divider. Therefore, it is *only* a check on the frequency divider and is *not* assurance that the internal time base frequency is accurate even though the digital display is correct to the last digit. The only way to check the time base is to use the meter to measure the output from a frequency standard whose accuracy exceeds that of the meter in question.

Exercise 15.3

Using two signal generators, apply the output of one to the input terminals and the output of the second to the external time-base terminals of a digital frequency meter. Set the meter for external time-base operation. Using the values of K determined in Exercise 15.2, vary the frequencies of both generators to demonstrate that the display is proportional to the frequency ratio f_I/f_T.

The digital display in terms of the periods of the input and time-base waveforms is given by KT_T/T_I. Thus if the unknown frequency is applied to the external time-base terminals and a known frequency is applied to the input terminals, the digital display will indicate the period of the unknown waveform in units of T_I/K. For convenience T_I should be chosen as some multiple of 10 such as 10 μs (100 kHz) or 1 μs (1 MHz).

Exercise 15.4

Connect two signal generators to the digital frequency meter as described in Exercise 15.3. With the meter set on internal time base, adjust T_I to 10 μs. Switch the meter to external time base and measure the period of the signal connected to the external time-base terminals for generator settings of 50, 100, and 200 kHz.

If the gate circuit is controlled by some external time control, then the digital frequency meter can be used as an event counter over long time intervals. For example, with the input derived from a photoelectric cell, the instrument may be used to count items passing on a conveyor belt for durations of minutes, hours, days, etc., up to the full capacity of the digital display. When used in this fashion, the gate may be controlled and the display reset manually by an operator or automatically, with a printed readout at the end of each sampling period.

15.3.2 analog frequency meters

An *analog frequency meter* is very similar to an AC voltmeter in that a waveform converter circuit is employed to convert the frequency of the waveform to a DC current which in turn deflects a d'Arsonval movement. If a *linear* RC or LC network is used in the waveform converter, then not only will the DC output be proportional to frequency, it will also depend on signal amplitude. This situation is undesirable in that it is impossible to differentiate between indicated output due to frequency or amplitude. In some situations it is practical to limit the amplitude to some known level and thus suppress output due to this cause. A second problem with LC waveform converters is that the frequency range for high accuracy is very limited.

The problems associated with an RC or LC converter circuit may be overcome with the arrangement shown in Figure 15.5. The input signal operates a pulse generator which produces an output pulse of amplitude A and duration δ whenever the input voltage has a positive-going zero crossing. Thus the output of the pulse generator will be periodic with the same period as the input waveform. The average value of the pulse generator output will be

$$\overline{V_{\text{out}}} = \frac{A\delta}{T} = (A\delta)f \tag{15.1}$$

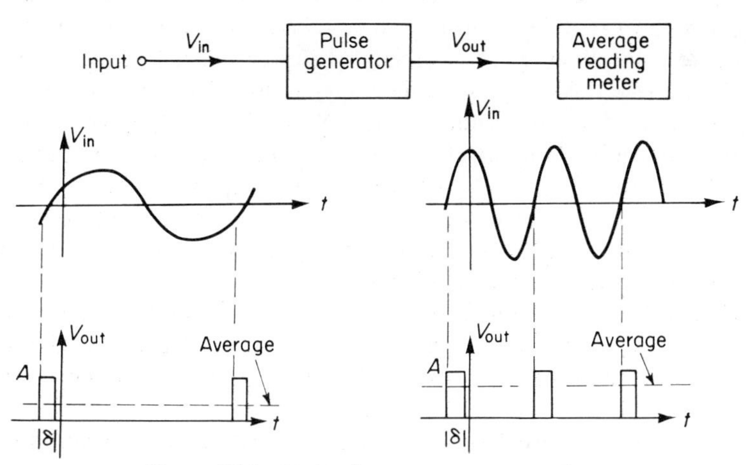

Figure 15.5. Analog frequency meter operation.

where T is the period and f is the frequency of the input waveform. Since the output pulses have an amplitude and duration independent of the input waveform, the product $A\delta$ is a constant and the average reading meter will provide a deflection directly proportional to the input frequency.

Exercise 15.5

Construct the circuit in Figure 15.6. The value of L is not critical.

Find the resonant frequency f_0 as indicated by a maximum signal on the oscilloscope. This should be near $1/2\pi\sqrt{LC}$. Plot the oscilloscope response to the circuit as a function of frequency between the two frequencies at which the response is 0.1 times the value at f_0. If the analog frequency meter has an expanded scale capability, use it for maximum accuracy.

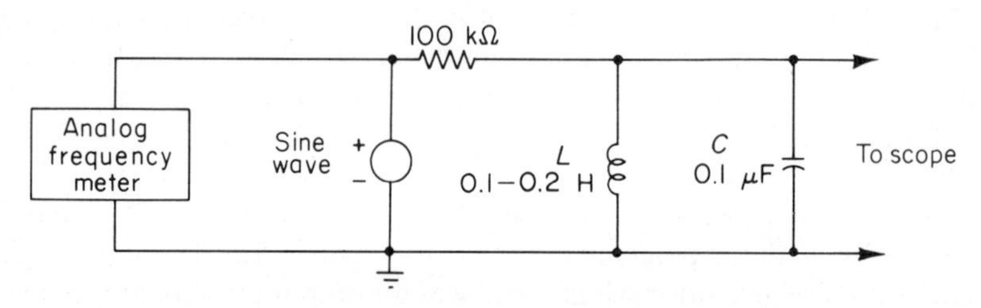

Figure 15.6. Analog frequency measurement of resonances.

15.4 harmonic analysis

In Section 15.1 the relationship between period and frequency of a periodic waveform were given. No requirement was placed on the amplitude variation of the periodic wave. Since most linear systems are described in terms of their performance in response to a sinusoidal waveform, it becomes useful to describe arbitrary periodic waveshapes in terms of equivalent sinusoidal components of different frequency and phase. Such a decomposition of a waveform is known as *harmonic analysis* or a *Fourier series* representation. Thus, if the response of a given linear system to a sinusoid is known as a function of frequency, then the response to the complex waveform can be calculated as the superposition of the responses of the individual sinewave components of the complex waveform.

While a detailed treatment of Fourier series is beyond the scope of this chapter, we will state some basic principles without proof, and present the results for some simple waveforms which are frequently encountered in the laboratory.

If a periodic waveform has a frequency f_1, then the sinusoidal components present in the waveform will be integer multiples of f_1. These integer multiple frequencies are known as the harmonic frequencies, nf_1 being called the nth harmonic. Thus the periodic waveform $g(t)$ may be expressed as

$$g(t) = A_0 + \sum_{n=1}^{\infty} A_n \sin (n\omega_1 t + \phi_n) \tag{15.2}$$

where $\omega_1 = 2\pi f_1$, A_n is the magnitude of the sinusoid of frequency nf_1, and ϕ_n is its phase. A_0 is simply the constant or DC component of $g(t)$. For a given waveform, A_n and ϕ_n may be calculated and their values serve to describe the frequency content or *spectrum* of the waveform.

As stated above, the calculation of A_n and ϕ_n will not be discussed. References 2 and 3 deal with this topic. However, two common periodic waveforms are shown in Figure 15.7 along with their expressions for A_n.

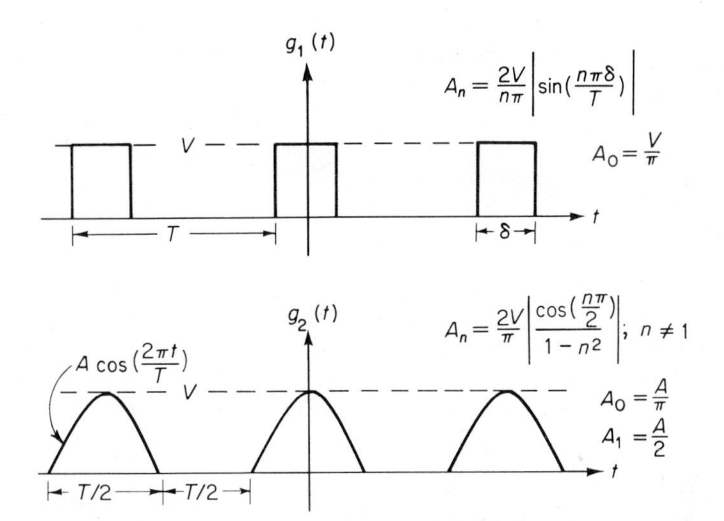

Figure 15.7. Two periodic waveforms and their harmonic amplitudes.

The simple pulse train $g_1(t)$ will in general contain all harmonics. However, for special cases certain harmonics will vanish. For example, if $\delta = T/2$, then A_n will be zero for all even values of n. This is the well-known result that a square wave contains only *odd* harmonics. Similarly, if $\delta = T/3$, then the 3rd, 6th, 9th, . . . harmonics will vanish.

15.4.1 wave analyzers

A wave analyzer is basically an AC voltmeter preceded by a very narrow-bandwidth, tunable, bandpass filter. Thus, the AC voltmeter will respond only to the narrow range of signal frequencies which can be passed by the filter. By changing the filter's bandpass frequency, the various amplitudes of the harmonic frequencies in the input waveform can be measured. Like a prism dividing a light beam into its component wavelengths, the waveform analyzer permits the measurement of the components of an electronic signal and thus experimentally determines the values of A_n. Since the response is not dependent on phase, the information relative to ϕ_n is lost. The bandwidth of the tunable filter usually ranges from 3 to 50 Hz for a wave analyzer with a 50-kHz range. Units with higher frequency capabilities have a correspondingly wider filter bandwidth.

Since it is an AC voltmeter, the waveform analyzer usually indicates the rms value of the sine wave component. In this case, the indicated value must be multiplied by $\sqrt{2}$ to obtain the value of A_n. For further discussion of this point, see Section 5.5.1.

Frequently, wave analyzers are coupled to graphic recorders to provide a written record of the frequency spectrum of the waveform under test. Ideally, this spectrum should be simply a series of spikes with amplitude A_n located at the harmonic frequencies. However, both the bandwidth of the filter and the response characteristics of the recorder will cause a broadening of the spectral lines. This effect is minimized by keeping the bandwidth small and the sweep speed slow.

Exercise 15.6

Use the circuit of Figure 15.8 to explore the frequency components in a pulse train. Set the period of the pulse train at 5 ms and the amplitude at 4 volts. Observe the waveform on the oscilloscope. Using the scope, set $\delta = T/2$ and then measure the second harmonic. Try reducing the second harmonic by adjusting δ without changing the period. Plot a frequency spectrum for some value of δ and compare it with that predicted by the expression for A_n in Figure 15.7. Be sure to correct for an rms indication on the analyzer.

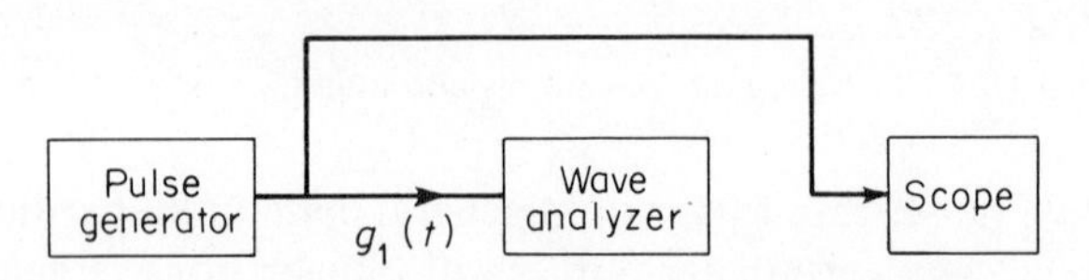

Figure 15.8. Harmonic waveform analysis.

15.4.2 *frequency response of nonlinear networks*

One property of a linear network is that the steady-state response contains no frequency components not present in the excitation signal. Conversely, a nonlinear network will always produce frequency components not present in the input signal. The magnitude of these components compared with the magnitude of the components present in the excitation signal depends on the degree of nonlinearity present.

If the input signal is a sinusoid, then the nonlinearity will produce additional frequency components at the output which are harmonically related to the input signal. An example of this is the half-wave rectifier shown in Figure 15.9.

With an input signal consisting only of a single frequency, the data in Figure 15.7 shows that the output V_0 contains DC (zero frequency), and all the even harmonics of the input frequency, as well as a component of the input frequency itself.

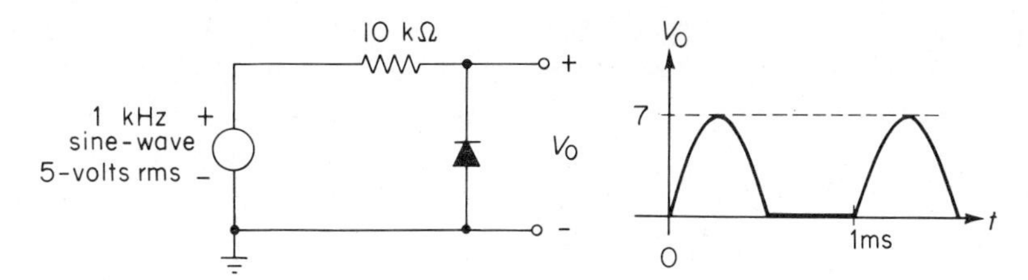

Figure 15.9. Half-wave rectifier.

Exercise 15.7

Using the circuit of Figure 15.9, measure the amplitude of the output frequency components at 1, 2, . . . , 10 kHz and compare them with the values predicted in Figure 15.7. Be sure to check the input waveform for harmonic content, and discuss how much error is introduced in your measurement by the presence of these harmonics at the input.

Since the presence of nonlinearities in a system produces harmonic frequencies at the output, the presence or absence of these frequencies is often taken as a measure of the degree of linearity of the system. One method of specification is to form the ratio of the output signal at the harmonic frequencies to the output signal at the fundamental frequency for a sinusoidal excitation. This ratio is called the *harmonic distortion* and is usually expressed as a percentage:

$$\text{Harmonic distortion } (\%) = \frac{\left[\sum_{n=2}^{\infty} A_n^2\right]^{1/2}}{A_1} \times 100 \tag{15.3}$$

The numerator is simply a calculation of the rms value of the output signal due to frequencies other than the fundamental.

Exercise 15.8

Using a wave analyzer, measure the harmonic distortion of an audio amplifier operating at its rated power output with a 1-kHz input sinusoid.

We have seen how a nonlinear network can produce output frequencies which were not present at the input. Since we have assumed that the input was a sinusoid, we found that additional frequency at the output were harmonics of the input. However, if more than one frequency is present at the input, then not only will there be harmonics of both frequencies at the output, but also sum and difference components as well. For example, if a 60-Hz and 1-kHz sinusoid were simultaneously applied to a nonlinear network, then we would expect the output to contain not only 60 Hz, 120 Hz, 180 Hz, . . . , 1 kHz, 2 kHz, 3 kHz, . . . , but also 940 Hz, 1.06 kHz, 880 Hz, 1.12 kHz, . . . , 1.94 kHz, 2.06 kHz, 1.88 kHz, 2.12 kHz, . . . , etc. Normally, the largest components would be expected at 60 Hz and 1 kHz, but depending on the nonlinearity other frequency components may be substantial. The production of these sum and difference components is often called *modulation* or *mixing*, and is frequently used to produce a desired frequency from two different ones.

Exercise 15.9

Construct the circuit of Figure 15.10. Using the wave analyzer, plot the frequency spectrum of the output. The outputs of the signal generators may be set after first shorting the diode. This will eliminate the cross-coupling and nonlinear response.

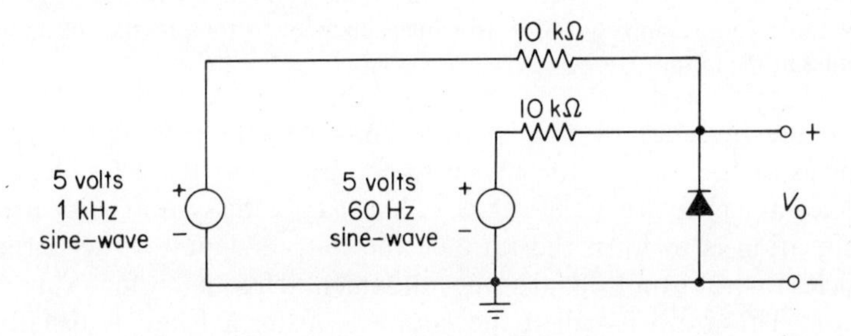

Figure 15.10. Frequency mixing in a nonlinear network.

The previous discussion of harmonic distortion was based on single-frequency excitation. If two frequencies are applied to a nonlinear amplifier, then sum and difference components will also result. This effect, called *intermodulation (IM) distortion*, is another measure of the perfection of a linear amplifier. To measure IM distortion, usually two frequencies f_1 and f_2 are chosen, with $f_1 < f_2$. Typically, f_1 may be 60 or 400 Hz, while f_2 will be 1-kHz or greater. These sinusoids are usually combined such that an input signal

$$v_i = 4A \sin \omega_1 t + A \sin \omega_2 t \tag{15.4}$$

is obtained. By means of a wave analyzer, the frequency components at the output are measured and the amount of power at frequencies other than at f_1, f_2, and their harmonics is a measure of IM distortion. A more detailed treatment of this effect is contained in References 4 through 6.

Exercise 15.10

Using the circuit of Figure 15.11, measure the frequency spectrum of the amplifier under full power. Observe the waveform to insure that clipping does not occur. If the waveform is clipped, reduce the input signal until the clipping is eliminated. Be sure to check the oscillator outputs for harmonic distortion components.

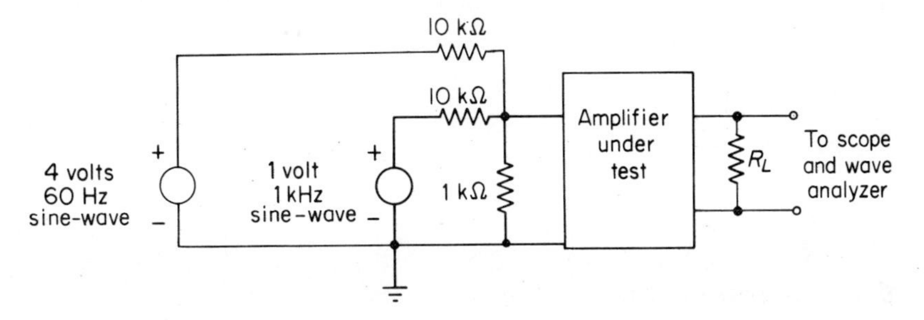

Figure 15.11. Testing an amplifier for IM distortion.

15.5 references

1. "Issue on Frequency Stability," *Proc. IEEE*, Vol. 54, No. 2, February, 1966.

2. Mason and Zimmermann, *Electronic Circuits, Signals, and Systems*, Chapter 6, John Wiley & Sons, Inc., New York, 1960.

3. Aseltine, *Transform Method in Linear System Analysis*, McGraw-Hill Book Company, New York.

4. R. S. Fine, "Intermodulation Analyzer for Audio Systems," *Audio Engineering*, Vol.,34, p. 11, July, 1950.

5. J. M. van Beuren, "Simplified Intermodulation Measurements," *Audio Engineering*, Vol. 34, p. 24, November, 1950.

6. E. W. Berth-Jones, "Intermodulation Testing," *Wireless World*, Vol. 57, p. 233, June, 1951.

16

operational amplifiers

16.1 introduction

In the past, the components available to the electrical engineer consisted of individual, discrete units such as resistors, capacitors, inductors, diodes, transformers, vacuum tubes, and transistors. Now various manufacturers offer complete assembled circuits which can be used as building blocks to construct electronic systems. The time and expense required for system development can often be dramatically reduced by using these types of circuits.

Available circuits can be divided into two broad categories: *digital*, in which the circuit input and output voltages assume only two distinct values of interest, and *linear* or *analog* circuits, in which input and output voltages vary continuously. These two classes of circuits can be further subdivided according to the technique used for circuit construction. *Discrete-component* circuits are collections of individual components, normally interconnected by printed wiring, with the final assembly molded or "potted" in epoxy to improve reliability. *Monolithic* integrated circuits are formed from single pieces of silicon by processing techniques similar to those used for transistor fabrication. Other integrated-circuit techniques combine thick- or thin-film passive components with individual transistor and diode chips.

Monolithic integrated-circuit technology has progressed to the point where the performance of such digital circuits is generally superior to that of discrete-component circuits. Conversely, superior linear circuits can normally be constructed by discrete-component techniques. However, in the case of linear circuits, integrated types are being used with increasing frequency since the performance, physical size, reliability, and price of these units compare favorably with discrete-component circuits and meet the requirements of many system designs. This chapter introduces an extremely versatile linear integrated circuit, the operational amplifier, and provides techniques for measuring its characteristics.

The term "operational amplifier" originated in the field of analog computation where these amplifiers are employed to achieve various mathematical "operations" such as addition and integration. Although these devices are now used in an extremely wide range of applications, the original terminology has remained.

16.1.1 basic performance characteristics of operational amplifiers

An operational amplifier is essentially a high-voltage gain (10^3–10^9) direct-coupled amplifier. Other desirable characteristics include:

1. High input impedance.
2. Wide bandwidth.
3. Low output impedance.

A circuit model which can be used to represent an operational amplifier along with a simplified designating symbol is shown in Figures 16.1a and 16.1b. The circuit model shown in Figure 16.1a is called a linear incremental model in that it provides relationships among incremental changes in terminal variables which occur when the amplifier is in its linear operating region. This model implies the following relationships:

$$I_i = \frac{V_1 - V_2}{Z_{in}} \tag{16.1}$$

$$V_o = A(V_1 - V_2) - Z_{out} I_o \tag{16.2}$$

As an illustration of the use of this model to calculate amplifier performance, consider the circuit shown in Figure 16.2. With the circuit model introduced in Figure 16.1a substituted into the circuit of Figure 16.2, and recalling the high Z_{in} and low Z_{out} characteristics of operational amplifiers, we have the approximate relationships $I_2 \approx 0$ and $V_o \approx -AV_2$. With the aid of these approximations, equations for the circuit of Figure 16.2 become:

$$I_2 = \frac{V_i - V_2}{Z_i} + \frac{V_o - V_2}{Z_f} = 0 \tag{16.3}$$

$$V_o = -AV_2 \tag{16.4}$$

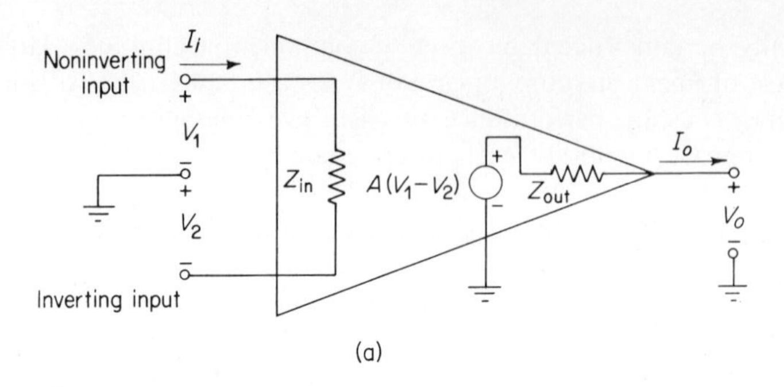

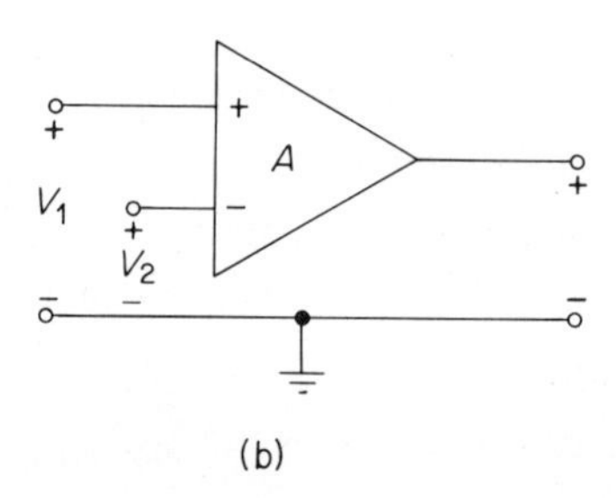

Figure 16.1. (a) Model for operational amplifier. (b) Simplified representation.

or

$$\frac{V_o}{V_i} = \frac{-AZ_f/(Z_i + Z_f)}{1 + AZ_i/(Z_i + Z_f)} \tag{16.5}$$

If the amplifier gain A is sufficiently large so that

$$\frac{AZ_i}{Z_i + Z_f} \gg 1, \tag{16.6}$$

then

$$\frac{V_o}{V_i} \approx -\frac{Z_f}{Z_i} \tag{16.7}$$

An alternate way of arriving at Equation (16.7) which results in greater physical insight into the circuit operation is as follows. Because of the high gain of the amplifier, the voltage required at the input of the amplifier (V_2 in Figure 16.2) is extremely small, in the millivolt to microvolt range or less. This small input voltage, coupled with the high input impedance of the amplifier, insures that the input current I_2 is also very small, and may

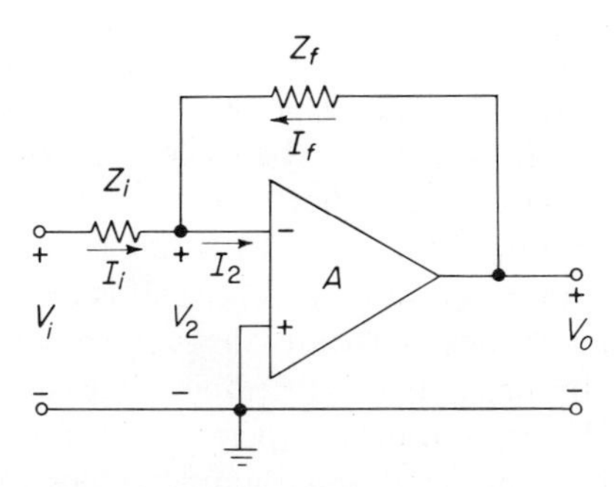

Figure 16.2. Amplifier with feedback.

be approximated as zero. Since V_2 is approximately zero, application of Kirchhoff's current law at the input node ($I_i + I_f = I_2 \approx 0$) then yields

$$\frac{V_i}{Z_i} = -\frac{V_o}{Z_f} \tag{16.8}$$

or

$$\frac{V_o}{V_i} = -\frac{Z_f}{Z_i} \tag{16.9}$$

This analysis is known as the *virtual ground* method, in that the potential V_2 at the input of the amplifier connected as shown in Figure 16.2 is essentially at ground potential. The concept of virtual ground may also be used to determine the input impedance at the terminal to which V_i is applied. Since $V_2 \approx 0$, the ratio $V_i/I_i \approx Z_i$.

The connection shown in Figure 16.2 is an example of *feedback* applied around an operational amplifier. The value of operational amplifiers as general-purpose analog building blocks stems from the fact that the high gain of the amplifier makes it possible to employ feedback in such a way that *the overall gain or transfer function of the amplifier with feedback is dependent primarily on the values of the feedback elements*, as shown by Equation (16.7) or (16.9). The stable, well-known parameters of passive feedback components thereby make possible accurate control of the transfer function *independent* of the properties of the active amplifier element.

16.2 circuit configuration

There are a number of integrated-circuit operational amplifiers available from several manufacturers. The performance (and cost) of these various amplifiers reflect the engineering compromises made during the amplifier design. An example of such an amplifier, which will be discussed in this assignment, is the Motorola Type MC1430.* The performance of this unit is representative of that which can be expected from several relatively inexpensive integrated-circuit amplifiers.

A schematic diagram of the MC1430 itself is shown in Figure 16.3. While a discussion of the operation of this circuit is beyond the scope of this chapter, the schematic does illustrate the complexity of circuits which can be constructed by present integrated-circuit technology. (Reference 1 contains a detailed description of the operation of this circuit.)

Most integrated-circuit operational-amplifier designs have several external compensation terminals available. It is possible to connect elements to these terminals and thus modify the performance of the amplifier as may be required for a specific application. Most frequently, external compensation is used to alter the gain of the amplifier as a function of frequency, since this characteristic determines whether a given feedback connection will be stable or if oscillations will occur.

The amplifier to be used in the exercises has been compensated in such a way as to insure stability in all suggested connections. It is important to realize that this type of compensation deteriorates the bandwidth of the amplifier in many of its connections, and some of the measurements, therefore, may not reflect the ultimate capability of the unit.

A special power supply which minimizes the chance of burnout if a wrong connection is accidentally made, and permits circuit operation from a single ungrounded power supply, has been designed for the amplifier. Components have also been connected to the input and output terminals of the amplifier to further reduce chances of accidental burnout.

It is convenient to mount the amplifier, together with compensation, special power supply, and protective elements on a printed-circuit card. This configuration reduces problems associated with excessive lead length, and permits connection of the external power source and feedback elements via an edge connector. The complete schematic for the amplifier is shown in Figure 16.4.

16.2.1 initial circuit connections and precautions

The operational amplifier used in these exercises is intended to operate from a *floating* 24-volt DC supply (see Figure 2.11). Connect the +12-volt

* There are a number of available operational amplifiers (both integrated-circuit and discrete-component) with performance characteristics superior to those of the MC1430. However certain performance limitations of the circuit actually enhance its value for instructional purposes, since straightforward measurements are possible. If a higher performance operational amplifier is used in place of the MC1430, the gain measurements as described in Sections 4.1 and 4.2, and the direct measurement of offset described in Section 4.3 may not be possible.

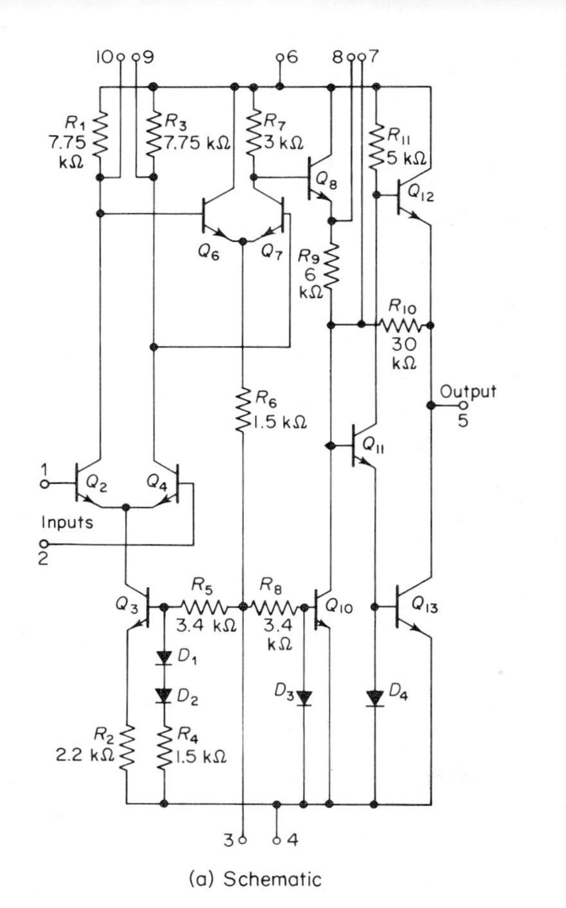

(a) Schematic

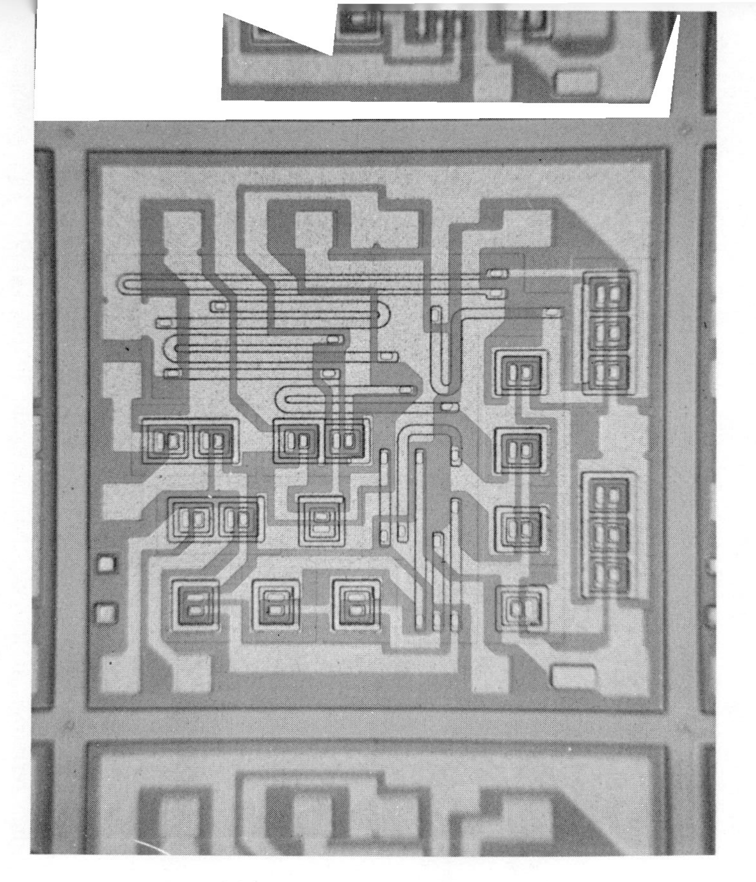

(b) Enlarged Photograph of Chip

Figure 16.3. MC1430 schematic and physical realization. Courtesy Semiconductor Products Division Motorola, Inc.

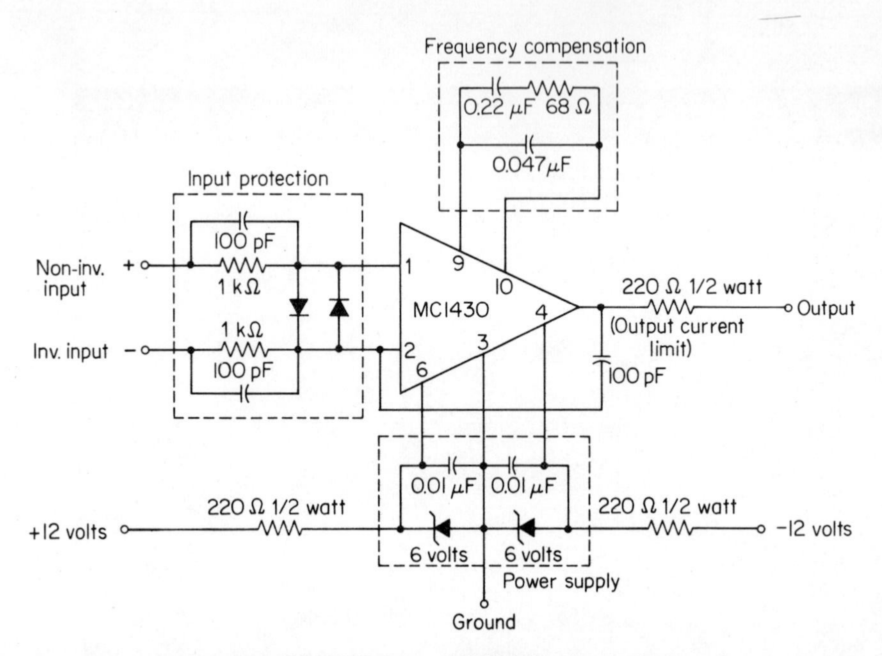

Figure 16.4. Circuit diagram of MC1430 printed circuit card.

amplifier terminal to the positive terminal of the supply and the −12-volt amplifier terminal to the negative terminal of the supply. The ground terminal of the amplifier on the printed-circuit card is used as the common terminal for all measurements. Adjust the supply to 24 volts. *It is assumed in all further circuits that the amplifier is connected to its power supply, although this connection is not specifically indicated.*

Even with the extremely conservative frequency compensation employed with this amplifier, it remains a high-frequency device. For this reason it is imperative that short leads be used to make all connections to the amplifier.

16.3 measurement of operational-amplifier characteristics

16.3.1 open-loop voltage gain

The analysis of Section 16.1 assumes that an operational amplifier has very high voltage gain. Direct measurement of the DC open-loop gain of the amplifier is complicated by amplifier voltage drift (see Section 16.3.3). However, the product of open-loop voltage gain and amplifier drift for the MC1430 is sufficiently low to permit direct gain measurement if care is taken.

Exercise 16.1

Connect the amplifier as shown in Figure 16.5. With the switch open, turn the fine adjust potentiometer to the middle of its range. Adjust the coarse potentiometer until the amplifier output is within ± 3 volts of ground potential. Then use the fine adjustment to set the output to approximately -2 volts.

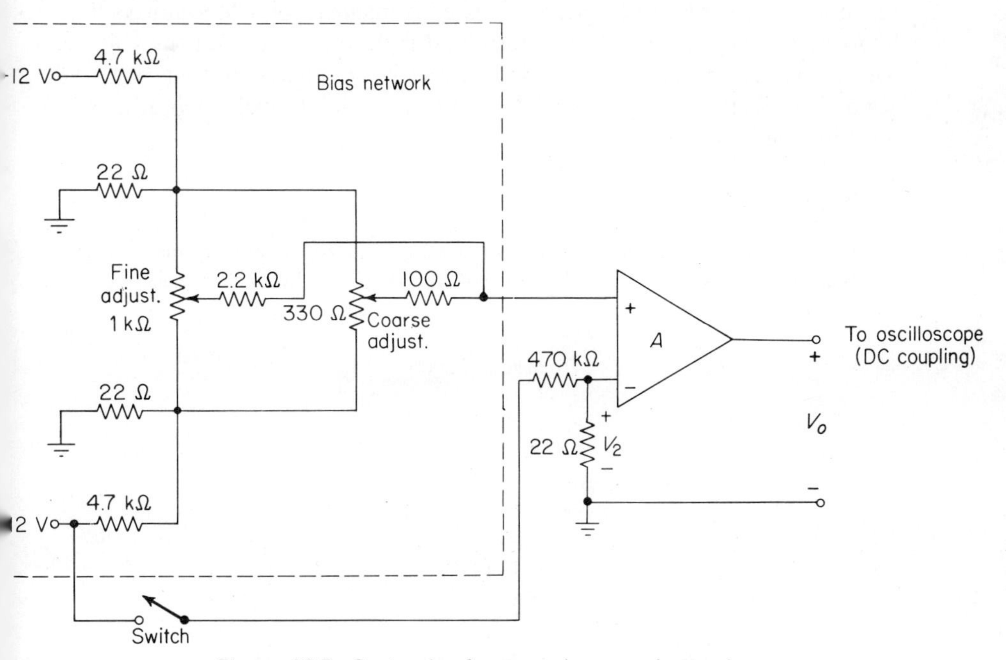

Figure 16.5. Connection for measuring open-loop gain.

With the switch open the input voltage V_2 is zero. When the switch is closed, an input voltage

$$V_2 = \frac{22 \text{ ohms}}{4.7 \times 10^5 \text{ ohms}} \times (-12) \text{ volts} \approx -0.5 \text{ mV}$$

is applied to the negative input terminal of the amplifier. Measure the *change* in output voltage V_o when the switch is closed. The ratio of this change to the applied input voltage is the DC open-loop gain. Open the switch to insure that drift has not caused the output with $V_2 = 0$ to change significantly from the original -2 volts.

16.3.2 gain vs. frequency

The gain of an operational amplifier is maximum at DC, and decreases as the frequency of the applied input signal increases. The nature of the gain vs. frequency characteristics of the amplifier determine the stability of the amplifier in a particular feedback connection. Compensation is normally used to modify these characteristics for specific applications.

The range of frequency over which the gain of an operational amplifier exceeds unity can extend from DC to more than 10 MHz. Similarly the gain can be as high as 10^9 at DC for some designs. Because of the wide range of both gain and frequency it is convenient to plot log gain vs. log frequency. Although not strictly correct, it is traditional to plot the voltage gain expressed in decibels on a linear scale vs. frequency plotted on a log scale. (See Sections 2.3.1 and 3.3.4 for review of the decibel.) A plot in this form is called a *Bode plot.*

Exercise 16.2

Connect the amplifier as shown in Figure 16.6. Adjust the potentiometer in the bias network to provide a DC output $V_o = 0$ volts. (Recheck the setting occasionally during these measurements.)

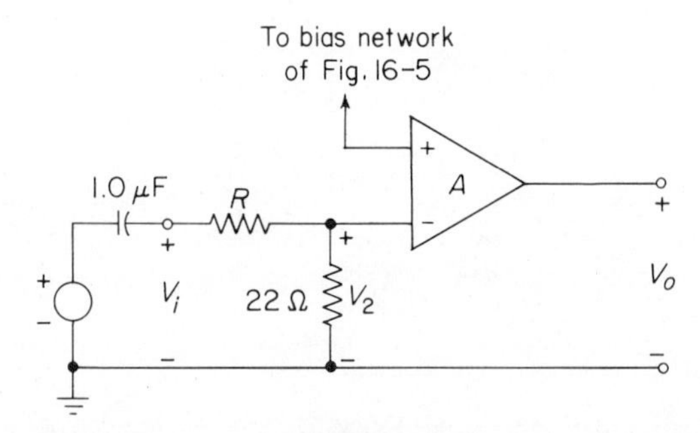

Figure 16.6. Connection for measuring gain vs. frequency.

Connect a signal generator with a lower frequency limit of 10 Hz or less to the circuit. Set the generator to a frequency of 10 Hz and select an initial value of $R = 100\ \text{k}\Omega$. Adjust the signal amplitude from the generator until an easily measured signal ($\sim$2 volts peak-to-peak) is obtained at the output of the amplifier.

Since V_i can be easily measured, the input signal V_2 is obtained by noting that

$$V_2 = \frac{V_i(22\ \text{ohms})}{R + 22\ \text{ohms}} \approx V_i\left(\frac{22\ \text{ohms}}{R}\right) \tag{16.10}$$

for $R \gg 22$ ohms. The gain at 10 Hz is equal to V_o/V_2.

Measure gain at frequencies of 10 Hz, 20 Hz, 50 Hz, 100 Hz, 200 Hz, etc., to at least 500 kHz. As the gain of the amplifier decreases with frequency, change the value of R as required to facilitate measurements. Reduce the magnitude of the output signal from 2 volts peak-to-peak if the output signal becomes nonsinusoidal at high frequencies. Plot gain in decibels vs. log frequency on 5-cycle semilog graph paper. The vertical scale should extend from 0 to at least 80 dB.

16.3.3 *input offset voltage*

The *input offset voltage* (also called *drift referred to the input*) is the voltage required between the input terminals ($V_2 - V_1$ in Figure 16.1) to set the output voltage to zero. This is a measure of the imbalance in the input circuit and would be zero in an ideal operational amplifier.

Exercise 16.3

Connect the amplifier as shown in Figure 16.6 with R an open circuit and adjust the potentiometer until the output voltage is zero. Measure the voltage applied to the positive input terminal of the amplifier with a direct-coupled oscilloscope. This voltage is the input offset voltage.

The direct measurement of input offset voltage described above is not possible for amplifiers with very low offset. It is possible to connect the amplifier in such a way that the input drift itself is amplified to permit convenient measurement. Consider the connection shown in Figure 16.7.

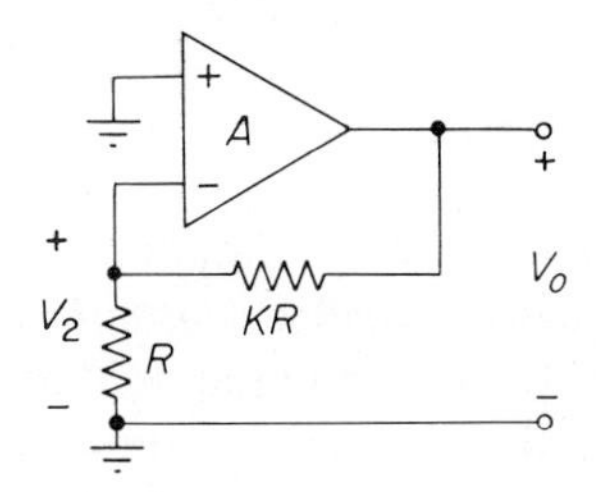

Figure 16.7. Alternative connection for measuring offset voltage.

Let the input offset voltage be equal to V_d. Then

$$V_o = -A(V_2 - V_d) \tag{16.11}$$

since a voltage of $V_2 = V_d$ is required at the input to force the output to zero. If the effects of amplifier input current are negligibly small, then

$$V_2 = \left(\frac{1}{1 + K}\right) V_o \tag{16.12}$$

Combining Equations (16.11) and (16.12) yields

$$V_o = -A\left[\left(\frac{1}{1 + K}\right) V_o - V_d\right] \tag{16.13}$$

or, solving for V_d

$$V_d = \left[\frac{1}{A} + \left(\frac{1}{1 + K}\right)\right] V_o. \tag{16.14}$$

If parameters are selected so that $A \gg 1 + K$, then

$$V_d \approx \left(\frac{1}{1 + K}\right) V_o \tag{16.15}$$

and the offset voltage may be determined without knowing the actual value of A.

Exercise 16.4

Connect the amplifier as shown in Figure 16.7 with $R = 22$ ohms and $KR = 2.2\ \text{k}\Omega$. Recalling the results of Section 16.3.1, is $A \gg 1 + K$? Determine the input offset voltage V_d by measuring V_o. Compare the result with that obtained in Exercise 16.3.

16.3.4 *input bias current*

The DC current required at the amplifier input terminal to establish operation in the linear region is called *bias current.* (This quantity is often defined as the average of the currents required at the two input terminals.) The bias current I_b required at the negative input terminal can be determined from the connection shown in Figure 16.8. If the input offset voltage is small compared with I_bR, then

$$I_b \approx \frac{V_o}{R}. \tag{16.16}$$

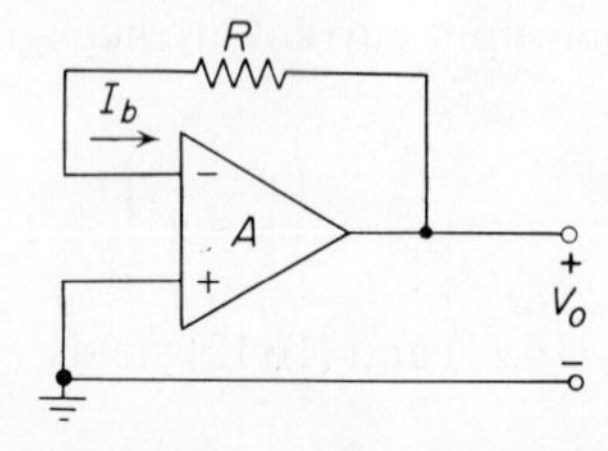

Figure 16.8. Connection for measuring bias current at the negative input terminal.

Exercise 16.5

Connect the amplifier as shown in Figure 16.8 with $R = 100\ \text{k}\Omega$. Determine the input bias current at the negative input terminal.

16.3.5 other characteristics

The operational amplifier is a complex circuit. Operating characteristics are dependent on the performance of many constituent components, and for this reason detailed specification of an amplifier is significantly more difficult than specification of a resistor or a transistor. The electrical characteristics of the MC1430 as supplied by the manufacturer are provided in Table 16.1. Definitions of these quantities, together with test circuits which can be used to measure them, are presented in Reference 2.

Table 16.1 / MC1430 Characteristics

Electrical Characteristics ($V_{CC} = +6$ Vdc, $V_{EE} = -6$ Vdc, $T_A = 25°$ C unless otherwise noted)

Characteristic Definitions*	Characteristic	Symbol	Min	Typ	Max	Unit
–3 dB; A_{VOL}; BW_{OL}	Open loop voltage gain	A_{VOL}	69 3000	74 5000	— —	dB
	Open loop bandwidth (no roll-off capacitance)	BW_{OL}	1.0	1.2	—	MHz
Z_{out}; Z_{in}	Output impedance (f = 20 Hz)	Z_{out}	—	25	50	ohms
	Input impedance (f = 20 Hz)	Z_{in}	5 k	15 k	—	ohms
+V; –V	Output voltage swing (1000 ohm load)	V_O	±4.0	±5.0	—	V_{peak}
V_i; V_O	Input common mode voltage swing	V_{ICM}	±2.0	±2.5	—	V_{peak}
V_i +V –V $A_{VCM} = \frac{V_O}{V_i}$; $CM_{Rej} = A_{VCM} - A_{VOL}$	Common mode rejection ratio	CM_{Rej}	65	75	—	dB
I_1; I_2	Input bias current $I_b = \frac{I_1 + I_2}{2}$	I_b	—	5	15	μA
I_1; I_2	Input offset current $I_{io} = I_1 - I_2$	I_{io}	—	0.4	4	μA
$\lvert V_d \rvert$; $V_O = 0$	Input offset voltage	$\lvert V_d \rvert$	—	2	10	mV
	DC power dissipation power supply = ±6 volts, $V_O = 0$	P_D	—	110	150	mV
	Input offset voltage +75° C 0° C	$\lvert V_d \rvert$	— —	3.0 3.0	12.0 11.0	mW

*All definitions imply linear operation.

16.4 closed-loop performance

As mentioned in Section 16.1, most operational-amplifier applications involve the use of negative feedback applied from the output to the inverting input of the amplifier. If the amplifier is ideal, its transfer function or gain is determined only by the feedback elements. In Section 16.3 some of the performance characteristics of an actual amplifier were measured. In this section we will see that in spite of departure from ideal characteristics, feedback can be successfully used with actual amplifiers, and further that the closed-loop performance of actual amplifiers can often be predicted by considering ideal models.

16.4.1 inverting amplifier

Consider the circuit connection shown in Figure 16.2. In Section 16.1.1 the ideal closed-loop gain of this connection was shown to be $-Z_f/Z_i$. One way to obtain frequency-independent amplification is to use resistors for both Z_i and Z_f. An operational amplifier connected with single resistive input and feedback elements is called an *inverting amplifier* since the ideal output is an inverted, scaled replica of the input.

If the amplifier is ideal, any combination of input and feedback resistors can be selected. However, departures from ideal operational-amplifier characteristics limit the range of element values which can be used. If we assume an input resistor value R_i and a feedback resistor value R_f, the analysis of Section 16.1.1 shows the gain to be approximately $-R_f/R_i$, provided that $AR_i/(R_f + R_i) \gg 1$. Thus, the ratio R_f/R_i must be limited to satisfy this inequality. Even if the ratio R_f/R_i is kept low, the output of the amplifier with feedback still differs from the ideal value because of input offset voltage and input bias current. Figure 16.9 shows an amplifier with an input offset voltage V_d and an input bias current I_b connected with resistive input and feedback elements. (It is assumed that $AR_i/(R_f + R_i) \gg 1$.)

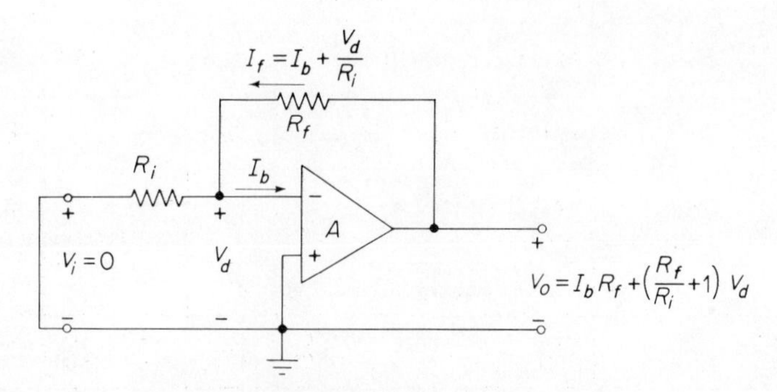

Figure 16.9. Determination of inverter errors.

The output voltage with $V_i = 0$ is

$$V_o = I_b R_f + \left(\frac{R_f}{R_i} + 1\right) V_d \tag{16.17}$$

This expression indicates that R_f/R_i must be kept low to minimize errors in V_o from V_d, and R_f must be kept low to minimize errors from I_b.

Exercise 16.6

Connect to amplifier as shown in Figure 16.9 with $R_f = 10\ \text{k}\Omega$ and $R_i = 3.3\ \text{k}\Omega$. Measure the output voltage with the input connected to ground. Compare this value with what you predict from your previous measurements of V_d and I_b. *Caution:* Be sure to observe the correct polarities of V_d and I_b.

Measure the closed-loop gain of the circuit with a 1-volt peak-to-peak, 1-kHz input signal V_i and compare it with $-R_f/R_i$.

16.4.2 summation

Consider the connection shown in Figure 16.10. If circuit parameters are chosen so that the amplifier behaves ideally, we can analyze the circuit by means of the virtual ground method described in Section 16.1. With

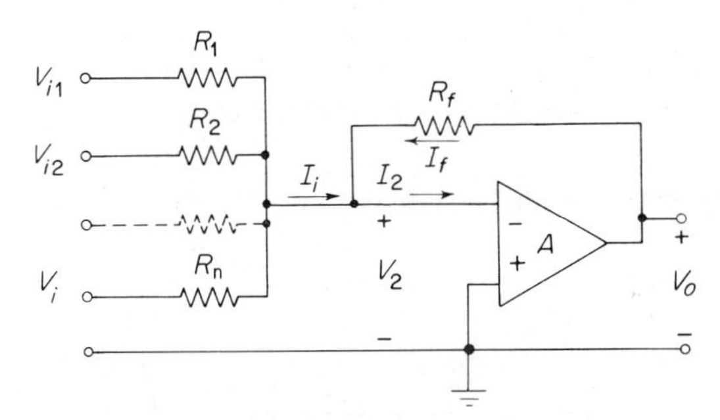

Figure 16.10. Summing amplifier.

$V_2 \approx 0$, then I_i is approximately given by

$$I_i \approx \frac{V_{i1}}{R_1} + \frac{V_{i2}}{R_2} + \dots \frac{V_i}{R_n} \tag{16.18}$$

Since $I_2 \approx 0$, then

$$I_i \approx -I_f \tag{16.19}$$

and therefore

$$V_o \simeq -\left(\frac{R_f}{R_1}\right) V_{i1} - \left(\frac{R_f}{R_2}\right) V_{i2} - \dots - \left(\frac{R_f}{R_n}\right) V_i \tag{16.20}$$

It can be seen that this connection yields an output which is a sum of the input voltages, weighted according to the resistance ratios.

Exercise 16.7

Construct a summing amplifier with $R_f = 10\ \text{k}\Omega$ and one 1-kΩ and one 3.3-kΩ input resistor. Verify the gain relationship predicted for this connection, Equation (16.20).

16.4.3 non-inverting amplifier

The connections introduced in Sections 16.4.1 and 16.4.2 have two common characteristics which may be undesirable in some applications. The circuits provide an inversion between input or inputs and output, and the incremental resistance at a voltage input terminal is fairly low. These limitations are circumvented by the connection shown in Figure 16.11. If

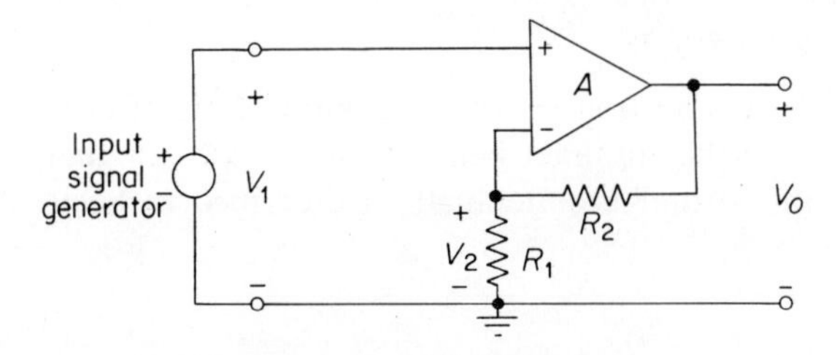

Figure 16.11. Non-inverting amplifier.

the current required at either input is negligibly small, then

$$V_o = A(V_1 - V_2) = A\left(V_1 - \frac{R_1}{R_1 + R_2} V_o\right) \tag{16.20}$$

$$\frac{V_o}{V_1} = \frac{A}{[1 + (AR_1)/(R_1 + R_2)]} \tag{16.21}$$

If

$$AR_1/(R_1 + R_2) \gg 1$$

then

$$\frac{V_o}{V_1} \approx \frac{R_1 + R_2}{R_1} \tag{16.22}$$

This circuit is called a *non-inverting amplifier* or in the special case where $R_1 = \infty$, $R_2 = 0$, a *follower.*

It is possible to calculate an input impedance for the connection of Figure 16.11 in terms of the impedance between the input terminals of the operational amplifier and the gain. Such calculations are usually optimistic because of the presence of an internal amplifier impedance (not included in our simplified model) from the noninverting input to ground. However, input impedances ranging from 1 to 100 MΩ are typically obtained at low frequencies.

Exercise 16.8

Construct the amplifier connection of Figure 16.11 with $R_2 = 10\,\text{k}\Omega$, $R_1 = 1\,\text{k}\Omega$. Measure the gain of the circuit and compare with the predicted result. Connect a 10-kΩ resistor in series with the noninverting input. Is the gain of the circuit noticeably altered? What does this imply about the input impedance of the circuit?

16.4.4 integration

Operational amplifiers are frequently used as integrators. In fact, the only linear operations normally performed in analog computation are summation (with inversion) and integration. It can be shown that the implementation of those two functions is sufficient to solve *any* linear, time-invariant, differential equation.

The analysis associated with Figure 16.2 shows that if Z_f is a capacitor C and Z_i is a resistor R, the "gain" expression for an ideal operational amplifier will be

$$v_o = -\frac{1}{RC}\int v_i\,dt \qquad (16.23)$$

A practical difficulty is encountered with the integrator connection, since even with no input signal, the output becomes

$$v_o = \frac{1}{C}\int i_b\,dt \qquad (16.24)$$

where i_b is the bias current required at the inverting input terminal of the amplifier. For example, with $i_b = 1\,\mu\text{A}$ and $C = 1\,\mu\text{F}$, the output voltage drift rate is 1 volt/sec with no input! For this reason, the requirement for accurate integration has led to the development of amplifiers with extremely low ($\sim 10^{-11}$ amp) input currents. At present, amplifiers with very low input currents are not available in integrated-circuit form.

Exercise 16.9

Construct the integrator circuit shown in Figure 16.12. The mercury-wetted relay is used to reset the capacitor voltage to zero 60 times a second. Adjust the 10-kΩ potentiometer until the integrator output drifts less than 0.1 volt between reset intervals. Connect a 1-volt DC source to the input terminal labeled v_i in Figure

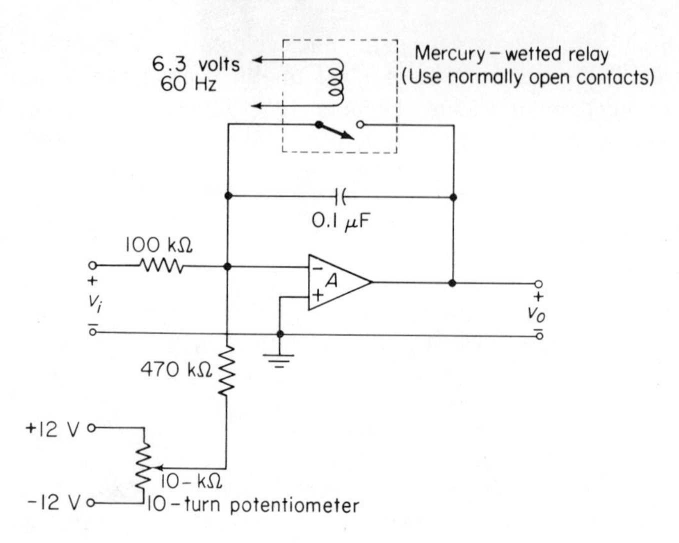

Figure 16.12. Integrator.

16.12. (The required 1-volt DC source can be realized with a battery and a resistive divider.) Measure the rate of change of output voltage with time during the period when the amplifier is integrating. Compare with predicted results.

16.4.5 other connections

The operational amplifier is, more than any other circuit, a general-purpose analog building block. New applications of this device frequently result from the imagination of the designer. Figures 16.13 through 16.16 show several additional connections which may be investigated.

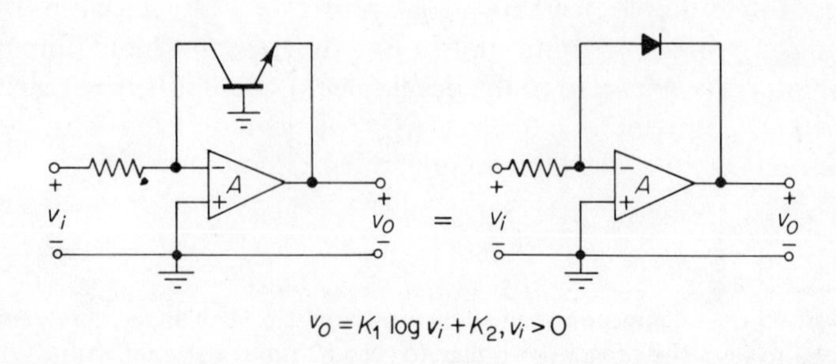

Figure 16.13. Log circuit.

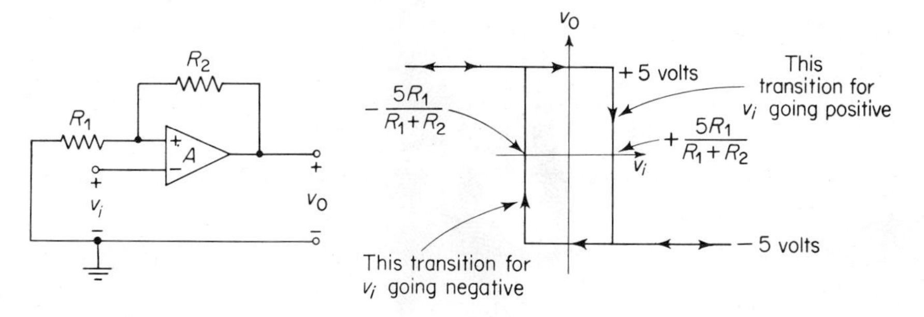

Figure 16.14. Schmitt trigger.

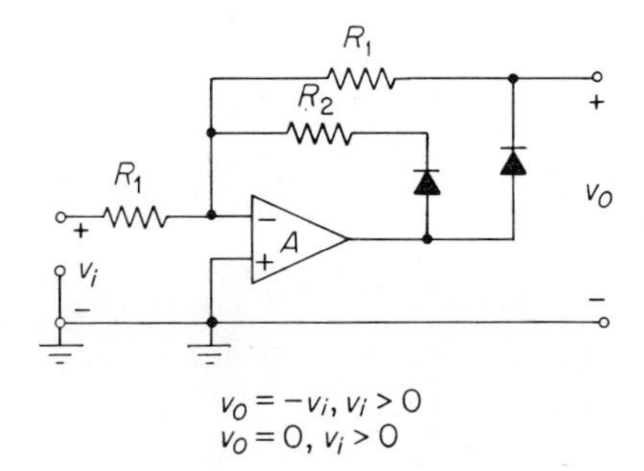

$v_O = -v_i,\ v_i > 0$
$v_O = 0,\ v_i > 0$

Figure 16.15. Precision rectifier.

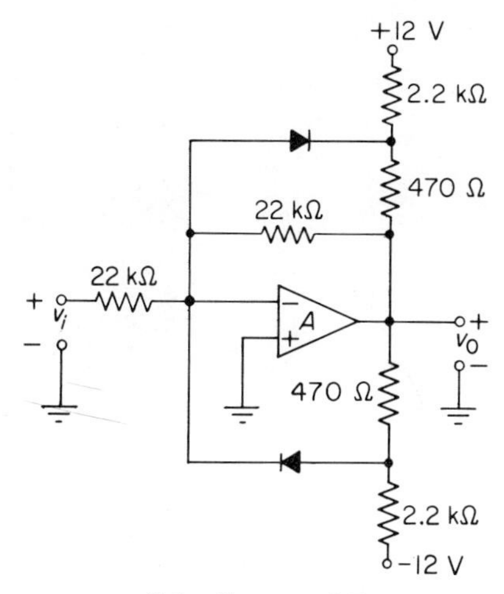

$v_O = 2.5$ volts, $v_i > +2.5$ volts
$v_O = -v_i$ 2.5 volts $> v_i > -2.5$ volts
$v_O = +2.5$ volts, $v_i < -2.5$ volts

Figure 16.16. Limiter.

16.5 references

1. *High Performance Integrated Operational Amplifiers*, Motorola Semiconductor Products Inc. Application Note AN-204.

2. *Handbook of Operational Amplifier Applications*, Burr-Brown Research Corporation, Tucson, Ariz.

3. *Applications Manual for Computing Amplifiers*, George A. Philbrick Researches, Inc., Nimrod Press, 1966, Boston.

4. *Fairchild Semiconductor Linear Integrated Circuits Applications Handbook*, James N. Giles, 1967, Fairchild Semiconductor.

17

digital integrated circuits

17.1 introduction

This chapter introduces a family of digital, monolithic integrated circuits and demonstrates some of their important characteristics. The comparisons between linear and digital circuits and several common fabrication techniques are discussed in Chapter 16.

An essential feature of a digital circuit is that the input and output voltages or currents assume only two distinct levels. Because all signals are two-leveled, or *binary*, the operation of such circuits is relatively immune to variations in component parameters, which, in turn, permits them to be fabricated easily and economically in monolithic integrated circuit form. A further advantage stems from the fact that *any* digital function (including those performed by the most sophisticated computing systems) can be realized by appropriately combining a very few individual types of building-block elements. Thus, the development of a complete digital family requires a minimal investment in the masks used to fabricate the monolithic circuits, resulting in significant advantages with respect to cost, size, and reliability as compared with the discrete-component realizations of the same circuit functions.

Integrated circuits reduce the task of designing a digital system to that of determining appropriate interconnections between building blocks. Once the interconnection diagram is complete, elements with guaranteed performance specifications are available to implement the diagram. In this way integrated circuits reduce design effort by a substantial margin over discrete-component realizations.

The logic described in this assignment is resistor-transistor logic (RTL). Although it has certain performance deficiencies compared with other logic types, RTL is the most economical logic type currently available. Further, all other available types follow the same interconnection rules as RTL, and therefore RTL circuits are perfectly acceptable vehicles for instruction. The three individual building blocks of the RTL family which will be studied are the NOR gate, the buffer, and the J-K flip-flop.

17.2 boolean algebra

The mathematics used to describe the operation of binary elements is called *Boolean algebra.* In this section we will introduce a few of the essential concepts of Boolean algebra by way of example. For a more detailed treatment of the use of Boolean algebra in digital system design see Reference 1.

17.2.1 example

Assume that you are approaching an intersection on foot and that traffic at the intersection is controlled by either a policeman or a traffic light. Local laws permit crossing with a green light, or crossing in any direction if a WALK light is lit, providing no officer is present. We further assume that we approach the intersection with the intent of crossing a particular street. We wish to develop a Boolean equation which will tell us whether or not we can legally cross the street of our choice without waiting for the assistance of the policeman.

Note that all variables in this problem have only two values. The light is not 75% green; it is either green or it is not. Similarly, the officer is either present or he is not. In order to state the problem in the notation of Boolean algebra, we assign variables to certain events. We assign the value of 1 to a variable if the condition it represents is true; otherwise we assign the value 0.

The assignment of variables for our problem:

The light is green in the direction we wish to cross $\Rightarrow A = 1$, or simply A. ($\Rightarrow$ means implies.)

The WALK light is lit $\Rightarrow B$.

Policeman present $\Rightarrow C$.

We are able to cross the street of our choice without delay $\Rightarrow D$.

Note that if a policeman is present, $C = 1$. Our original problem statement indicates that we can cross without delay only if $C = 0$. This statement is equivalent to the statement $\overline{C} = 1$. (Read "C complement equals

1".) With this notation we can reduce our problem statement to "D if A or B, providing $\bar{C}$."

17.2.2 OR *and* AND *operations*

We now define the OR operation for three variables as follows:

$$X = Y \text{ OR } Z \tag{17.1}$$

meaning

$$X = 1 \text{ if } Y = 1$$
$$\text{OR if } Z = 1$$
$$\text{OR if both } Y \text{ and } Z = 1$$

We use an arithmetic + sign as shorthand notation for the OR operation. Thus (17.1) becomes

$$X = Y + Z \tag{17.2}$$

The manipulation rules which insure equivalence between (17.1) and (17.2) are

$$0 + 0 = 0$$
$$1 + 0 = 0 + 1 = 1 + 1 = 1$$

Similarly, we denote the AND operation with a dot as illustrated below.

$$X = Y \cdot Z \tag{17.3}$$

which means

$$X = 1 \text{ if } Y = 1$$
$$\text{AND if } Z = 1$$

The accompanying manipulation rules are

$$1 \cdot 1 = 1$$
$$1 \cdot 0 = 0 \cdot 1 = 0 \cdot 0 = 0$$

Parentheses may be used to give precedence to operations as in ordinary algebra, and in cases where parentheses are used the dot indicating AND is sometimes omitted. If there are no parentheses, the AND functions are evaluated first.

With the aid of these rules, we can state our original example as

D	$=$	$(A$	$+$	$B)$		$\bar{C}$
cross street with no delay		light green	OR	walk light	AND	no policeman

or, if we prefer to omit the dot,

$$D = (A + B)\overline{C} \tag{17.4}$$

This function is equivalent to

$$D = A \cdot \overline{C} + B \cdot \overline{C} \tag{17.5}$$

Exercise 17.1

Verify the equivalence of (17.4) and (17.5) and that they correctly state the example by checking all possible combinations.

The implementation of our Boolean function into a digital system design proceeds with the development of a *logic flow diagram.* This diagram combines the three basic digital building blocks in such a way that the digital system behaves according to the desired Boolean algebraic problem statement. The three building blocks, along with defining relations, are shown in Figure 17.1. One possible logic flow diagram for our example, assuming that only the C variable (not its complement) is available as an input, is shown in Figure 17.2.

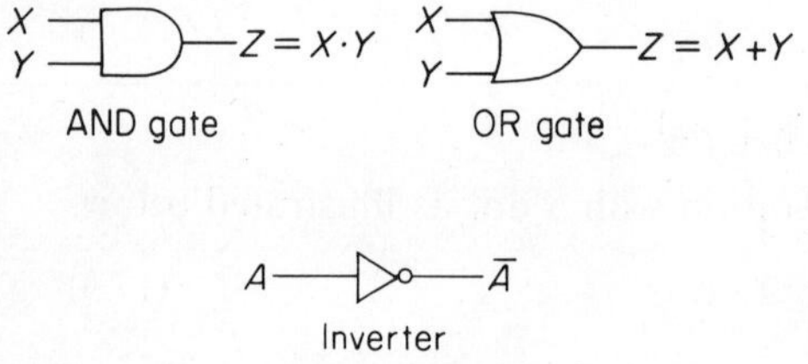

Figure 17.1. Elements used in logic flow diagrams.

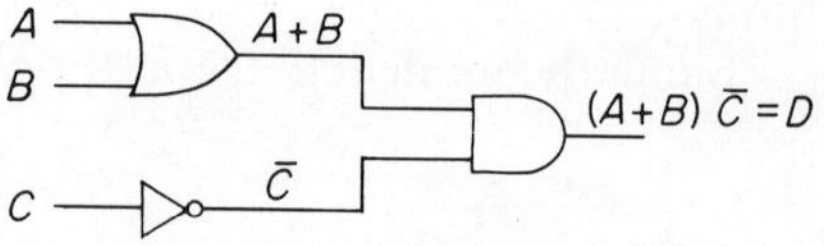

Figure 17.2. Logic flow diagram for example.

17.3 NOR *logic*

Once the flow diagram is completed we are in a position to implement our function by using integrated circuit realizations of our building blocks. It is important to note that while the variables in flow diagrams are 1's or 0's, the variables in the actual system are voltage or current levels. In the logic form introduced in this section, a variable is assumed to be a 1 if its value is greater than +1 volt, and it is assumed to be a 0 if its value is less than +0.4 volt. Note that there is nothing special about this association of levels with logical variables; it is a consequence of the physical circuit used for logic function realization. This is an example of *positive logic*, since the voltage which represents a 1 is more positive than that which represents a 0.

In the RTL family, the basic logic building blocks are constructed with only resistors and transistors. An inverter is shown in Figure 17.3. (The resistor values indicated in this figure are typical ones for RTL.) The transistor is used as a switch between the output and ground in this application, and whether the switch is open or closed depends on the voltage applied to the input of the circuit. If the input voltage is less than approximately +0.4 volt ("low"), the transistor switch is off (or nonconducting), and therefore the output voltage is close to the supply voltage. If the input voltage exceeds 1 volt ("high"), there is sufficient base current to turn the transistor switch on and to force its collector potential (the output voltage) to less than +0.3 volt. Thus this circuit can be used as an inverter, since a 1 input causes a 0 output and vice versa.

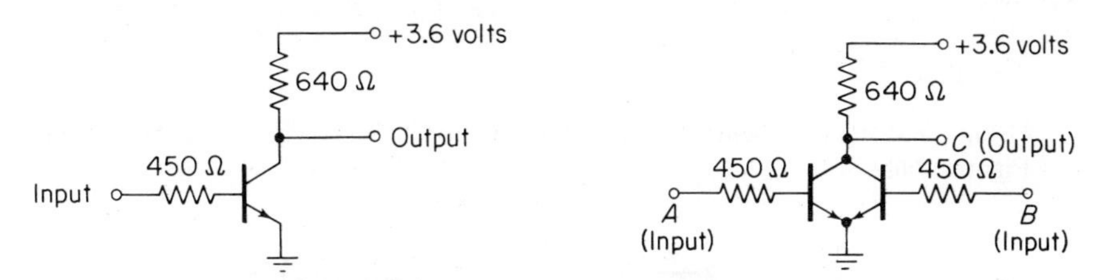

Figure 17.3. Inverter.

Figure 17.4. NOR gate circuit.

If we add one additional transistor and resistor to the inverter, as shown in Figure 17.4, we form an element which can be used for all the operations discussed to this point. Note that this circuit can also be used as an inverter. If we ground the A input, for example, the circuit behaves as if the left-hand transistor is not present and functions as the inverter described previously. However, if a positive signal is applied to either input, the output will be near ground. We have previously defined the OR combination of two inputs as being a 1 if either input is a 1. Note that the operation realized with the circuit of Figure 17.4 is the complement of the OR operation. The output is low for either input high, or

$$C = \overline{A + B} \tag{17.6}$$

For this reason such circuits are called "NOT OR" gates or simply NOR gates. Viewed alternatively, recognize that the output is a 1 only if both inputs are 0. Both inputs 0 implies that the complements of both inputs are 1's. Therefore, we see that we can also write the input-output relationship as

$$C = \bar{A} \cdot \bar{B} \tag{17.7}$$

The equivalence between Equations (17.6) and (17.7)

$$\overline{A + B} = \bar{A} \cdot \bar{B} \tag{17.8}$$

is called deMorgan's theorem. Figure 17.5 shows the symbol used to

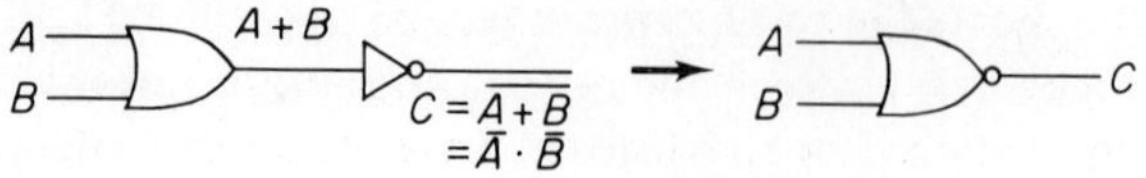

Figure 17.5. Symbol for NOR gate.

represent a NOR gate and illustrates its evolution from an OR gate followed by an inverter.

The NOR gate is a "universal" element, since we can use combinations of NOR gates to perform the three basic operations of AND, OR, and inversion. Figure 17.6 illustrates the connections used to form OR and AND gates from NOR gates.

Exercise 17.2

Using deMorgan's theorem, verify the relationship shown at the output of Figure 17.6b.

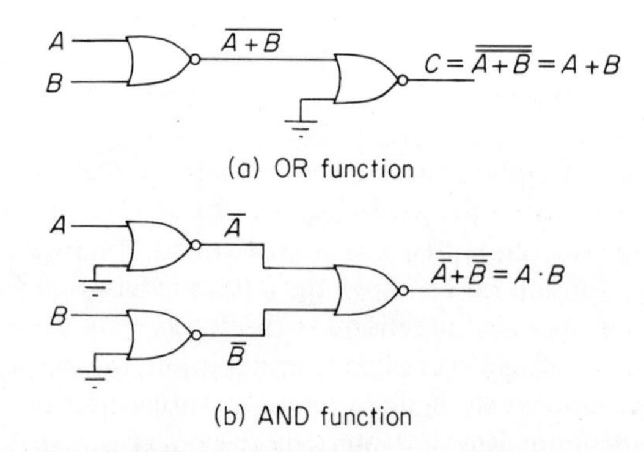

(a) OR function

(b) AND function

Figure 17.6. NOR synthesis of AND and OR operations.

17.4 *applications and properties of* NOR *gates*

In previous sections, it has been shown that combinations of NOR gates can be used to implement AND, OR, and inversion operations. The mechanization of these functions, as well as several other useful connections which combine two or more NOR gates, are discussed in this section.

17.4.1 *implementation of logic flow diagrams*

Exercise 17.3

Connect NOR gates as shown in Figures 17.6a and 17.6b to synthesize OR and AND gates. Use the positive supply voltage to provide a 1 input, and ground as a 0 input.

Measure and record the output and intermediate signals for all four possible combinations of inputs and compare them with the predicted relationships. *Note:* As seen in Figure 17.4, each logic element requires power supply and ground connections *in addition* to the signal inputs. These connections are not shown on logic flow diagrams, but must be made when constructing the actual circuit.

Exercise 17.4

Use the elements developed in Exercise 17.3 to implement the logic flow diagram for the traffic-light problem discussed in Section 17.2.1. Measure the output for all possible combinations of input variables and compare them with the predictions of Equation (17.5).

17.4.2 the set-reset flip-flop

It can be shown that combinations of NOR gates assembled as indicated in earlier sections can be used to generate *any* single-valued Boolean function which is dependent only on the instantaneous value of the independent variables. However, the limitation to functions dependent on instantaneous values of signals is a severe one. Such a system has no storage or memory, which many of the more interesting digital operations require.

It is possible to connect NOR gates in such a way that storage is obtained. Consider the connection of two NOR gates, each operating as an inverter, shown in Figure 17.7a. This connection, which simply repeats the value of the input, seems of no value. However, consider what happens if the input and output are connected as shown in Figure 17.7b. We have now introduced an additional constraint, and as a result it is possible to maintain $B = A = 1$ *or* $B = A = 0$ with no external input! The circuit is capable of supplying its own "holding" signal, and has two possible states. If the unused inputs are ungrounded, the state of the combination can be changed by external signals. Further, a given input will be "remembered" until a new input is applied. This type of connection is one example of a *bistable multivibrator* or *flip-flop.* The more conventional way of drawing a set-reset flip-flop, together with its designating symbol, is shown in Figure 17.8.

The rules concerning the operation of this circuit can be summarized as follows:

If set = 1 and reset = 0, then the output marked (0) is forced low and the output marked (1) is forced high.

If reset = 1 and set = 0, (0) goes high and (1) goes low.

If set = 0 and reset = 0, the circuit maintains its state indefinitely.

If set = 1 and reset = 1, both outputs are low. If both inputs return to 0 simultaneously, the resultant state is indeterminant. This combination is generally avoided.

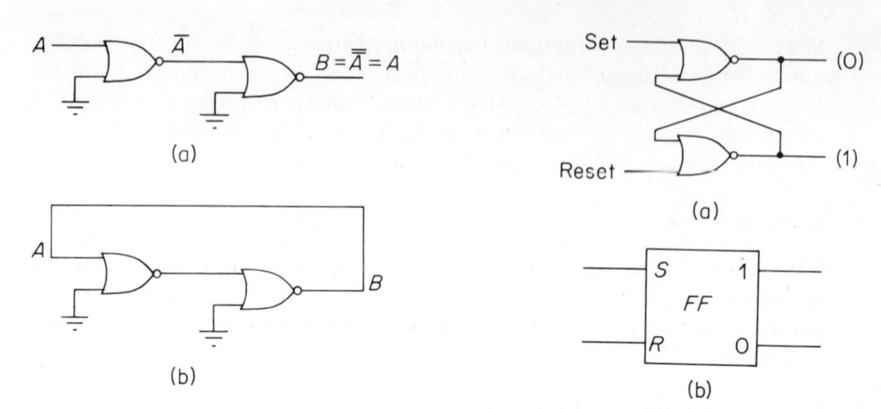

Figure 17.7. Circuit with two stable states.

Figure 17.8. Set-reset flip-flop and symbol.

Exercise 17.5

Construct a set-reset flip-flop using two NOR gates. Verify the behavior summarized above.

17.4.3 timing circuits

The flip-flop illustrates one application of *feedback* around digital elements. Another connection which exploits feedback is shown in Figure 17.9. Assume that the input is low and has been low for a long period of time. If R is not too large, then the current through R keeps the output of gate 2 low. Since both inputs to gate 1 are then low, its output is near the supply voltage. In this state the voltage at input B of gate 2 is typically $+0.7$ to $+1.2$ volts, depending on the value of R. Because of the difference between the output voltage of gate 1 and the input voltage of gate 2, the capacitor C is charged to approximately 2 to 3 volts with the indicated polarity. Now assume that the input V_i suddenly becomes a 1. This forces the output of gate 1 to near ground potential. Because the voltage across C cannot charge instantaneously, input B of gate 2 is forced negative. Since both inputs to gate 2 are now 0, its output V_o goes high, and this high signal fed back to gate 1 keeps the output A of gate 1 low even if the input signal returns to ground. The capacitor then starts to charge through resistor R, and, at a time determined primarily by the RC product, the output of gate 2 switches low. If it is assumed that V_i is low at this time, the capacitor recharges to its original voltage. If V_i has not yet become low, capacitor voltage recovery is delayed until some later time. In either case, the output V_o remains low until the recovery is completed and the input is "triggered" a second time.

The important feature of this circuit is that it generates an output pulse with a start synchronized to the input pulse, but with a time duration in-

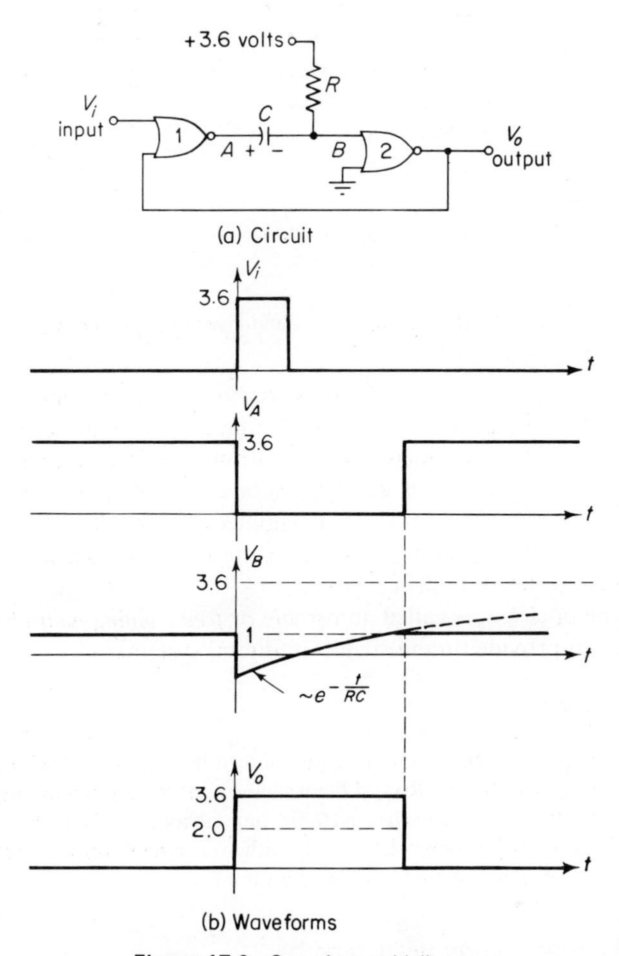

Figure 17.9. One-shot multivibrator.

dependent of the width of the input pulse. Such a circuit is called a *monostable* or *one-shot multivibrator*.

Exercise 17.6

Construct the circuit shown in Figure 17.10. (The diodes are used to protect the input of the first gate from excessive signal levels.) Select $R = 1.5\ \text{k}\Omega, C = 0.0047\ \mu\text{F}$. Record the waveforms at the input and output of each gate in your notebook. Repeat these measurements with $R = 2.7\ \text{k}\Omega$, $C = 0.01\ \mu\text{F}$. Relate the width of the output pulse to the RC product.

It is also possible to use capacitive feedback around two NOR gates to form a free-running pulse generator, as shown in Figure 17.11. The

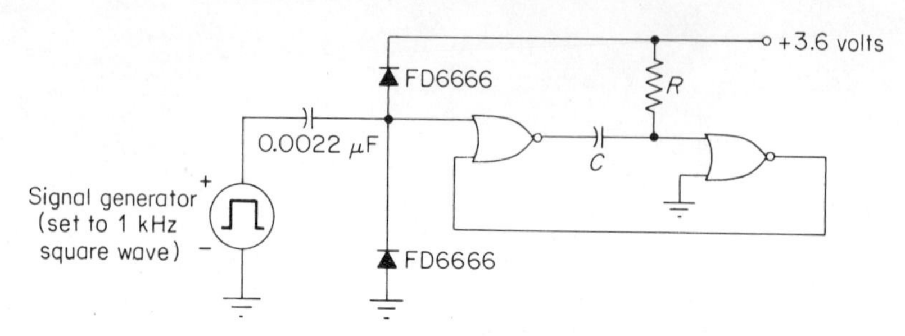

Figure 17.10. One-shot multivibrator with trigger generator.

operation of this circuit is as follows. Assume that the output of gate 1 has been high and has just switched low. Since the voltage across capacitor C_1 cannot change instantaneously, the output of gate 2 switches high as in the monostable circuit. This state is maintained until the voltage at the right-hand side of C_1 charges to approximately $+0.8$ volt, at which time the output of gate 2 switches low. This new state is maintained while capacitor C_2 charges, and the cycle then repeats.

This type of circuit is called an *astable* or *free-running multivibrator*, and can be used to provide timing signals in digital systems.

Exercise 17.7

Construct the free-running multivibrator shown in Figure 17.11 with $R = 4.7\ \text{k}\Omega$ and $C_1 = C_2 = 0.001\ \mu\text{F}$. Record the waveforms at the input and output of each gate. (*Note:* It is possible to have a stable, time-invariant state with the output of both gates low. If you observe this state with your circuit, turn the power supply off and then on again to start astable operation.)

17.4.4 propagation delay time

The output of a gate or inverter requires some time interval to change state after an input is switched. The definition of this *propagation delay time* t_{pd} is illustrated in Figure 17.12. Although the delay times are not

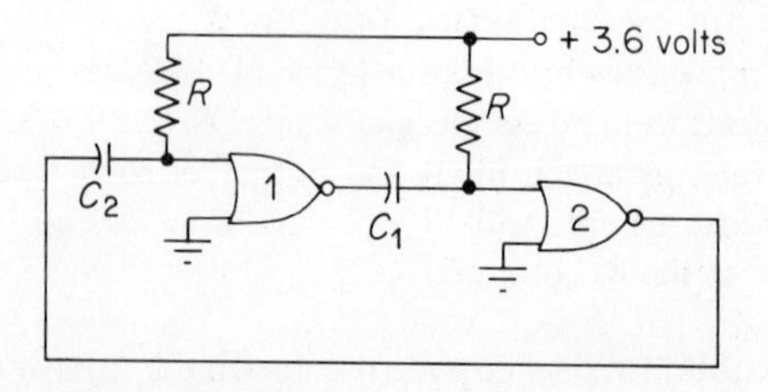

Figure 17.11. Free-running multivibrator.

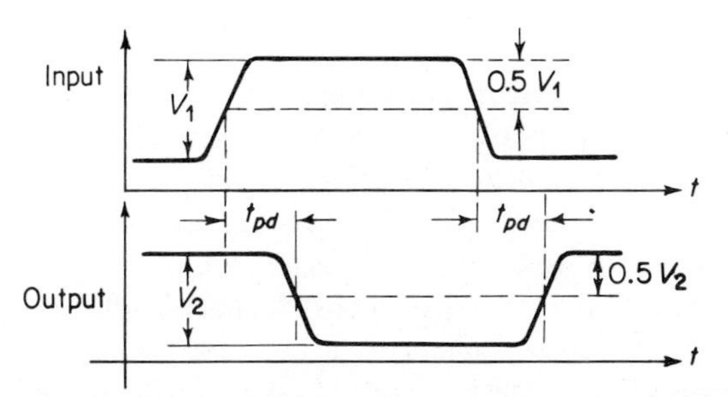

Figure 17.12. Illustration of propagation delay time for an inverter.

necessarily equal for positive and negative transitions, it is assumed that this is the case to simplify the following discussion.

The propagation delay time can be measured with the connection shown in Figure 17.13. Note that in contrast to the flip-flop, no stable combinations of output states exist since an odd number of inverters are connected in a loop. Assume that the signal at the output of one of the gates has just switched negative. This signal propagates around the loop and, after a time equal to $5t_{pd}$, causes the output of the gate in question to switch positive. After a further delay of $5t_{pd}$, the output of the gate again switches negative. Therefore, a square wave with a frequency equal to $1/10t_{pd}$ is present at the output of each gate.

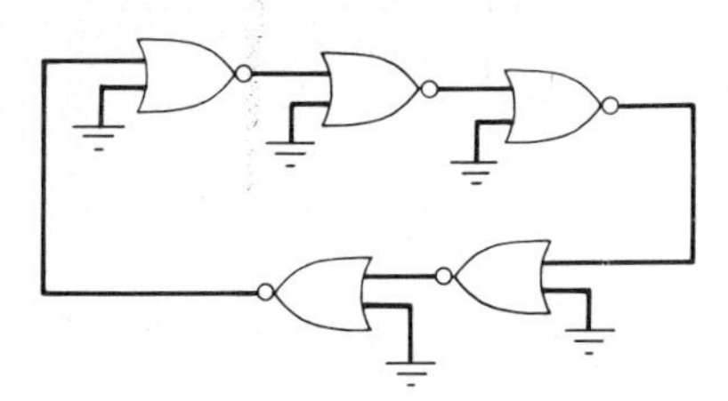

Figure 17.13. Connection for measuring propagation delay time.

Exercise 17.8

Use the connection shown in Figure 17.13 to measure the propagation delay of a NOR gate connected as an inverter. It is necessary to use a wide-bandwidth oscilloscope for this measurement since t_{pd} is typically 20 ns.

17.5 other digital circuits

It is possible to synthesize any required digital function using NOR gates. However, the amount of circuitry required to implement many functions can be reduced if two additional elements, a buffer and a J-K flip-flop, are used in addition to the NOR gate.

17.5.1 buffer

The NOR gate described in preceding sections cannot supply unlimited output current. For this reason signals are deteriorated when many subsequent inputs are connected to the output of a single gate. The term *fan-out* is used to specify the number of inputs which one element can drive before its performance becomes unacceptable. Determination of fan-out is complicated by the fact that not all digital circuits have identical values for input loading or output drive parameters. One convenient method to describe the problem is to specify a loading factor for inputs and outputs of all elements in a family of logic. For example, the output of the NOR gate may have a rated capacity of 16, and each NOR gate input represents a load of 3. Therefore, the output of one NOR gate can drive the inputs of up to five other NOR gates before it is loaded to the point where errors may be introduced.

If we wish to drive ten gate inputs from a single output, we must provide additional output capability. A buffer element, which is effectively an inverter with a high output current capability, can be used to provide the signal required in high load situations. The buffer element included in the RTL family may have a drive capability as high as 80, or more than five times as great as that of the NOR gate. The symbol for a buffer is shown in Figure 17.14.

The additional drive capacity of a buffer can be used to operate indicator lamps directly from digital signals and this element is used at times to provide a visual readout. It also permits driving large capacitive loads with good switching speed.

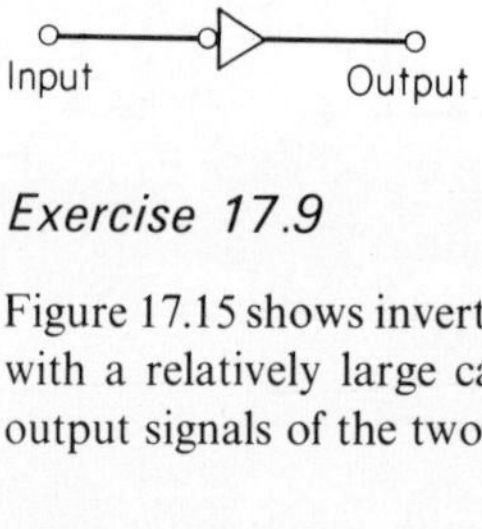

Input Output

Figure 17.14. Symbol for inverting buffer.

Exercise 17.9

Figure 17.15 shows inverters realized with a NOR gate and with a buffer, each loaded with a relatively large capacitor and driven with the same signal. Compare the output signals of the two elements.

17.5.2 J-K flip-flop

The third member of the RTL family to be introduced is the *J-K flip-flop*. Like the set-reset flip-flop, this element can be used to store information, but differs in that it has three types of inputs:

1. A *forcing* input. A positive voltage applied to such an input causes the unit to assume a known state, independent of the signals applied to other inputs. The RTL element usually provides only one such input, called *preset* (*P*).

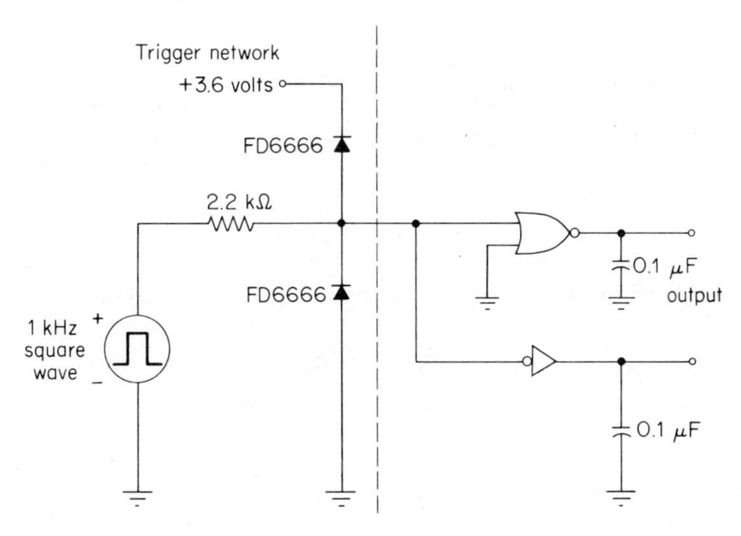

Figure 17.15. Capacitively loaded inverters.

2. A *set* input and a *clear* input (S and C) which are used in conjunction with the trigger.
3. A *trigger* input (T). If the forcing input (P) is low, the output can change state only when the trigger goes from high to low. The final state is a function of S and C just prior to the trigger transition.
   If S and C are high, there is no change in output.
   If S and C are low, the outputs change state or complement.
   If S and C are complements ($S = \bar{C}$), then output 1 assumes the value of S and output 0 assumes the value of C.

The symbol for the J-K flip-flop is shown in Figure 17.16.

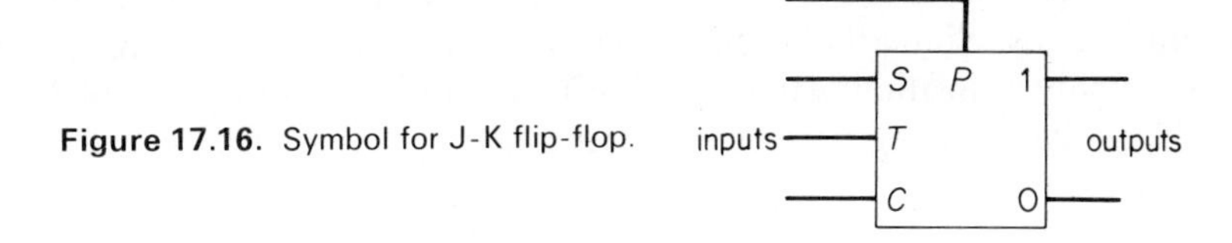

Figure 17.16. Symbol for J-K flip-flop.

Flip-flops are frequently used to construct counters. Figure 17.17 illustrates a 2-bit binary counter. Assume that both flip-flops are in the 0 state (the 1 output is low). The first negative transition applied to the T input of flip-flop A causes it to switch to the 1 state. Since a negative-to-positive transition is applied to the trigger of flip-flop B, its state does not change. The second negative transition applied to the T input of A causes

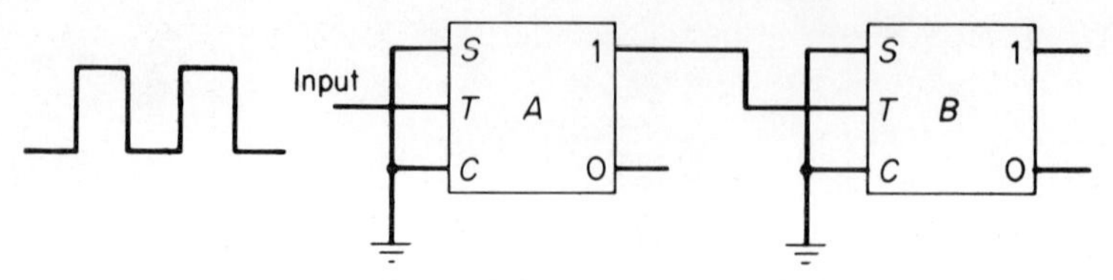

Figure 17.17. Counter.

both flip-flops to change state. A possible sequence of output states is:

A	B
0	0
1	0
0	1
1	1
0	0

Repeats in groups of four.

Exercise 17.10

Construct a 2-bit binary counter. Use the square-wave generator and resistor-diode network shown in Figure 17.15 to trigger the first flip-flop. Observe the waveforms at the output of both flip-flops, and record the time relationships between these signals. *Hint:* In order to obtain meaningful time relationship measurements, your oscilloscope should be triggered externally from an event which occurs only once per sequence, such as a 0 to 1 or 1 to 0 transition of flip-flop B.

J-K flip-flops are also frequently used in *shift registers.* A shift register is an element which sequentially transfers data from one storage unit to another. A 3-bit shift register connection is shown in Figure 17.18. Transfer of data occurs when a negative transition is applied to the shift input line as follows. Since the S and C inputs of flip-flop A are driven with complementary signals, the state of A following a shift transition is controlled by these input signals. Similarly flip-flop B assumes the state of A immediately prior to the shift transition, while flip-flop C assumes the prior state of B. It is evident that shift registers of any length can be easily constructed using the J-K flip-flop.

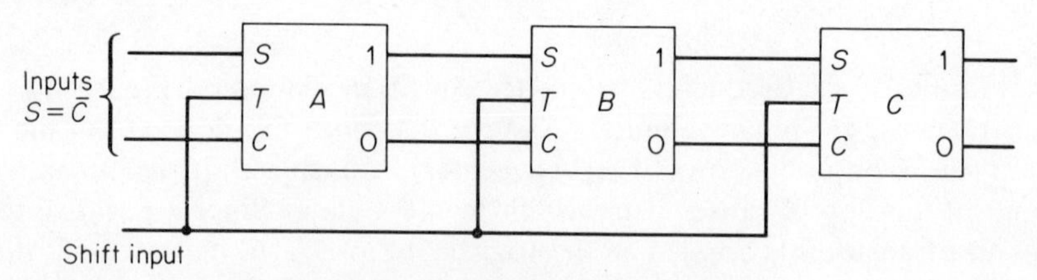

Figure 17.18. Three-bit shift register.

We have seen earlier in this section that counters can be constructed with J-K flip-flops. In many cases the outputs of counters are *decoded* with gates. For example, a particular requirement might be to provide a 1 output whenever a 2-bit counter is in the 11 or the 00 state. If we call the desired output C and the outputs of the two flip-flops A and B, we can write this condition as

$$C = (A \cdot B) + (\bar{A} \cdot \bar{B}) \tag{17.9}$$

and implement the required function using gates.

Exercise 17.11

Using the outputs of the 2-bit counter constructed in Exercise 17.10 (Figure 17.17), design and construct a decoding gate network that will realize the Boolean function of Equation (17.9). Assume a flip-flop is in the 1 state when the 1 output is high and the 0 output is low. Verify the operation of your circuit experimentally.

Exercise 17.12

Figure 17.19 shows an example of decoding gates connected to a 2-bit counter. While the counter itself has only four possible states, the inputs to the gates have eight possible distinct combinations, since the input to the first flip-flop (which may be either high or low while the counter is in each of its states) is also supplied to the gates. Predict the values of A and B for all eight possible combinations of input variables.

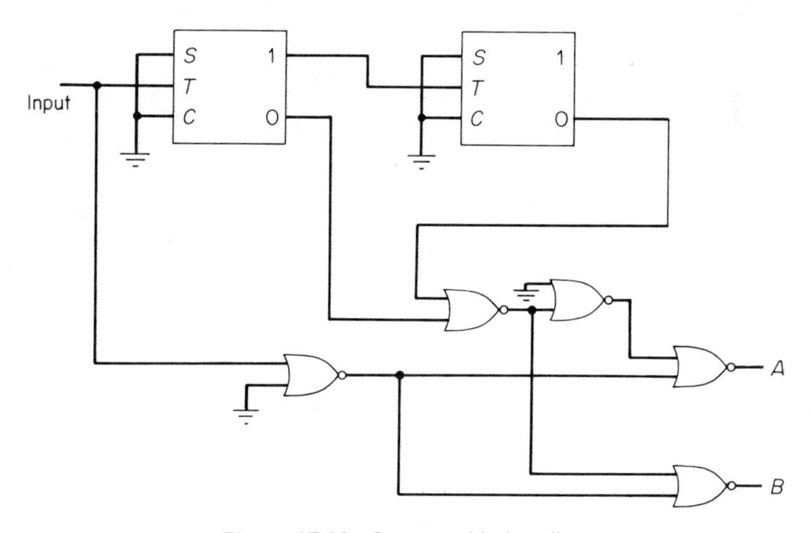

Figure 17.19. Counter with decoding.

Investigate the performance of the counter and decoder shown in Figure 17.19. Use the trigger network shown in Figure 17.15 to provide input signals. Record the A and B output signals and compare with predicted results. (*Note:* In order to obtain a stable oscilloscope display, trigger your oscilloscope from either the high-to-low or low-to-high transition of signal A.)

17.6 reference

1. J. N. Harris et al., *Digital Transistor Circuits*, John Wiley & Sons, Inc., New York, 1966.

18

RF impedance measurements

18.1 introduction

In the audio-frequency region, the actual behavior of passive components, particularly resistors and capacitors, often very closely approximates the expected behavior of an ideal device. However as operating frequencies are extended into the radio-frequency (RF) region, which we will loosely term the frequencies from 1 to 100 MHz, component behavior often departs significantly from the idealized description. This chapter introduces component behavior in the RF region, including some graphical techniques for constructing circuit models. Also several instruments used for component characterization at radio frequencies will be described.

18.2 parasitic effects in components

The circuit equations previously given to describe a component's terminal behavior are actually approximate solutions to the more general electromagnetic field description of the component. At audio frequencies the approximation may be excellent. However, as the operating frequency is increased, a point is reached where the approximate circuit equations no longer provide an adequate description of the component's behavior. Thus, a resistor may begin to behave more like a capacitor or an inductor at high frequencies, depending on the detailed solution to the electromagnetic field equations. Such behavior is often termed *stray capacitance* or *stray inductance.* Because these effects are usually undesirable and serve to limit the high-frequency performance of components, they are also called *parasitic effects.*

18.2.1 lead inductance

One important parasitic effect is that of *lead inductance.* At low frequencies, short lead lengths are desirable to minimize noise pickup. At high frequencies, lead length can have a strong effect on circuit performance through the self-inductance of the lead wire. The origin of this inductance is the stored energy in the magnetic field surrounding the leads which is produced by the terminal current. This situation in a composition resistor along with a possible circuit model, is shown in Figure 18.1.

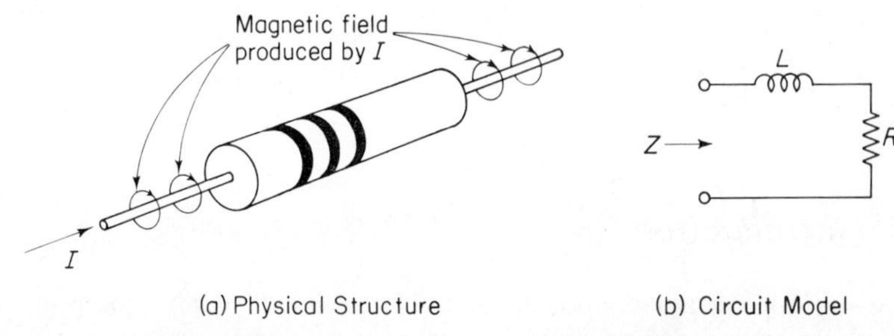

Figure 18.1. Lead inductance in a composition resistor.

18.2.2 stray capacitance

A second parasitic effect which becomes very important in the RF region is stray capacitance. The origin of the stray capacitance is the stored energy in the electric fields surrounding the component which are produced by the terminal voltages. Figure 18.2 illustrates this problem in a composition resistor, along with a possible circuit model.

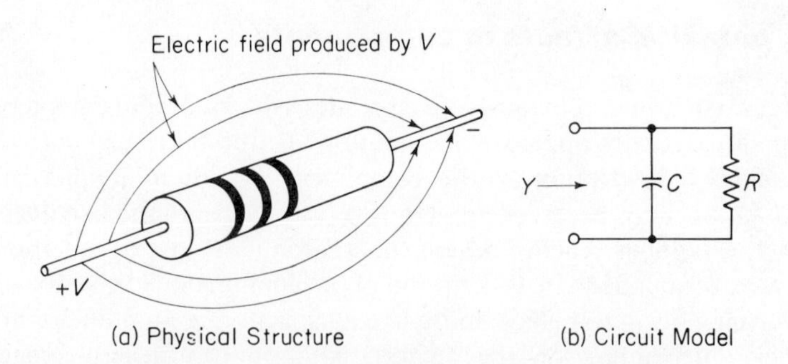

Figure 18.2. Stray capacitance in a composition resistor.

18.2.3 parasitic effects in reactive components

In the case of reactive components, which are primarily energy storage devices in the first place, the parasitic effects of lead inductance and stray capacitance limit the useful operating frequency through *self-resonance.* For example, if the stray shunt capacitance of an inductor L is C_p, then at the frequency $\omega_0 = 1/\sqrt{LC_p}$, the inductor will resonate with C_p alone, and the impedance at its terminals will appear to be a pure resistance (due to the losses in the windings and core material). Above ω_0, the terminal behavior will be that of a capacitor. Clearly, if the component is to be called an "inductor," then the maximum useful frequency is limited by ω_0.

A similar situation occurs with the lead inductance of a capacitor. At some frequency, the capacitor and the lead inductance will form a series resonant circuit. Above this resonant frequency the element will no longer perform its intended function as a "capacitor."

18.3 circuit models for parasitic effects

In general, both the lead inductance and stray capacitance will affect the performance of a component in the RF range, and the simple models, such as those in Figures 18.1 and 18.2, will not provide a complete circuit description. Usually, however, one effect is dominant as the component performance departs from the ideal behavior, and a simple model is then sufficient for circuit calculations over a limited frequency range. Theoretical calculation of the parasitic effects is extremely difficult; the practical approach is to measure the component behavior and then fit the data to a suitable circuit model.

One useful technique of deriving a circuit model from measured data is to examine the behavior of the magnitude of the impedance or admittance of the component under test as a function of frequency. For example, the impedance of the resistor in Figure 18.1, including the lead inductance, is

$$Z = R + j\omega L \tag{18.1}$$

and

$$|Z| = [R^2 + \omega^2 L^2]^{1/2} \tag{18.2}$$

If we plot (18.2) on log-log coordinates, then there will be a low frequency region where $\omega \ll R/L$, and hence $|Z| \simeq R$, the ideal component behavior; and a high frequency region where $\omega \gg R/L$ and hence $|Z| \simeq \omega L$, a linearly increasing function of ω. This behavior is shown in Figure 18.3, along with the corresponding result for $|Y| = 1/|Z|$.

If the measured values of $|Z|$ or $|Y|$ of a resistor under test are plotted as a function of ω on log-log coordinates, then the equivalent circuit values of

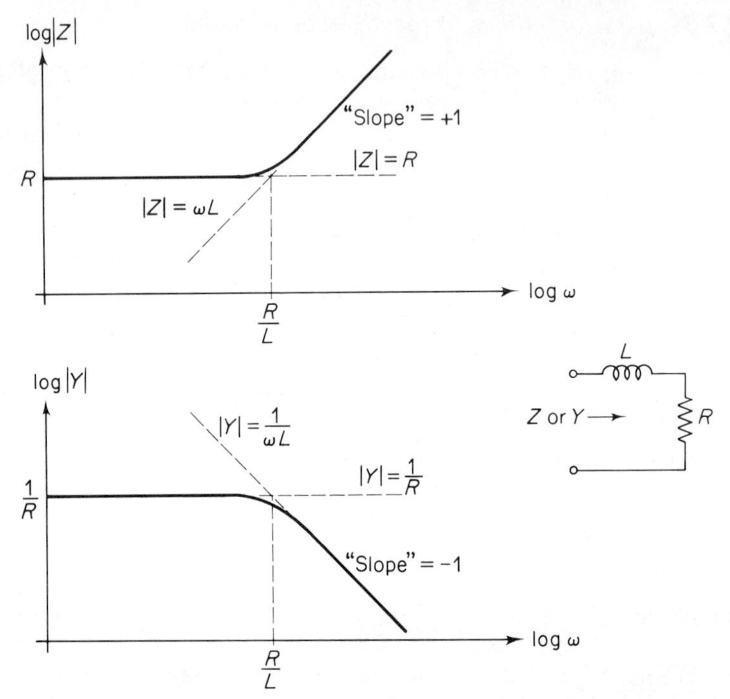

Figure 18.3. Variation of $|Z|$ and $|Y|$ vs. ω for a series R–L circuit.

R and L may be directly determined from the straight-line segments of the actual data. Notice that intersection of the straight-line segments occurs at $\omega = R/L$, independent of whether $|Z|$ or $|Y|$ is plotted.

If we consider the case where stray capacitance is the dominant effect, as illustrated in Figure 18.2, then a dual analysis yields the variations of $|Z|$ and $|Y|$ as shown in Figure 18.4. In this case, the intersection of the low and high frequency asymptotes occurs at $\omega = 1/RC$. As in the previous example, by fitting straight-line segments to the data points, the circuit model parameters can be determined.

The final circuit model that we will consider is that of a self-resonant coil, as shown in Figure 18.5. At low frequencies, the magnitude of the impedance behaves like an ideal inductor, while at high frequencies the stray capacitance is dominant and $|Z| = 1/\omega C$. At the self-resonant frequency, $\omega_0 = 1/\sqrt{LC}$, the inductive and capacitive reactance cancel one another and the impedance is pure real (0° phase angle) and equal to equivalent resistance associated with the coil losses. Once more, by fitting straight-line segments to the low and high frequency data points and by noting the self-resonant frequency, the parameters of the circuit model can be determined.

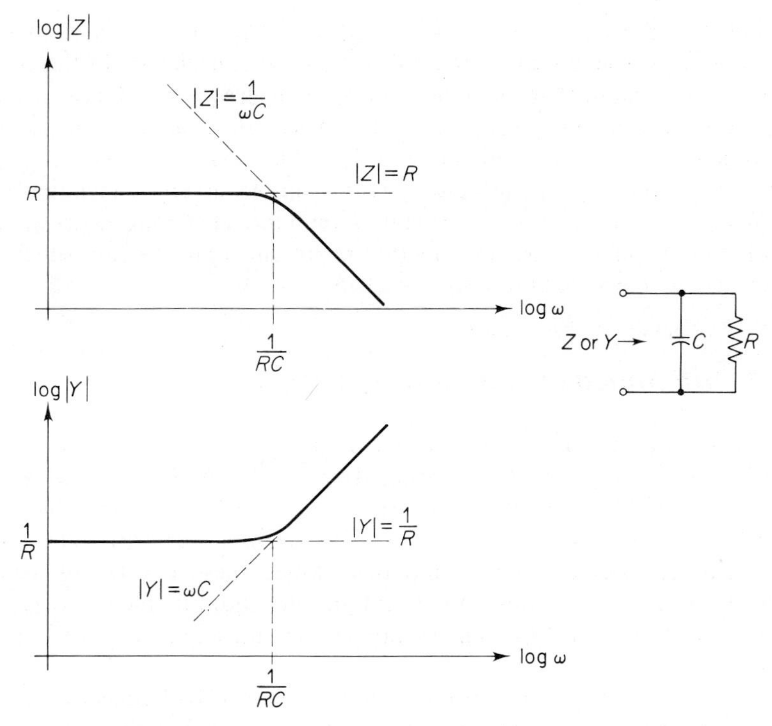

Figure 18.4. Variation of $|Z|$ and $|Y|$ vs. ω for a parallel R–C circuit.

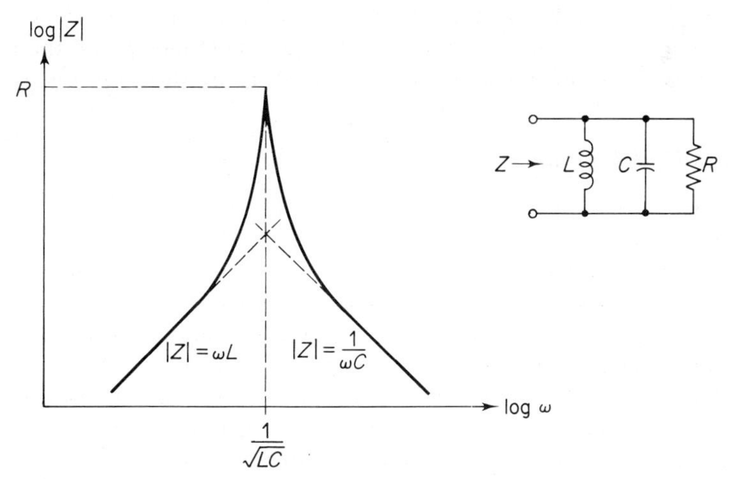

Figure 18.5. Variation of $|Z|$ vs. ω for a lossy self-resonant coil.

Before leaving the question of circuit-model parameter determination, the issue of frequency range of validity must be considered. In the simple models proposed here, it was assumed that as the frequency of measurement was increased, a range was encountered where the magnitude of Z or Y varied as either ω or $1/\omega$. Ultimately, as the test frequency is further increased, the data points may well deviate from this simple, assumed variation with frequency. At this point of deviation the circuit model no longer corresponds to the actual physical device and the maximum frequency for which the particular model is valid has been reached.

18.4 RF impedance measurements

In performing impedance or admittance measurements in the RF region, there is a distinct advantage in testing at or near the frequency at which the component will be used. In Section 18.2 we have seen how lead inductance, distributed capacitance, and winding and core losses can drastically alter the terminal behavior of a component. Since these effects are usually difficult or even impossible to detect in the audio-frequency region, a measurement made at 1 kHz and extrapolated to the RF region is of doubtful value.

It is also desirable to perform measurements over the frequency range of interest to observe the impedance or admittance variation. In this way, the complex behavior of the parasitic effects can often be represented by a simple circuit model as described in Section 18.3. It is important to note that a single frequency measurement is not sufficient to make the proper choice of circuit model.

There are three methods in common use by which the RF properties of components may be measured: an RF bridge, the Q meter, and the vector impedance meter. Examples of these instruments are shown in Figure 18.6.

18.4.1 RF bridges

An AC bridge for measurements in the RF region is basically the same as the AC bridge used at audio frequencies. (See Section 8.4.) That is, a sinusoid at the test frequency is applied to a bridge circuit, in which the unknown impedance forms one of the bridge arms. Adjustable elements in the remaining arms are then varied to produce a null output on a detector. The adjustable elements are calibrated to read an equivalent impedance or admittance at the test frequency when a null on the detector is obtained. In many RF bridges the test oscillator and detector are integral parts of the equipment. In others, only a single test frequency is internally available, with provision for a variable-frequency test signal to be applied externally.

A few instruments require both an external test signal and external detector. In Table 18.1 the major performance characteristics of several RF bridges are listed.

There are always two main adjustment controls on an AC bridge: one to balance the real part of the bridge equation and one to balance the imaginary part. The calibration of these main balance controls in terms of the equivalent circuit measured at the bridge terminals depends upon the detailed bridge circuit. One bridge may indicate an equivalent series impedance $R + jX$, while another may yield the equivalent admittance $G + jB$. Furthermore, a bridge yielding equivalent admittance may actually indicate $1/G$ in ohms rather than G in mhos. Finally, the reactive component may be calibrated directly in equivalent capacitance or inductance at the test frequency, or in terms of ohms of reactance or mhos of susceptance at some specified test frequency. In the latter case a conversion factor must usually be applied to the reactance or susceptance reading as the test frequency is varied.

Most RF bridges can only achieve a balance for a restricted range of values of the real and imaginary part of the complex impedance under test. It is therefore important to estimate the expected result before attempting a measurement, in order to be reasonably confident that the particular bridge is capable of the required measurement. When a measurement is attempted beyond the capability of a bridge, either little detector variation is observed or the minimum reading occurs at an end of scale setting on one or both of the main balance controls. When a true balance is achieved, the null indication is normally very sharp.

Because RF bridges are called upon to operate over a wide frequency range, some controls must be included to cancel out the parasitic effects within the bridge circuit itself. These controls are often referred to as *initial balance* or *zero balance* adjustments. This initial balance is simply accomplished by selecting a test frequency, setting the main balance controls to specified positions (usually reactance equal to zero and resistance equal to either zero or infinity) and adjusting the initial balance controls to obtain a null on the detector.

Exercise 18.1

Connect a 47-ohm, $\frac{1}{2}$-watt composition resistor to an RF bridge using the maximum possible lead length. Measure the impedance as a function of frequency over as wide a range as possible. Plot the magnitude of the impedance on log-log coordinates, and construct a circuit model for the resistor. How large is the lead inductance?

Exercise 18.2

Reconnect the resistor from the previous exercise to the bridge with the minimum possible lead length, cutting the excess lead length from the resistor. Repeat the

GR Model 1606-B

GR Model 1606-A

Figure 18.6. Examples of *RF* impedance measurement instruments.

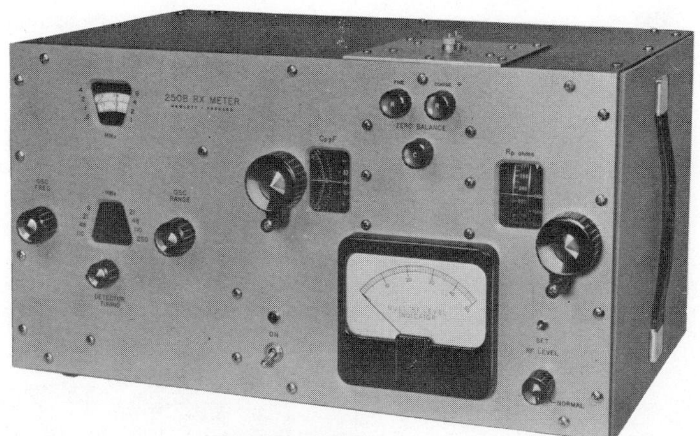

HP Model 250B

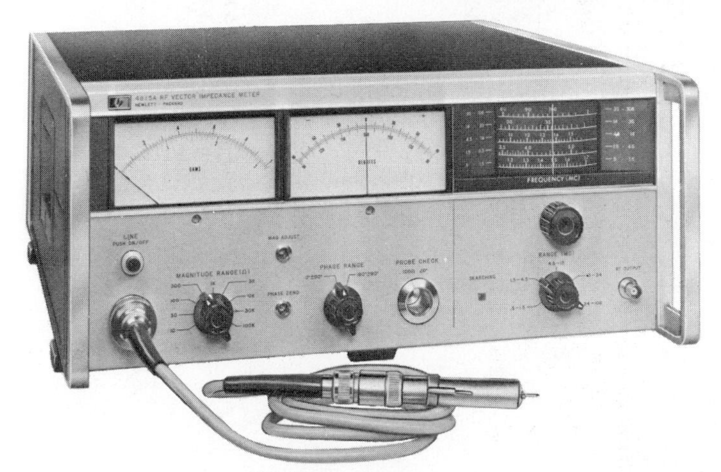

HP Model 4815A

HP Model 260A

Table 18.1 / Characteristics of RF Impedance Instruments

Model	Frequency	Basic Accuracy	Equivalent Circuit Readout	Measurement Range	Features
Hewlett-Packard 250B RX Meter	0.5–250 MHz internal	±2%	R_p parallel C_p	15 ohms $< R_p <$ 100 kΩ −100 pF $< C_p <$ 20 pF	self-contained bridge
Hewlett-Packard 260A Q Meter	0.05–50 MHz internal 1–50 kHz external	±2%	L series resistor, Q	0.09 μH $< L <$ 130 mH 10 $< Q <$ 625	self-contained bridge
General Radio Type 1606-B	0.4–60 MHz external	±2%	$R + jX$	0 $< R <$ 1 kΩ −5 kΩ $< X <$ 5 kΩ at 1 MHz	external generator and detector required
Hewlett-Packard 4815A Vector Impedance Meter	0.5–108 MHz internal	±4%	$\|Z\|, \angle Z$	1 Ω $< \|Z\| <$ 100 kΩ 0° $< \angle Z <$ 360°	automatic tuning, direct reading

impedance measurements vs. frequency and reevaluate your previous circuit model. How large is the stray capacitance?

Exercise 18.3

Measure the impedance of a 50-ohm wirewound resistor as a function of frequency with an RF bridge and compare your results with the data obtained on the composition resistor.

The useful reactance range of an RF bridge may be extended by the use of auxiliary coils or capacitors connected to the bridge terminals. In effect, the auxiliary component is chosen to cancel a portion of the unknown reactance such that the remaining portion can be measured by the bridge. The actual value is then determined by combining the bridge indication with the reactance of the auxiliary component. For example, a bridge capable of a 100-pF maximum capacitance is to be used to measure a 200-pF capacitor at 10 MHz. If a 2-μH coil is connected in parallel with the 200-pF capacitor, then the equivalent reactance of the parallel combination at 10 MHz is 217 ohms capacitive, which is equivalent to a 73-pF capacitor, well within the bridge capability of 100 pF maximum; of course, the auxiliary coil must have very low loss in order not to affect the resistive reading of the bridge. Many bridge manufacturers provide accessory coils and capacitors of very high quality specifically for range extension applications.

Exercise 18.4

Select a test frequency for which a 10-μH RF choke is beyond the reactance range of the RF bridge. Calculate the range of values for an auxiliary capacitor which will bring the choke within the measurement range of the bridge at the selected test frequency. Consider the cases of both series and parallel connection of the auxiliary capacitor. Using a capacitor within the calculated range, measure the 10-μH choke at the selected test frequency.

18.4.2 Q meter

The basic operation of the Q meter has been described in Section 9.3.3. The commercial version, such as the Hewlett-Packard 190A or 260A, is a self-contained instrument consisting of the signal generator, the series-tuned circuit, and the high-impedance voltmeter. In addition to the measurement of inductor Q, the instrument is also capable of measuring the self-resonant frequency, capacitance, effective RF resistance, inductance and capacitance of resistors, and frequency response and coupling coefficient of IF transformers.

Exercise 18.5

Using no. 22 copper wire construct a 10-turn coil 0.5 in. in diameter and 1 in. long. Using the Q meter, determine the inductance and Q in the frequency range of 20 to 50 MHz. Compare the measured inductance with the value calculated from the empirical formula for a single layer coil

$$L = \frac{n^2 r^2}{9r + 10l} \mu\text{H} \tag{18.3}$$

where n is the number of turns, r the radius of the coil, and l the length of the coil in inches. (If a Q meter is not available, this exercise may be performed on an RF bridge.)

18.4.3 vector impedance meter

The RF vector impedance meter is a relatively new development in the field of impedance measurement. By means of a single probe, a component is excited by a test signal and the resulting response sampled. The response signal is detected and processed to yield the magnitude and phase angle of the unknown impedance directly on individual front-panel meters. The test frequency is easily selected by a front-panel control and instrument requires no initial balance or detector tuning adjustments as the frequency is changed. With this instrument, rapid measurements of $|Z|$ and $|Y|$ can be made over the frequency range of interest. The Hewlett-Packard Model 4815A RF Vector Impedance Meter is capable of measurement of $|Z|$ from 1 ohm to 100 kΩ over the frequency range of 0.5 to 100 MHz. Within this range the instrument is unrivaled for speed and ease of measurement.

Exercise 18.6

Use the vector impedance meter to measure the magnitude and phase of a 10-μH RF choke over the frequency range of 1 to 100 MHz. Determine the self-resonant frequency. Plot $|Z|$ on log-log coordinates and determine an equivalent circuit model as shown in Figure 18.5. (If a vector impedance meter is not available, this exercise may be performed on a Q meter or an RF bridge.)

19

coaxial cables

19.1 introduction

At frequencies below 1 MHz, normal laboratory practice usually neglects the effects of connecting cables between components within a circuit and between the circuit and the measurement equipment. As the frequency of operation is increased to the radio-frequency (RF) region, roughly 1 to 100 MHz, connecting cables can have a marked effect on the circuit under test. In Chapter 14 the behavior of a coaxial cable was studied in terms of pulse excitation. The purpose of this chapter is to explore some of the basic properties and circuit performance characteristics of coaxial cables under steady-state sinusoidal excitation.

19.2 coaxial cables

Generally speaking, a coaxial cable is composed of a center conductor surrounded by a concentric conducting shell. Either or both of the conductors may be solid or stranded, depending on both the electrical and mechanical application. The two conductors are separated by a dielectric such as air, polyethylene, or teflon. Sizes range from less than 0.1 in. to several inches in diameter. Cables are usually designated by their military nomenclature, which is RG-*xy*/*U*; the "*xy*" being a particular number which specifies the cable. For general laboratory use, a small, flexible cable, such as RG-58/U, is usually employed. This cable, which will be considered as an example in this chapter, has a stranded center conductor, a polyethylene dielectric, and a single braided outer conductor.

19.2.1 input impedance of a coaxial cable

We begin by considering the simple cable system shown in Figure 19.1. At DC and very low frequencies, the input impedance Z_i would be equal to Z_L, assuming that there are no losses in the conductors or dielectric. This is simply the expected behavior of a pair of "wires" from a circuit theory point of view. However, as the frequency is increased, a point is reached where the electric currents are no longer "confined" to the wire. Rather, energy is transmitted along the cable by means of wave propagation guided by the conductors. At the point where this mechanism becomes important, the circuit theory concept of "wires" is no longer valid, and a field theory solution to the problem must be obtained.

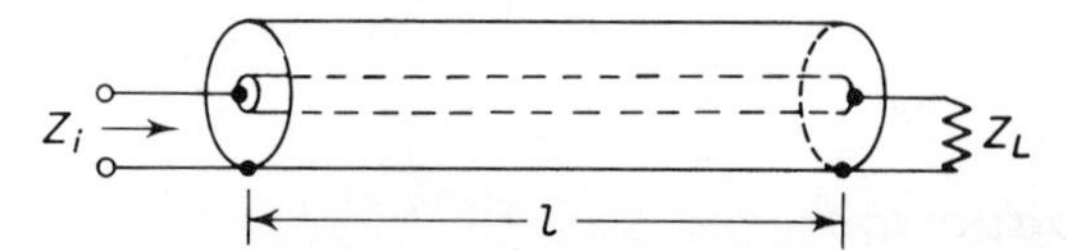

Figure 19.1. Coaxial cable circuit.

Fortunately in the case of coaxial cables it remains meaningful to use the circuit concept of a lumped impedance at the cable terminals. Thus for sinusoidal excitations, the input impedance Z_i is still the ratio of the complex voltage and current at the input terminals. However, the relationship between Z_i and Z_L depends upon the wave propagation between the input and output terminals. It is beyond the scope of this chapter to derive the theoretical relationship. Several references which present this material are listed at the end of the assignment. Rather, the theoretical result will be presented and investigated experimentally.

For the case of a lossless cable under sinusoidal excitation the relationship between Z_L and Z_i for a cable of length l is

$$Z_i = Z_0\left(\frac{Z_L + jZ_0 \tan \beta l}{Z_0 + jZ_L \tan \beta l}\right) \tag{19.1}$$

where Z_0 is the *characteristic impedance* and β is the *propagation constant*, both of which are properties of the cable alone. If we introduce the phase velocity from elementary wave theory,

$$v_p = f\lambda \tag{19.2}$$

where f is the frequency and λ is the wavelength, the propagation constant is given by

$$\beta = \frac{\omega}{v_p} = \frac{2\pi}{\lambda} \tag{19.3}$$

For a lossless coaxial cable, the phase velocity v_p is also the speed at which an electric signal will propagate along the cable, i.e., the velocity of light, and was previously studied in Section 14.3.3.

One final point must be made before we examine the result of (19.1) in detail. The phase velocity (19.2) refers to the medium in which the electromagnetic wave is moving. For a coaxial cable this is the dielectric. The wavelength in the dielectric is shorter than the wavelength in free space, since for a dielectric such as polyethylene $\varepsilon > \varepsilon_0$, $\mu \simeq \mu_0$. In free space, the phase velocity is given by

$$v_{p0} = \frac{1}{\sqrt{\mu_0 \varepsilon_0}} = 3 \times 10^8 \text{ m/sec} \tag{19.4}$$

In polyethylene, $\varepsilon = 2.25\varepsilon_0$ and $\mu \simeq \mu_0$; thus for the cable

$$v_p = \frac{1}{\sqrt{2.25}} v_{p0} = 0.68 v_{p0} \tag{19.5}$$

Equation (19.5) shows that in the cable the propagation velocity will be only 68% of the velocity of light in free space, and hence (19.2) shows that for a given frequency, λ in the cable must be shorter than λ in free space.

From Equation (19.3), we find the product

$$\beta l = 2\pi \left[\frac{l}{\lambda} \right] \tag{19.6}$$

It is common practice to refer to the length of a cable in terms of its wavelength rather than the actual physical length. When l is measured in terms of wavelength, it is often called the *electrical length* of the cable. Thus for a cable that is "$\lambda/4$ long," $l = \lambda/4$ and $\beta l = \pi/2$. For the remainder of this chapter we will describe cables by their electrical length unless specifically noted otherwise.

19.2.2 characteristic impedance

There are several physical interpretations for the characteristic impedance of a cable. In Section 14.3.3 the characteristic impedance was defined in terms of a line of infinite length. Here we will define it as that load impedance which results in the condition $Z_i = Z_L$, *independent* of the length of the cable. Notice that the length dependence in (19.1) is removed when $Z_L = Z_0$, and under this condition $Z_i = Z_0$. For lossless cables Z_0 is a positive real number, usually in the range of 50 to 75 ohms. RG-58/U cable has a nominal characteristic impedance of 53.5 ohms.

In the usual situation where it is desired to deliver power to a load at the end of a coaxial cable, termination of the cable in its characteristic impedance is the only way a constant, real impedance at the input of the cable can be achieved over a range of frequencies (changing electrical length).

19.2.3 capacitance and inductance of a cable

For electrical lengths less than about 0.1 λ and an open-circuit load ($Z_L = \infty$), the input impedance in Figure 19.1 will be that of a capacitor. This is not unexpected, since the physical structure is simply two conductors separated by a dielectric. Calculation of Z_i from (19.1) for this case yields

$$Z_i = \frac{1}{j\omega(l/v_p Z_0)} \tag{19.7}$$

in which the approximation $\tan \beta l \simeq \beta l$ for small values of βl (0.6 or less) has been used. The total value of capacitance depends linearly on the physical length l; hence it is convenient to define a capacitance per unit length C for the cable, which depends only on its intrinsic parameters. From (19.7)

$$C = \frac{1}{v_p Z_0} \tag{19.8}$$

Evaluation of C for the nominal parameters of RG-58/U yields $C = 91.5$ pF/m, or about 1 pF/cm. Occasionally when a small capacitance is required in a circuit, an appropriate length of coaxial cable is employed.

In a similar fashion, if Z_L in Figure 19.1 is replaced by a short circuit ($Z_L = 0$), then the physical structure is that of a one-turn inductor. Again for lengths less than about 0.1 λ, (19.1) yields

$$Z_i = j\omega\left(\frac{Z_0 l}{v_p}\right) \tag{19.9}$$

In this case, the inductance per unit length L is given by

$$L = \frac{Z_0}{v_p} \tag{19.10}$$

which is 0.26 μH/m in RG-58/U.

Equations (19.8) and (19.10) may be combined to express the phase velocity and characteristic impedance in terms of L and C:

$$v_p = \frac{1}{\sqrt{LC}} \tag{19.11}$$

$$Z_0 = \sqrt{\frac{L}{C}} \tag{19.12}$$

Recalling from (19.5) that the phase velocity depends only on the properties of the dielectric, (19.11) shows that the product of the inductance and capacitance per unit length is a function only of the dielectric material and is independent of the cross-sectional dimensions of the cable.

19.3 impedance calculations—the Smith chart

For a given cable of known characteristic impedance, propagation velocity (or L and C), and length, the input impedance may be calculated for any given frequency and load impedance (real or complex) by means of (19.1). In general, this is a tedious algebraic process. However, there exists a graphical aid for such calculations known as the Smith chart, shown in Figure 19.2.

Before describing the use of the Smith chart for these calculations, we must first introduce the normalized impedance or admittance. Direct calculation shows that (19.1) may be rewritten as

$$Z_{ni} = \frac{Z_{nL} + j \tan \beta l}{1 + jZ_{nL} \tan \beta l} \tag{19.13}$$

where $Z_{ni} = Z_i/Z_0$ and $Z_{nL} = Z_L/Z_0$ are the *normalized* input and load impedances respectively. Similarly, we find on an admittance basis

$$Y_{ni} = \frac{Y_{nL} + j \tan \beta l}{1 + jY_{nL} \tan \beta l} \tag{19.14}$$

where $Y_i = 1/Z_i$, $Y_{ni} = Y_i/Y_0$, $Y_0 = 1/Z_0$, etc. The parameter Y_0 is called the *characteristic admittance* of the coaxial cable. The functional equivalence of (19.13) and (19.14) is a very useful property since a graphical technique which works for (19.13) will be identical for (19.14).

Although elegant mathematical derivations of the properties of the Smith chart coordinate system exist, we will accept the chart as a graphical tool and confine our attention to its practical applications. Since the measurements to be made subsequently are usually more convenient on an

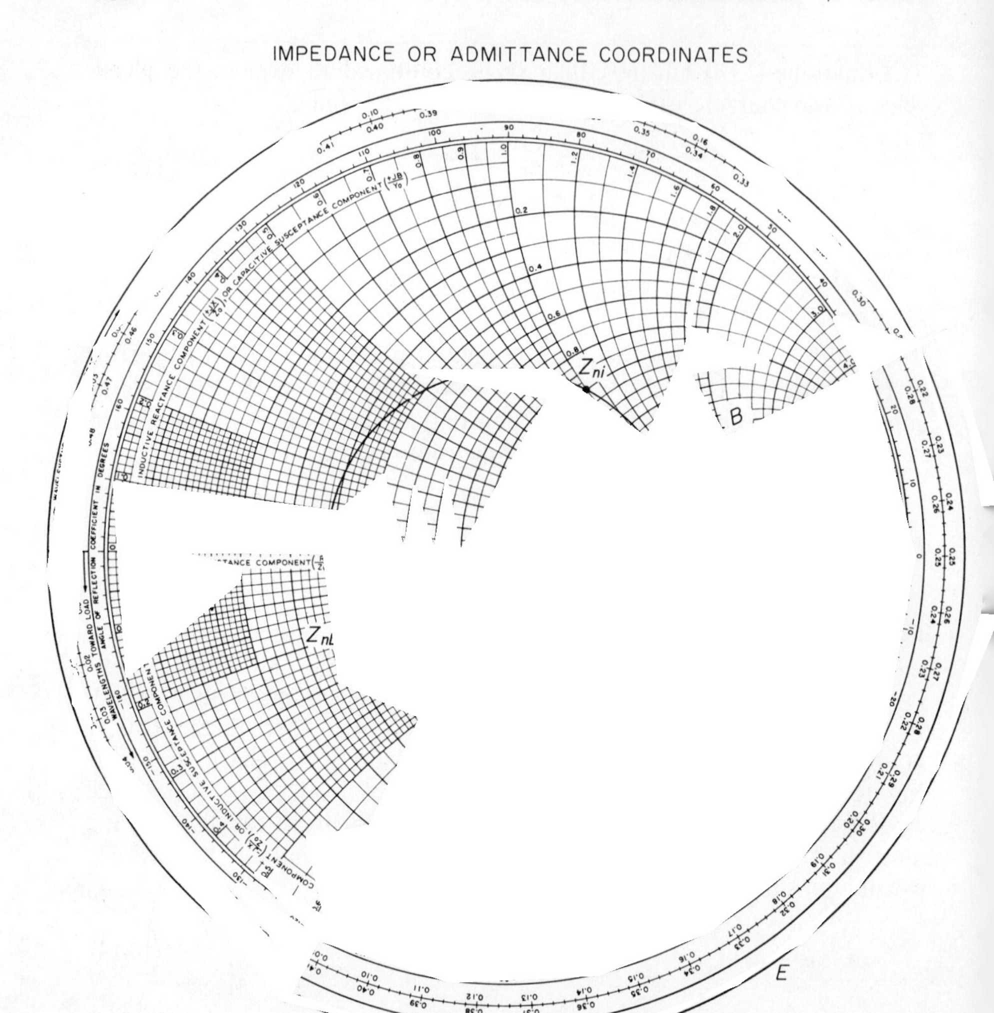

Figure 19.2. Example of Smith chart calculation.

admittance basis, we will use that basis for our examples. Calculations on an impedance basis are identical.

The basic description of the Smith chart is as follows:

1. Contained in a circle centered at $Y_n = 1 + j0$, is an orthogonal coordinate system in which any value of a passive normalized admittance

$$Y_n = \frac{G}{Y_0} + j\frac{B}{Y_0} = G_n + jB_n \tag{19.15}$$

may be located.

2. Loci of constant G_n are circles tangent to the right-hand edge of the circumference.

3. Loci of constant B_n are circles orthogonal to the circles of constant G_n, with the positive values of B_n in the upper half of the chart and negative values of B_n in the lower half of the chart.

4. The circumference of the chart is scaled in angular displacement of radii from the center of the chart such that one complete revolution is equivalent to an electrical length of $\lambda/2$.

The graphical solution to Equation (19.14) is accomplished as follows:

A. Convert the value of Y_L to normalized form by dividing by Y_0 (or multiply by Z_0).

B. Locate the point Y_{nL} in the chart and draw a radius from the center of the chart, through Y_{nL}, to the circumference.

C. Calculate the electrical length of the cable.

D. If the electrical length is greater than $\lambda/2$, subtract $\lambda/2$ from the length. Repeat, if necessary, until the length lies in the range $0 \leq l \leq \lambda/2$.

E. Starting at the radius constructed in (B) move along the circumference in a *clockwise* direction (toward the generator in chart nomenclature) by the electrical length calculated in (C) or (D).

F. Draw a radius from the center of the chart through the point on the circumference located in (E).

G. With the point $Y_n = 1 + j0$ as its center, draw a circle through the point Y_{nL}, located in Step B.

H. The intersection of the circle of (G) and the radius (F) is the admittance Y_{ni}.

I. To find Y_i multiply Y_{ni} by Y_0 (or divide by Z_0).

19.3.1 example on use of the Smith chart

The following example will illustrate the step-by-step procedure outlined above.

A 50-ohm coaxial cable, 2 meters long, with a phase velocity $v_p = 0.5v_{p0}$ is terminated by the parallel combination of a 25-ohm resistor and a 210-pF capacitor. Find the input admittance Y_i of the cable at 15 MHz.

A.
$$Y_L = \tfrac{1}{25} + j2\pi \times 15 \times 10^6 \times 210 \times 10^{-12}$$
$$= 40 + j19.8 \text{ mmhos}$$

$$Y_0 = \frac{1}{50} = 20 \text{ mmhos}$$

$$\therefore \quad Y_{nL} \simeq 2 + j1$$

B. See Figure 19.2.

C. From Equation (19.2)

$$\lambda = \frac{0.5 \times 3 \times 10^8}{15 \times 10^6} = 10 \text{ m}$$

$$\therefore \quad \text{electrical length} = \tfrac{2}{10}\lambda = 0.2\,\lambda$$

D. Since the length is in the range $0 \leq l \leq \lambda/2$, it is not necessary to subtract $\lambda/2$ from the length.

E, F, G, H. See Figure 19.2.

I. From Figure 19.2, we have

$$Y_{ni} = 0.5 - j0.5$$

and

$$Y_i = 20\, Y_{ni} = 10 - j10 \text{ mmhos}$$

Before leaving this example, an additional graphical property of the Smith chart should be pointed out. Any complex quantity $a + jb$ can be converted to its reciprocal in rectangular form by simply rotating the point by 180 degrees about the center of the chart. This is particularly useful for conversion between Z's and Y's. This is illustrated in Figure 19.2 where the impedances Z_{ni} and Z_{nL} are shown. From these points we have, after removal of the normalization,

$$Z_i = 50 + j50 \text{ ohms}$$

$$Z_L = 20 - j10 \text{ ohms}$$

19.3.2 other uses of the Smith chart

The Smith chart is by no means limited to the solution of (19.13) or (19.14). Since all values of an impedance or admittance with a positive real part are contained within a finite circle, it makes a useful tool for displaying the variation of an impedance or admittance as a function of frequency, and the loci of such plots can be used to construct circuit models in the same fashion that magnitude of impedance plots were employed in Section 18.3.

19.4 impedance measurements on coaxial cables

The frequency range in which we will experimentally explore the behavior of a coaxial cable is 1 to 100 MHz. Thus, we will be continuing the use of the equipment discussed in Chapter 18, and the following discussion will presume familiarity with that material.

If the measurement equipment can be fitted with a coaxial cable connector, such as a type BNC, it should be used. This will reduce the stray reactive effects present when the cable is connected to a set of binding posts.

In providing a resistive load for the coaxial cable, a 50-ohm termination must be used. This device is designed for minimum reactance and is fitted with a coaxial connector. Some types of terminations (Tektronix 011-149, for example) are provided with both a male and female BNC connector to permit a through connection, as shown schematically in Figure 19.3. If a through termination is not available, one can be made from a normal termination and a T adapter.

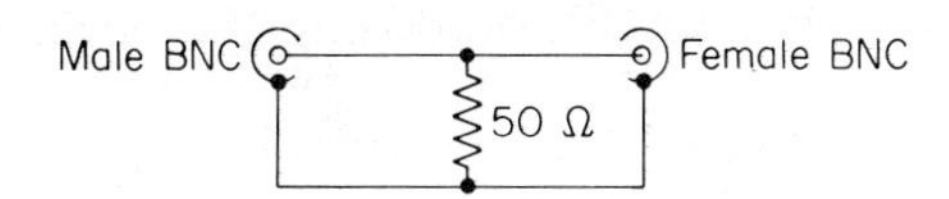

Figure 19.3. Schematic of a through termination.

If a bridge is to be used for the cable measurements, it would normally be required that an initial rebalance be made whenever the test frequency is changed. The following exercise will provide an estimate of the inherent error resulting from not initially rebalancing after each change of frequency. The magnitude of the error can then be compared with the actual measurement and a decision made whether or not a rebalance is necessary. In most of the following exercises, the rebalance will be unnecessary.

Exercise 19.1

With an open-circuit input, initially balance the bridge at 10 MHz. Without changing the initial balance controls, measure the equivalent impedance of the open-circuit input at 1 MHz and 100 MHz, by adjusting the main balance controls for a detector null.

Connect the 50-ohm termination to the bridge and measure the equivalent impedance at 10 MHz. If the reactive part is not zero, it can be compensated by the following procedure. Without changing the real part indicator, set the reactive part indicator to zero. Then rebalance the bridge using the initial balance controls. This corrects for the small reactance of the termination. Finally, without changing the initial balance controls, measure the impedance of the 50-ohm load at 1 MHz and 100 MHz.

These measurements at 1 and 100 MHz provide a measure of the parasitic effects at the low- and high-frequency ends of the range, and hence an estimate of the error introduced by not rebalancing the bridge as the frequency is changed. If these errors are small compared with the actual impedance being measured, they can be neglected for the purposes of these exercises.

19.4.1 measurement of C of coaxial cable

The capacitance per unit length of a cable can be measured by using an open-circuited piece of cable which has a length much less than a wavelength, as shown by (19.7).

Exercise 19.2

Obtain a short length of open-circuited RG-58/U and measure the total capacitance at several frequencies from 1 to 100 MHz. Divide the measured capacitance by the physical length of the cable to obtain the capacitance per unit length. If the capacitance varies as a function of frequency, which value should you use for C?

If you wish to correct for a BNC connector on the cable, it adds about 1 cm to the total physical length and adds approximately 0.8 pF to the measurement.

19.4.2 measurements on terminated cables

If the electrical length of a coaxial cable is equal to an odd number of quarter-wavelengths, then $\tan \beta l \to \infty$ and (19.1) shows that $Z_i = Z_0^2/Z_L$. Similarly, if the electrical length equals an even number of quarter-wavelengths, then $\tan \beta l = 0$ and $Z_i = Z_L$. The ability of an odd number of quarter-wavelengths to convert a load impedance into its reciprocal scaled by Z_0^2 is often called the impedance transformation property of a quarter-wave cable.

Exercise 19.3

Balance the bridge at 10 MHz with the 50-ohm load as described in Exercise 19.1 and shown in Figure 19.4a. Remove the load and reconnect it by means of the length of RG-58/U cable about 1.5 meters long as shown in Figure 19.4b. Measure the input admittance at several frequencies from 1 to 100 MHz and plot your data on a Smith chart.

Reverse the connection; that is, using a through termination, connect the load to the bridge input terminals and the open-circuit cable to the load (see Figure 19.4c). Repeat the measurements and plot the data on the Smith chart. Add a second 50-ohm load at the open-circuited end of the cable as illustrated in Figure 19.4d. Check the admittance at several frequencies and enter the data on the Smith chart. What conclusions can you draw about terminations at the source and load ends of a coaxial cable?

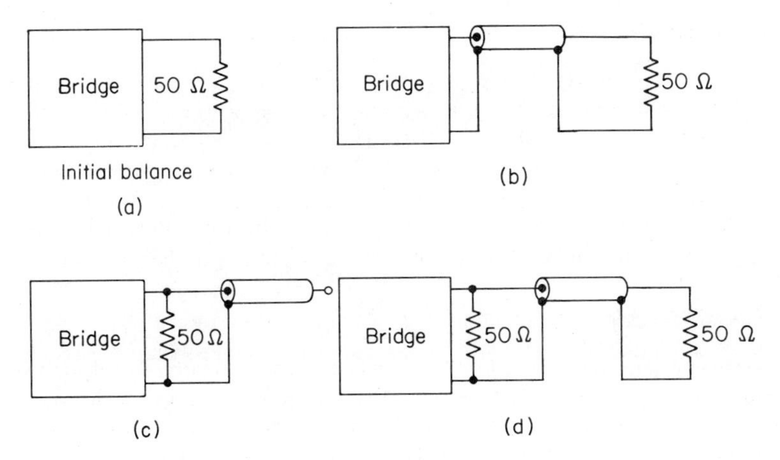

Figure 19.4. Measurements on terminated cable.

Exercise 19.4

Balance the bridge at 10 MHz with the 50-ohm load at the input. Construct a 25-ohm load by using two 50-ohm loads plus a T connector. Attach one end of the 1.5-meter length of RG-58/U cable to the bridge and terminate it with the 25-ohm load as shown in Figure 19.5. Measure the input admittance at several frequencies and plot the data on the Smith chart. The theoretical locus of the points should be a circle passing through the point $2.0 + j0$ with its center at the center of the chart. This is because the only variable parameter is the electrical length of the cable since the load impedance is assumed to be independent of frequency. Using the impedance transformation properties of quarter-wavelength cables described above, find the frequencies at which the electrical length is a multiple of $\lambda/4$.

19.4.3 measurement of open and shorted cable

From Equation (19.1) it is seen that the input impedance (or admittance) is purely reactive if Z_L is a short- or open-circuit. On the Smith chart, this corresponds to a locus at the outer circumference of the chart.

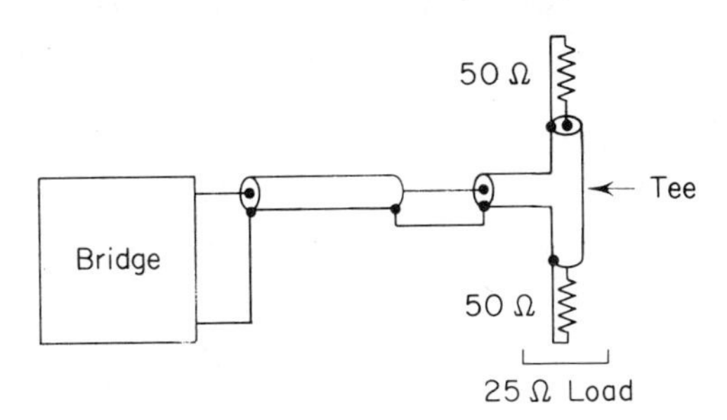

Figure 19.5. Measurements of terminated cables.

Exercise 19.5

Using a length of RG-58/U about 1.5 meters long, measure the input admittance with an open- and a short-circuit load in the range of 10 to 100 MHz. Plot the measured admittances on Smith charts and label the frequency of each measurement. (*Note:* Most bridges are not capable of measuring all values of impedance presented by the cable in this frequency range. Rather, data can only be obtained in certain frequency ranges. Estimate these ranges by using (19.1) and the known impedance ranges of the bridge, and begin measurements at these frequencies.)

Exercise 19.6

Using the impedance transformation properties of a quarter-wavelength cable described in Section 19.4.2 and the data from the preceding exercise to determine the frequencies at which the cable is an integral number of quarter-wavelengths long, calculate the phase velocity from (19.2). For example, if the electrical length is $\lambda/4$ at $f = 30$ MHz, then the wavelength is $\lambda = 4 \times 1.5$ meters or 6 meters, and $v_p = 6 \times 30 \times 10^6 = 1.8 \times 10^8$ m/sec. Compare the measured value of v_p with the prediction of (19.5). Also compare the value of v_p with that obtained for the velocity of pulse propagation in Chapter 14.

From the measured values of v_p and C, use (19.8) to calculate Z_0, and (19.10) to find L. Compare your value of Z_0 with the nominal value for RG-58/U of 53.5 ohms.

19.5 references

1. T. Moreno, *Microwave Transmission Design Data*, Chapters 3 and 5, Dover Publications, Inc., New York, 1958.

2. Adler, Chu, and Fano, *Electromagnetic Energy Transmission and Radiation*, Chapters 3 and 9, John Wiley & Sons, Inc., New York, 1960.

20

thermal measurements and heat sinks

20.1 introduction

Probably the most important nonelectrical parameter that affects the performance of an electronic component is its temperature. At low-temperature extremes, a device may fail to function because the change in temperature has significantly altered the electrical characteristics of the materials from which it is made. At elevated temperatures, the effects of temperature can result in complete destruction of the device. In applications such as amplifiers and regulated power supplies, where significant amounts of electric power are involved, roughly 1 watt or more, one must be concerned with the effects of elevated temperature on components resulting from internal power dissipation as well as the effects due to a high ambient temperature. Finally, it is axiomatic that the useful life of a component is inversely proportional to its operating temperature. Thus reliable performance of a complete system demands close attention to the problems of heat dissipation in the design phase.

In this chapter we will study several techniques for temperature measurement. Also, we will explore the use of heat dissipators to extend the power-handling capabilities of semiconductor devices.

20.2 temperature measurement

The familiar bulb thermometer, which depends upon the thermal expansion of mercury or alcohol in a capillary tube, is not well suited for temperature measurement of electronic components. Generally, the small size of the components and their shapes make it difficult to provide a good thermal connection between, say, a transistor and the thermometer bulb. Also, the slow response time of the conventional thermometer makes its use tedious.

An alternate approach to this problem is to employ techniques where the effect of temperature on a device produces a known effect that can be measured electrically. Such devices are known as *thermal transducers* because they translate temperature into an electric signal. Since the transducer can be made very small, it is capable of monitoring temperature in a localized region. Also, the corresponding low thermal mass makes possible rapid response to temperature changes.

20.2.1 thermistors

The thermistor has already been described in Section 7.3.5. It is basically a resistor which has a large negative temperature coefficient of resistance, and therefore temperature can be measured by simply measuring the resistance of a thermistor. The useful temperature range for these devices is from $-50°$ to $+300°C$.

Thermistors are manufactured in the form of beads as small as 0.010 in. in diameter, rods, washers, and in mountings such as glass rods, hypodermic needles, and a wide range of threaded studs. Thus there is a size and shape for nearly every application. The range of available resistance values at 25°C—10 ohms to 10 MΩ—is to some extent a function of the physical size of the thermistor, and the nominal tolerance is $\pm 20\%$. Although some manufacturers will provide precision calibrated thermistors, accurate to 1% or better for temperature measurement applications, in most cases calibration is required before the thermistor can be used as an accurate temperature transducer.

A calibration can be carried out in an oven or a refrigerated box as described in Exercise 7.4. However, this method is time consuming and an alternate procedure will be described here. The basic technique is to employ a series of *isothermal* baths. An isothermal bath is one whose temperature is determined by an equilibrium between two thermodynamic phases, such as a mixture of ice and water at 0°C and boiling water at 100°C. If we measure the resistance of a thermistor placed first in an ice-water mixture and then in a boiling-water bath, the resistance at 0° and 100°C will be determined. These two points plus the measured resistance at room temperature will establish a line on a $\log R$ vs. $1/T$ ($°K^{-1}$) plot (see Sections 7.3.5 and 7.4.2), from which intermediate resistance-temperature values may be interpolated. Extension of the range beyond the 0° and 100°C points may be done with some uncertainty as to accuracy.

Exercise 20.1

Using isothermal baths, calibrate a thermistor encapsulated in a threaded screw mount by measuring the resistance at 0°C (ice-water mixture), room temperature, and 100°C (boiling water). Choose a thermistor with a resistance of about 5 kΩ at 25°C. Be sure to stir the ice-water mixture constantly to insure that an equilibrium is maintained. If available, a vacuum dewar or other insulated container should be used for the ice-water mixture. The resistance measurement can be made with a multitester, digital ohmmeter, VTVM with a resistance scale, or a resistance bridge. From the three data points, draw an interpolation graph on $\log R$ vs. $1/T$ (°K^{-1}) coordinates.

20.2.2 resistance thermometers

Resistance thermometers, constructed from platinum wire, are capable of reproducing a resistance change with temperature on the order of 0.1 ppm, thus making them an extremely precise temperature transducer. While their useful range runs from −250° to +1200°C, they are used as the interpolation standard from the isothermal baths of the oxygen boiling point, −183°C, to the antimony freezing point, +630.5°C. Since these transducers are relatively expensive, they are normally employed only where the required accuracy and stability can justify the cost.

In less critical applications, other materials such as nickel may be used to construct resistance thermometers. However, not being noble metals they are subject to errors through corrosion and impurities present in the raw material.

20.2.3 thermocouples

When two dissimilar metals are joined and their junctions are at different temperatures, then a voltage proportional to the temperature difference is produced, as shown in Figure 20.1. The overall arrangement is called a *thermocouple.* Thus if T_1 is held at a known temperature, measurement of the voltage V will yield the unknown temperature T_2. The junction at T_1 is called the *reference junction,* since T_2 is measured with respect to T_1. The production of this voltage proportional to the temperature difference is known as the Seebeck effect, and the proportionality constant α is termed the *thermoelectric power.*

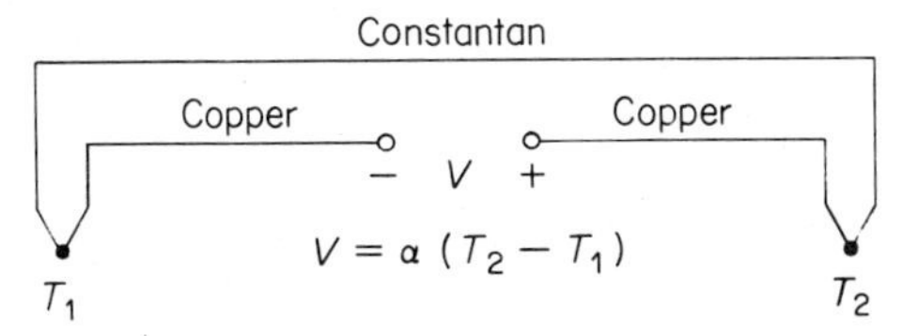

Figure 20.1. Basic thermocouple circuit.

The materials commonly used to form thermocouple junctions include iron, copper, platinum, and several alloys: platinum-rhodium (90% Pt-10% Rh), chromel (90% Ni-10% Cr), alumel (94% Ni-2% Al-3% Mn-1% Si), and constantan (60% Cu-40% Ni). These materials are used in the pairs which yield the largest values of thermoelectric power. The most common pairs along with their useful temperature ranges and values of α are given in Table 20.1. Also listed is the standard letter code used to identify given pairs. The most common thermocouple pairs are copper-constantan and chromel-alumel. The platinum pairs are expensive and used only where other materials are not suitable. The thermoelectric power listed in Table 20.1 is valid only in the range 0° to 100°C. Actually α is a mild function of temperature and may vary as much as 10% over the entire temperature range of the thermocouple pair. For this reason the actual thermocouple voltages as a function of temperature have been tabulated. This tabulation for copper-constantan is given in Table 20.2. As is standard practice, temperature of the reference junction T_1 is assumed to be 0°C, which is easily established in the laboratory with an ice-water mixture isothermal bath.

Table 20.1 / Characteristics of Thermocouples

Thermocouple Pair	Letter Code	Thermoelectric Power μV/°C (0°–100°C)	Useful Temperature Range (°C)*
Copper-Constantan	T	42.4	−200 to +300
Iron-Constantan	J	52.8	−200 to +1300
Chromel-Constantan	E	63	0 to +1100
Chromel-Alumel	K	41	−200 to +1200
Platinum-Platinum/Rhodium(10)	S	64.3	0 to +1450

* Depends considerably on particular application.

For most laboratory applications, thermocouple junctions must be fabricated from thermocouple wire by the user. Thermocouple wire is supplied in sizes from No. 8 to No. 28 AWG, and the choice of size depends primarily on considerations of mechanical strength, heat loss through the wire, and speed of response, as the output voltage is not influenced by the wire size. In the smaller sizes, *duplex* thermocouple wire, which consists of two color-coded wires of different thermocouple materials enclosed in a single jacket, is often convenient. Table 20.3 lists some common duplex thermocouple wires with their color codes.

The fabrication of the thermocouple junction is usually accomplished by fusing the two wires by means of a flame or electric weld. In the flame method, one simply places the two-wire ends in contact and heats them with a fine oxyacetylene or oxyhydrogen flame until a molten ball two-to-four times the wire diameter is formed between the wires. The flame is then removed and the wires are thus fused together by the alloy ball. Occasionally, the application of a brazing flux prior to heating will aid in this fusion process. Formation of a junction can also be accomplished electrically by spot welding the two wires or by melting and fusing them by means of an electric arc or resistance heater.

If the thermocouple is to be used at temperatures below about 200°C, then the junction can be formed by simply soldering the wires together. The presence of the solder in the junction may contribute an error of a degree or two in the thermocouple reading, hence this method should be avoided when maximum accuracy is required. A solder composed of 50% tin and 50% lead without a flux core is suitable for most work. Also, a zinc chloride flux such as those commercially available for soldering stainless steel* should be used. After the junction has been soldered, wash it thoroughly with warm water to remove all traces of the corrosive flux.

Exercise 20.2

Construct a copper-constantan thermocouple at one end of a 3-foot length of No. 24 or No. 28 AWG duplex wire. Remove 1 inch of insulation from each end of the duplex wire and clean the wires with steel wool or emery paper. On the end where the junction is to be formed, twist and trim the wires as shown in Figure 20.2. If a gas torch or electrical resistance heater is available, use it to fuse the junction. Otherwise, apply a small amount of zinc chloride flux and solder the junction with 50-50 lead-tin solder.

Since the thermoelectric power of thermocouple junctions is so small, only a few millivolts can be obtained for each 100°C of temperature difference. Thus, accurate temperature measurements require a sensitive voltmeter. In most instances a potentiometer (see Section 13.2.3) is used. Some potentiometers are furnished with interchangeable scales which permit direct readout of the temperature for whatever thermocouple pair is being used. Also, sensitive digital voltmeters are very convenient since they automatically track voltage, and hence temperature, changes. An X-Y recorder also may have sufficient sensitivity for use with a thermocouple and can be used to make a permanent record of a temperature variation.

Two practical temperature measurement schemes using thermocouples are shown in Figure 20.3. In Figure 20.3a, an ice-water mixture is used to establish the temperature of the reference junction. This bath should be stirred frequently to insure equilibrium between the ice and water. In Figure 20.3b, the ice bath has been replaced by an equivalent voltage. This method

* For example, Stay-Clean manufactured by J. W. Harris Company, 433 W. Ninth Street, Cincinnati, Ohio.

Table 20.2 / Copper vs. Constantan Thermocouple

Degrees Centigrade — Reference Junction 0°C

°C	0	1	2	3	4	5	6	7	8	9
					millivolts					
−190	−5.379	−5.395	−5.411	—	—	—	—	—	—	—
−180	−5.205	−5.223	−5.241	−5.258	−5.276	−5.294	−5.311	−5.328	−5.345	−5.362
−170	−5.018	−5.037	−5.056	−5.075	−5.094	−5.113	−5.132	−5.150	−5.169	−5.187
−160	−4.817	−4.838	−4.858	−4.878	−4.899	−4.919	−4.939	−4.959	−4.978	−4.908
−150	−4.603	−4.625	−4.647	−4.669	−4.690	−4.712	−4.733	−4.754	−4.775	−4.796
−140	−4.377	−4.400	−4.423	−4.446	−4.469	−4.492	−4.514	−4.537	−4.559	−4.581
−130	−4.138	−4.162	−4.187	−4.211	−4.235	−4.259	−4.283	−4.307	−4.330	−4.354
−120	−3.887	−3.912	−3.938	−3.964	−3.989	−4.014	−4.039	−4.064	−4.089	−4.114
−110	−3.624	−3.651	−3.678	−3.704	−3.730	−3.757	−3.783	−3.809	−3.835	−3.861
−100	−3.349	−3.377	−3.405	−3.432	−3.460	−3.488	−3.515	−3.542	−3.570	−3.597
−90	−3.062	−3.091	−3.120	−3.149	−3.178	−3.207	−3.235	−3.264	−3.292	−3.320
−80	−2.764	−2.794	−2.824	−2.854	−2.884	−2.914	−2.944	−2.974	−3.003	−3.033
−70	−2.455	−2.486	−2.518	−2.549	−2.580	−2.611	−2.642	−2.672	−2.703	−2.733
−60	−2.135	−2.167	−2.200	−2.232	−2.264	−2.296	−2.328	−2.360	−2.392	−2.423
−50	−1.804	−1.838	−1.871	−1.905	−1.938	−1.971	−2.004	−2.037	−2.070	−2.103
−40	−1.463	−1.498	−1.532	−1.567	−1.601	−1.635	−1.669	−1.703	−1.737	−1.771
−30	−1.112	−1.148	−1.183	−1.218	−1.254	−1.289	−1.324	−1.359	−1.394	−1.429
−20	−0.751	−0.788	−0.824	−0.860	−0.897	−0.933	−0.969	−1.005	−1.041	−1.076
−10	−0.380	−0.417	−0.455	−0.492	−0.530	−0.567	−0.604	−0.641	−0.678	−0.714
(−)0	−0.000	−0.038	−0.077	−0.115	−0.153	−0.191	−0.229	−0.267	−0.305	−0.343
(+)0	0.000	0.038	0.077	0.116	0.154	0.193	0.232	0.271	0.311	0.350
10	0.389	0.429	0.468	0.508	0.547	0.587	0.627	0.667	0.707	0.747
20	0.787	0.827	0.868	0.908	0.949	0.990	1.030	1.071	1.112	1.153
30	1.194	1.235	1.277	1.318	1.360	1.401	1.443	1.485	1.526	1.568
40	1.610	1.652	1.694	1.737	1.779	1.821	1.864	1.907	1.949	1.992

50	2.035	2.078	2.121	2.164	2.207	2.250	2.293	2.336	2.380	2.423
60	2.467	2.511	2.555	2.599	2.643	2.687	2.731	2.775	2.820	2.864
70	2.908	2.953	2.997	3.042	3.087	3.132	3.177	3.222	3.267	3.312
80	3.357	3.402	3.448	3.493	3.539	3.584	3.630	3.676	3.722	3.767
90	3.813	3.859	3.906	3.952	3.998	4.044	4.091	4.138	4.184	4.230
100	4.277	4.324	4.371	4.418	4.465	4.512	4.559	4.606	4.654	4.701
110	4.749	4.796	4.843	4.891	4.939	4.987	5.035	5.083	5.131	5.179
120	5.227	5.275	5.323	5.372	5.420	5.469	5.518	5.566	5.615	5.663
130	5.712	5.761	5.810	5.859	5.908	5.957	6.007	6.056	6.105	6.155
140	6.204	6.254	6.303	6.353	6.403	6.453	6.503	6.553	6.603	6.653
150	6.703	6.753	6.803	6.853	6.904	6.954	7.004	7.055	7.106	7.157
160	7.208	7.258	7.309	7.360	7.411	7.462	7.513	7.565	7.616	7.667
170	7.719	7.770	7.822	7.874	7.926	7.978	8.029	8.080	8.132	8.184
180	8.236	8.288	8.340	8.392	8.445	8.497	8.549	8.601	8.654	8.707
190	8.759	8.812	8.864	8.917	8.970	9.023	9.076	9.129	9.182	9.235
200	9.288	9.341	9.394	9.448	9.501	9.555	9.608	9.662	9.715	9.769
210	9.823	9.877	9.931	9.985	10.039	10.093	10.147	10.201	10.255	10.309
220	10.363	10.417	10.471	10.526	10.580	10.635	10.689	10.744	10.799	10.854
230	10.909	10.963	11.018	11.073	11.128	11.183	11.238	11.293	11.348	11.403
240	11.459	11.514	11.569	11.624	11.680	11.735	11.791	11.847	11.903	11.959
250	12.015	12.071	12.126	12.182	12.238	12.294	12.350	12.406	12.462	12.518
260	12.575	12.631	12.688	12.744	12.800	12.857	12.913	12.970	13.027	13.083
270	13.140	13.197	13.254	13.311	13.368	13.425	13.482	13.539	13.596	13.653
280	13.710	13.768	13.825	13.882	13.939	13.997	14.055	14.112	14.170	14.227
290	14.285	14.343	14.400	14.458	14.515	14.573	14.631	14.689	14.747	14.805
300	14.864	14.922	14.980	15.038	15.096	15.155	15.213	15.271	15.330	15.388
310	15.447	15.506	15.564	15.623	15.681	15.740	15.799	15.858	15.917	15.976
320	16.035	16.094	16.153	16.212	16.271	16.330	16.389	16.449	16.508	16.567
330	16.626	16.685	16.745	16.804	16.864	16.924	16.983	17.043	17.102	17.162
340	17.222	17.281	17.341	17.401	17.461	17.521	17.581	17.641	17.701	17.761
350	17.821	17.881	17.941	18.002	18.062	18.123	18.183	18.243	18.304	18.364
360	18.425	18.485	18.546	18.607	18.667	18.727	18.788	18.849	18.910	18.971
370	19.032	19.093	19.154	19.215	19.276	19.337	19.398	19.459	19.520	19.581
380	19.642	19.704	19.765	19.827	19.888	19.949	20.011	20.072	20.134	20.195
390	20.257	20.318	20.380	20.442	20.504	20.565	20.627	20.688	20.750	20.812

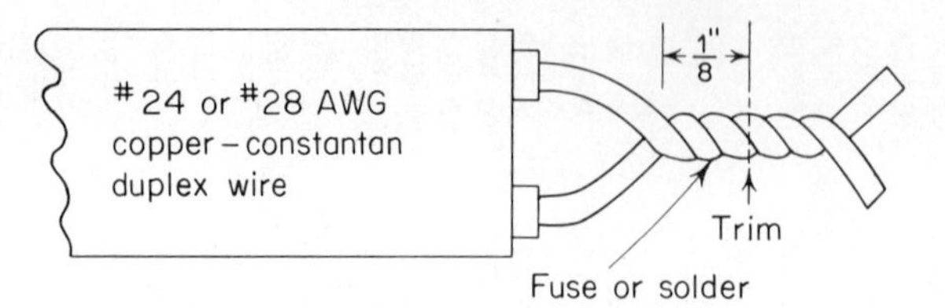

Figure 20.2. Thermocouple construction.

Table 20.3 / Duplex Thermocouple Wire

Thermocouple Pair*	Color Code
CHROMEL-alumel	YELLOW-red
CHROMEL-constantan	PURPLE-red
COPPER-constantan	BLUE-red
IRON-constantan	WHITE-red

* The polarity of the capitalized material is positive with respect to the material in lower-case type when the measuring junction exceeds the temperature of the reference junction.

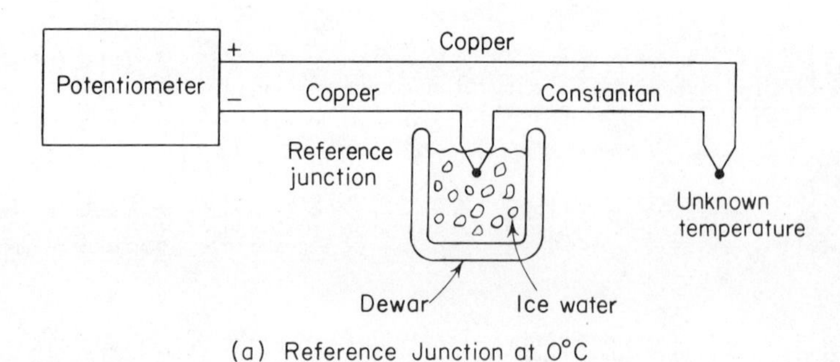

(a) Reference Junction at 0°C

Potentiometer
Reference junction
Copper
Constantan
Unknown temperature

(b) Electrical Compensation

Figure 20.3. Thermocouple measurement schemes.

is known as electrical compensation. When a potentiometer is used to measure the thermocouple voltage, electrical compensation may be achieved by use of the reference junction control on the potentiometer. This control introduces a small adjustable voltage in series with the thermocouple which is adjusted to substitute for the reference junction. To electrically compensate a junction, first establish the temperature of the junction accurately. This may be by means of an isothermal bath or simply the room temperature. Next, look up the thermocouple voltage corresponding to the known junction temperature and set the potentiometer at that voltage. Finally, depress the tap key and adjust the reference junction control for zero deflection of the detector. This completes the compensation procedure. Note that in both schemes in Figure 20.3 the measured voltage reverses sign as the unknown temperature goes through 0°C.

If a DVM is used to measure a thermocouple voltage without a reference junction, a correction factor must be employed. As above, the first step is to establish the temperature of the junction and to look up the corresponding thermocouple voltage. Next subtract the correct voltage from the DVM reading. This difference is the correction factor that must be subtracted from the DVM reading before using the thermocouple table to determine the actual junction temperature.

Exercise 20.3

Connect the thermocouple previously constructed to a potentiometer and adjust the reference junction at room temperature as described above. Alternatively, if a DVM or potentiometer without internal reference junction compensation is used, determine the voltage correction factor. Check the thermocouple operation by measuring the temperature of ice-water mixture and a boiling-water bath.

20.2.4 other temperature indicators

There are two other methods of temperature measurement that are frequently employed and that deserve mention here. In the high temperature range from 775° to 4200°C, temperatures can be quickly determined with an optical pyrometer. In this instrument, the infrared radiation from the surface whose temperature is to be measured is compared with the radiation from the filament of a calibrated lamp. The radiation of the lamp is in turn controlled by the filament current which is calibrated in terms of temperature. In operation, the surface to be measured is observed through the pyrometer telescope, which also shows the superimposed image of the filament. By adjusting the filament current such that the filament image "disappears" the radiation from the measured surface and the filament are balanced. The filament current control is then read directly in terms of temperature. This technique is particularly useful for measurement of

Figure 20.4. Optical pyrometers. Courtesy Leeds & Northrup Company, North Wales, Pa.

furnace temperatures at a distance from the extremely hot surface. Figure 20.4 shows a manual and an automatically adjusted optical pyrometer.

The second technique that is particularly useful for monitoring maximum temperatures of components is the use of temperature-indicating paints, crayons, and films. These materials have the property that they undergo a distinct change in their original color when a specified temperature is reached. In crayon form, a temperature range of 65° to 670°C is available, while paints covering 40° to 1350°C can be supplied. The accuracy is nominally ± 5°C at all temperatures. Also available are self-adhering plastic films that indicate whether the surface to which they are applied has exceeded a specified temperature. These films are available in 10°C steps from 40° to 300°C. The indicating crayons and plastic films are very convenient for checking the maximum temperature of semiconductor devices. They can be directly applied to the case of the component and thus the maximum operating temperature monitored directly in the actual circuit. The paints are useful when the temperature distribution over large surfaces is to be monitored.

20.3 thermal ratings of components

Most components are supplied by their manufacturers with one or more temperature and related power dissipation ratings. Virtually every device will have an upper and lower storage and operating temperature which is usually determined by the limits of thermal expansion and contraction and by the thermal limits of decomposition. Devices which generate significant amounts of heat internally, notably resistors and semiconductor devices, are also governed by power dissipation limits, which are related to the maximum permissible internal temperature at which the component can reliably operate. In the next section we will develop a simple model which describes the relationship between the maximum permissible temperatures and the internal power generation. We will, for the sake of clarity, consider the specific case of a transistor. However, the design techniques and conclusions are identical for power resistors and other semiconductor devices such as zener diodes, silicon controlled rectifiers, and power rectifiers.

20.3.1 thermal characteristics of transistors

As power is dissipated within the internal structure of a transistor, the temperature in that region increases until the heat flow through the transistor structure and outer case to the ambient surroundings just balances the internal power dissipation. The establishment of this steady-state balance is an example of the law of conservation of energy, since the power dissipation within a device is the rate of energy input (electrical) which in the steady-state must equal the rate of energy removal (thermal). The maximum power-dissipation rating of the transistor is then simply that rate of energy input which produces the maximum permissible internal operating temperature. Since the internal temperature will depend on the ambient temperature, the power-dissipation rating of a component is meaningless without some statement regarding ambient conditions.

Because the heat flow within a transistor is primarily by conduction, it is proportional to the temperature difference between the internal structure and the transistor outer case:

$$T_J = T_C + \theta_{JC} P_D \tag{20.1}$$

where T_J is the internal temperature (junction temperature in a transistor), T_C is the case temperature, and P_D is the internal power dissipation. The parameter θ_{JC} is called the *thermal resistance* with dimensions °C/watt, and the subscripts indicate that it relates the junction and case temperatures to the thermal heat flow. For a given transistor, $T_{J(\max)}$ and θ_{JC} are specified parameters. In silicon transistors $T_{J(\max)}$ is normally 200°C, while in germanium devices $T_{J(\max)}$ is usually 100°C. The value of θ_{JC} depends on the mechanical size of the structure and ranges from 0.5°C/watt in high-power units to 500°C/watt in miniature transistors intended for low-power applications.

Equation (20.1) also relates the maximum internal power dissipation to the corresponding maximum case and junction temperatures. This information is frequently presented in graphical form, known as the *power-temperature derating curve*, shown in Figure 20.5. The slope of this curve is called the *derating factor*, which is the reciprocal of θ_{JC}. Some data sheets give the derating factor rather than the thermal resistance. The maximum power dissipation in Figure 20.5 is shown constant below 25°C rather than the continued increase predicted by (20.1). This is a statement of the maximum permissible temperature difference between the junction and case which is governed by differential thermal expansion considerations rather than the absolute junction temperature.

The transistor described in Figure 20.5 would be called a 100-watt transistor in the advertising literature. But this dissipation level can only be achieved if the case temperature is held at 25°C or less—not an insignificant problem. For example, at a case temperature of 112°C the maximum

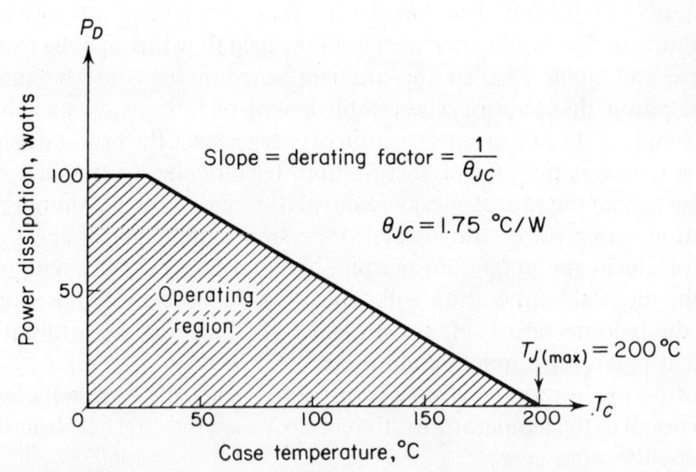

Figure 20.5. Typical power-temperature derating curve for a silicon power transistor.

dissipation is reduced to 50 watts. Thus, high-power dissipation levels require careful attention to the cooling of the transistor case.

For a transistor case to dissipate power to its surroundings, the case temperature must increase above the ambient temperature T_A. Since the heat rejection process involves conduction, convection, and radiation, it is in general a nonlinear function of the case-to-ambient temperature differential. However, in most practical situations, a linear approximation to the actual function is sufficiently accurate. Thus, we can relate the case temperature to the ambient temperature and power dissipation as follows:

$$T_C = T_A + \theta_{CA} P_D \tag{20.2}$$

where θ_{CA} is the thermal resistance between the transistor case and the ambient. For a transistor mounted in air, θ_{CA} is mainly a function of the case surface area, and typical values for some common transistor cases are given in Table 20.4. By combining (20.1) and (20.2) we obtain the relationship between T_J, T_A, and P_D,

$$T_J = T_A + (\theta_{JC} + \theta_{CA})P_D = \theta_{JA} P_D \tag{20.3}$$

where θ_{JA} is the total thermal resistance between the internal structure and the ambient, given by the sum of θ_{JC} and θ_{CA}.

Table 20.4 / Case to Ambient (Free Air) Thermal Resistance

Transistor Case	θ_{CA} (°C/watt)
TO-3	33
TO-5	125–200
TO-18	250–350

In a low-power transistor, θ_{JC} and θ_{CA} are usually the same order of magnitude. However, in a high-power transistor θ_{CA} may be much larger than θ_{JC}. For example, if the 100-watt silicon transistor, whose derating curve was shown in Figure 20.5, were mounted in a TO-3 case, then

$$\theta_{JA} = 1.75 + 33 = 34.75° \text{ C/watt} \tag{20.4}$$

for the transistor in free air. Thus the maximum power dissipation in free air for $T_A = 25°\text{C}$ becomes, from (20.3),

$$P_D = \frac{200 - 25}{34.75} = 5 \text{ watts!!} \tag{20.5}$$

Thus we see that the 100-watt transistor has a much reduced power dissipation capacity when simply mounted in free air.

Exercise 20.4

From the thermal characteristics listed on the data sheet for a 2N5067 power transistor and the data in Table 20.4, determine the maximum power dissipation of the transistor in free air. Calculate the free air case temperature T_C when the transistor is operated at 80% of the maximum power dissipation.

Exercise 20.5

Solder the copper-constantan thermocouple previously constructed into the slot of a 6-32 × $\frac{1}{2}$ in. round-head brass screw. Mount this screw in one hole of a 2N5067 transistor which is in free air. Connect the thermocouple to a voltmeter and compensate for the reference junction. Using the circuit of Figure 20.6, adjust the collector current so that the power dissipation ($P_D = I_C V_{CE}$) is 80% of the maximum value. Measure the case temperature and compare it with the calculation of the previous exercise.

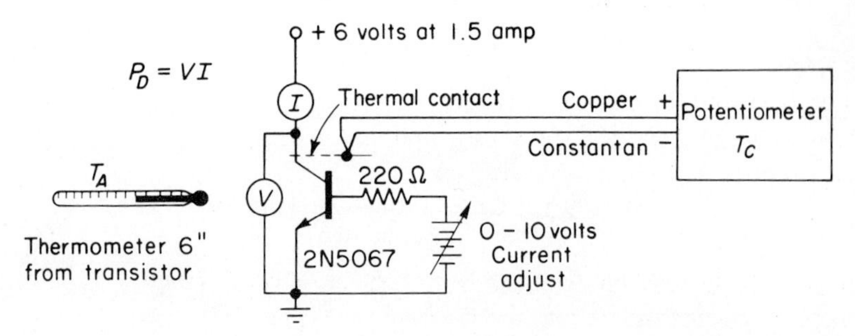

Figure 20.6. Measurement of transistor case temperature.

Exercise 20.6

Calculate the thermal resistance to ambient and maximum internal operating temperature for a 1-watt carbon composition resistor using the power derating curve shown in Figure 7.3.

20.4 heat sinks

We have seen in Section 20.3.1 that the major factor limiting transistor power dissipation capability is the high thermal resistance from the component case to the ambient surroundings. In order to increase the power-handling capability, the component case must be mounted in intimate thermal contact with a *heat sink.* A heat sink is simply a device with an improved heat-transfer capability and hence a low thermal resistance to the ambient surrounding. The heat sink may consist of a metal chassis, a finned structure with or without forced-air cooling, or even a liquid-cooled structure for maximum heat transfer in a minimum volume. The available shapes and sizes vary widely, with some typical examples illustrated in Figure 20.7.

The choice of a particular heat sink for a specific application begins by calculating from Equation (20.3) the required value for the net thermal resistance from the transistor case to the ambient, θ_{CA}, in terms of the required power dissipation P_D, the thermal characteristics of the transistor $T_{J(\max)}$ and θ_{JC}, and the maximum ambient temperature T_A. Note that the maximum expected ambient temperature must be used for this calculation in order that the maximum permissible junction temperature not be exceeded under the worst case ambient conditions. If the required net value of θ_{CA} is greater than about 2°C/watt, a forced air heat sink is indicated. For thermal resistances less than 0.2°C/watt a liquid-cooled heat exchanger will normally be required, although it probably would be better to consider an alternate transistor with a higher dissipation capability or a parallel combination of transistors in order to divide the power dissipation requirement among several devices.

Although there are heat sinks specifically designed for forced-air cooling, an air flow directed at a free-convection heat sink will substantially increase the heat dissipation capacity. This is illustrated in Figure 20.8, which shows the thermal characteristics of a Motorola MS-10 heat sink. We thus see that a small fan providing general ventilation for an instrument may improve free-convection heat-sink performance to the point that a specialized forced-air system is not required. It will also be noticed in Figure 20.8 that at the higher temperatures the heat flow improves somewhat over that calculated from the nominal θ_{SA} of 3°C/watt. This nonlinearity in the thermal characteristic is due to the effects of convection and radiation on the overall heat transfer from sink to ambient. However, for design purposes the low-temperature nominal value of θ_{SA} will provide a conservative calculation of junction temperature. In more demanding situations, the initial calculations can be corrected with the actual graphical data.

20.4.1 heat sink mounting considerations

The mechanical mounting of a transistor to a heat sink and of the heat sink to the chassis both influence the net heat rejection capabilities of the complete assembly. For convection-cooled heat sinks, mounting should provide a free air flow and be as far removed as possible from other heat-generating components. Also, mounting with cooling fins in a vertical plane will usually improve the heat rejection capacity through improved convection.

The method of mounting a transistor to a heat sink is an important factor in the net heat-rejection capability of the assembly. The manufacturer's specification for thermal resistance of a heat sink, θ_{SA}, is always

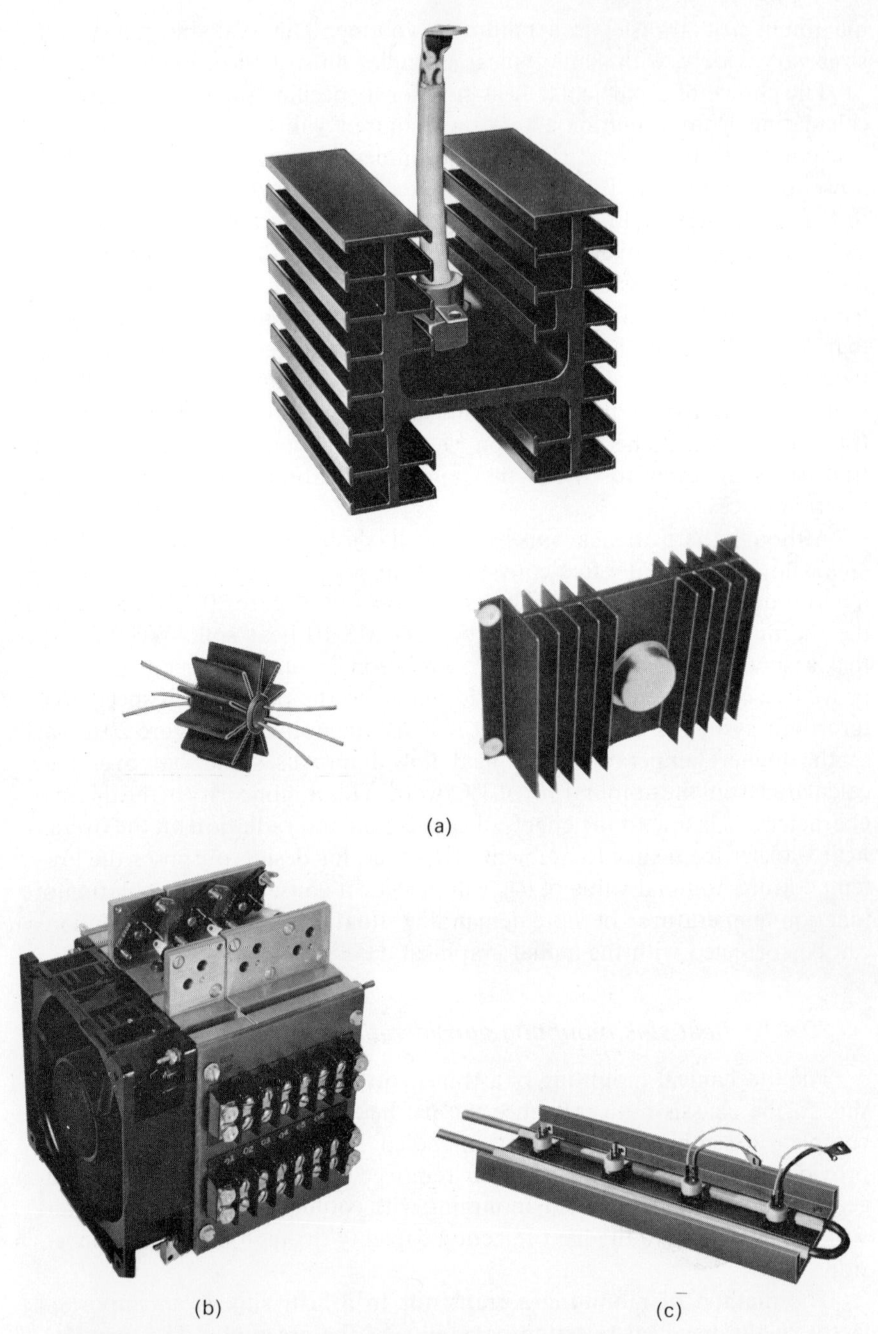

Figure 20.7. Typical heat sinks. (a) Natural convection. (b) Forced air. (c) Liquid cooled.

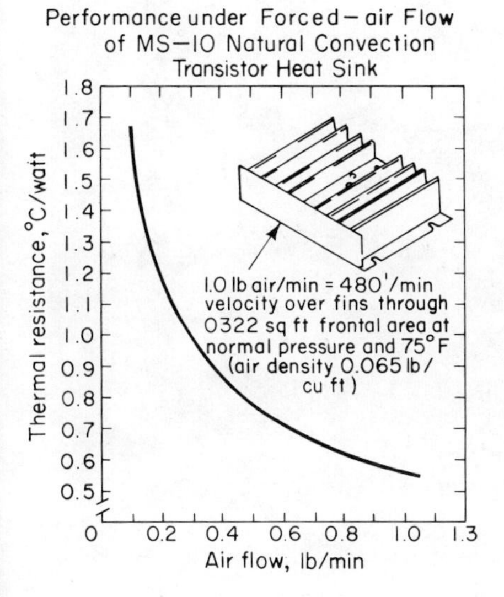

Specifications

Material	Aluminum alloy
Finish	Black
Total surface area	65 sq.in.(approx.)
Thermal resistance	3°C/watt

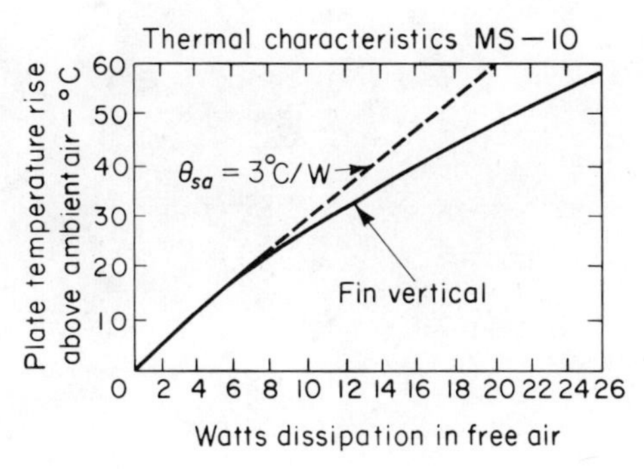

Figure 20.8. Typical thermal characteristics of a Motorola MC-10 heat sink.

measured between the heat-sink mounting surface and ambient. It does not include the additional component of thermal resistance θ_{CS} which exists at the case-to-heat sink interface. This additional component of thermal resistance, sometimes called the *thermal contact resistance*, is added to θ_{SA} to obtain the net thermal resistance from case-to-ambient:

$$\theta_{CA} = \theta_{CS} + \theta_{SA} \tag{20.6}$$

The specific value of θ_{CS} depends strongly on the properties of the mating surfaces, particularly smoothness and flatness, but is nominally 0.2°C/watt for a power transistor. In order to improve the thermal conduction at this interface, a thin coating of silicone lubricant, called *thermal joint* or *heat sink compound*, is applied to the mating surfaces, as shown in Figure 20.9. Typical examples of thermal joint compound are DC4 Silicone Lubricant and No. 340 Compound, manufactured by Dow Corning Co., and Thermacote, manufactured by the Thermalloy Company.

Since the minimum value of thermal contact resistance depends on intimate contact between case and sink, a secure mounting is important. However, overtightening of mounting screws and studs will often warp the mating surfaces and hence increase the contact resistance through decreased intimate contact area. Many specification sheets will include the maximum torque to be used in mounting a device in order to prevent mating-surface distortion.

Most power transistors have the collector junction electrically connected to the case in order to minimize θ_{JC}. Thus, the transistor case is at the

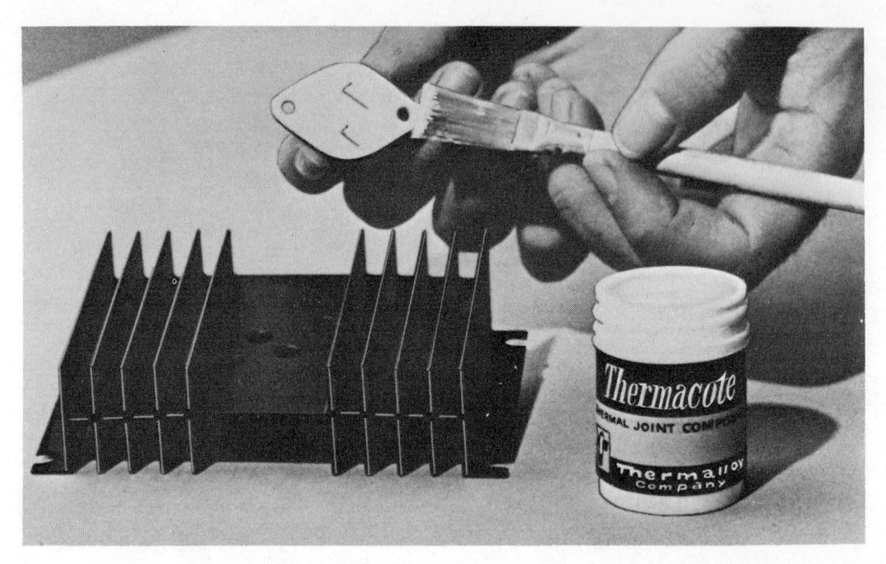

Figure 20.9. Application of thermal joint compound.

collector potential, which is not normally at zero volts or ground. Some electrical insulation is therefore necessary either between the heat sink and the chassis or between the transistor case and the heat sink. If the heat sink is protected by an enclosure, the fact that it may not be at ground potential may be of little consequence. However, heat sinks are often mounted on external surfaces in order to improve convection cooling. To prevent accidental short circuits in this situation the transistor itself should be insulated from the heat sink. Also, insulation of individual transistors may be required if more than one device is mounted on a single heat sink. Insulation from heat sinks is accomplished by placing an insulating washer between the transistor case and heat sink and by using insulating sleeves and washers with the mounting screws. This technique has the disadvantage that it may substantially increase the effective thermal contact resistance between the transistor case and heat sink. Typical insulating washer materials, along with their nominal values of θ_{CS}, are listed in Table 20.5. As in the case of no insulating washer, the use of a thermal joint compound will improve the heat transfer from case to sink. The choice of insulating washer is influenced not only by thermal resistance considerations, but also by dielectric strength, high frequency losses, and cost. In low-voltage applications, aluminum heat sinks with anodized coatings can provide sufficient insulation without the need for an insulating washer.

Exercise 20.7

Mount a 2N 5067 power transistor on a Motorola MS-10 (or equivalent) power transistor heat sink using the thermocouple-screw assembly from Exercise 20.5 for

Table 20.5 / Power Transistor Insulating Washers

Washer Material	Thermal Contact Resistance θ_{CS} (°C/watt) Dry Joint	With Silicone Compound
Teflon-glass cloth	1.45	0.80
Mica	0.80	0.40
Beryllium oxide	0.70	0.50
Anodized aluminum	0.40	0.35
No insulating washer	0.20	0.10

one mounting screw. Also attach the thermistor, previously calibrated, in one of the two holes near the edge of the heat-sink mounting surface. Use a thin coating of thermal joint compound, but do not use an insulating washer. (*Note:* For an accurate measurement of the thermal resistance of the heat sink, the temperature of the sink should actually be measured at the case-to-sink interface, not some distance away from the case.) From the thermal characteristics of the transistor, heat sink, and the nominal contact resistance, calculate the maximum permissible power dissipation for a 25°C ambient.

Using the circuit of Figure 20.10, adjust the power dissipation to 80% of the calculated maximum permissible value and measure the steady-state case temperature with the thermocouple. Monitor the heat sink temperature with the thermistor.

Compare the measured temperatures with calculated predictions and discuss the reasons for disagreement. Calculate the net thermal resistance from case-to-ambient from the measurement of T_C, T_A, and P_D.

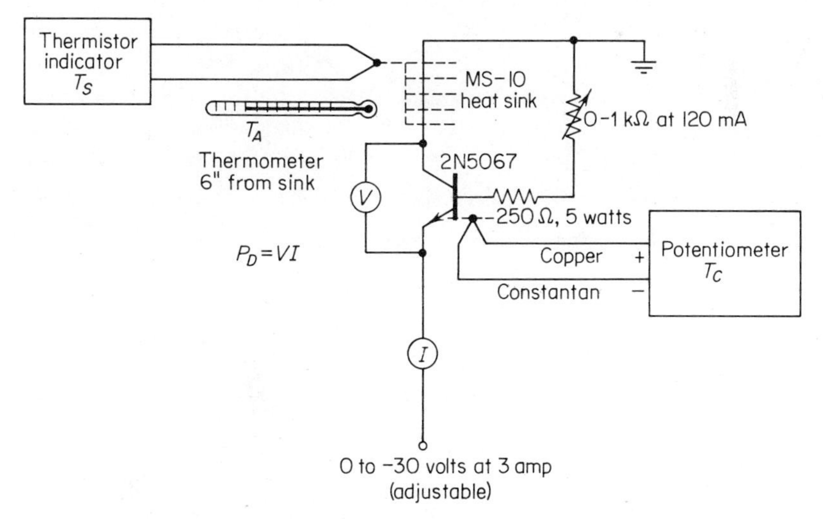

Figure 20.10. Evaluation of free-convection heat sink performance.

Exercise 20.8

With the transistor mounted as in the previous exercise and operating at 80% of the maximum power dissipation for free convection, direct the air flow from a fan onto the heat-sink assembly. From the new steady-state value of T_C, calculate net thermal resistance from case-to-ambient in the presence of the forced-air cooling. Compare this with the free-convection value obtained in the previous exercise.

20.5 references

1. D. M. Considine, *Process Instruments and Controls Handbook*, Sec. 2, pp. 5–20, McGraw-Hill Book Company, New York, 1957.
2. *Temperature Measurement Thermocouples*, American Standard Association Bulletin C96.1—1964.
3. D. I. Finch, *General Principles of Thermoelectric Thermometry*, Technical Publication D1.1000, Leeds and Northrup Company, 1966.
4. R. D. Thornton, et al., *Characteristics and Limitations of Transistors*, SEEC, Vol. 4, Chapter 2, John Wiley & Sons, Inc., New York, 1966.
5. *The Semiconductor Data Book*, "Selecting Commercial Power Transistor Heat Sinks," 3rd ed. pp. 16-17 to 16-22, Motorola Semiconductor Products, Inc., 1968.
6. *SCR Manual*, "Cooling the Power Semiconductor," 4th ed., pp. 341–365, General Electric Co., 1967.

21

basic characteristics of semiconductor devices

21.1 introduction

Since the invention of the bipolar transistor in 1948, the development of a wide variety of devices based on the electronic properties of semiconducting solids has touched virtually every area of electrical and electronic engineering. In this chapter we will continue the investigation of semiconductor device characteristics to include field-effect transistors, unijunction transistors, and silicon controlled rectifiers.

It is not the intention here to discuss the applications aspect of semiconductor devices other than perhaps to provide an example of their use. Rather, the basic terminal characteristics and circuit limitations will be studied as a basis for future design work. For those interested in additional exploration of circuit applications, the references provide a rich source of practical information.

21.2 semiconductor diodes

An ideal diode is a lossless circuit element that conducts current in one direction with no voltage drop and maintains a voltage with no current flow in the opposite direction. This ideal behavior is illustrated in Figure 21.1a. Semiconductor diodes approach the ideal diode behavior in that they conduct high currents in the *forward direction*, $V > 0$, with about 1 volt or less between the terminals, while in the *reverse direction*, $V < 0$, they will support as much as 1000 volts with currents in the microampere range.

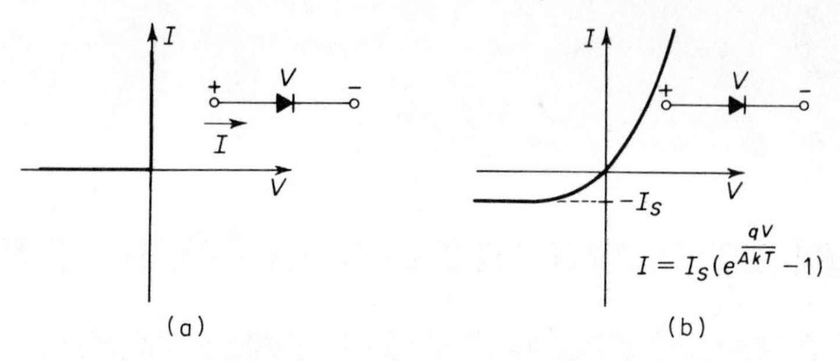

Figure 21.1. Diode characteristics. (a) Ideal diode. (b) Semiconductor diode.

Analysis of the physics of a semiconductor diode yields the following equation as an approximation to the current through the diode:

$$I = I_S[e^{(qV/AkT)} - 1] \tag{21.1}$$

where

I_S = saturation current of the diode.
q = electronic charge, 1.6×10^{-19} coulomb.
k = Boltzmann's constant, 1.38×10^{-23} joule/°K.
T = diode temperature, °K.
V = diode voltage.
A = a constant on the order of unity, characteristic of a given diode.

The volt-ampere characteristic of a semiconductor diode is shown in Figure 21.1b.

Exercise 21.1

Using a curve tracer (Section 6.4), examine the V-I characteristic of a 1N 4003 (or equivalent) semiconductor diode. Use a scale of 0.2 volt/division to observe the exponential nature of the characteristic near the origin. Can you measure I_S?

The values of I_S and A for a semiconductor diode are best measured by examining the V-I characteristic in the forward direction. For values of V such that $e^{(qV/AkT)} \gg 1$, we can approximate Equation (21.1) by

$$I \simeq I_S e^{(qV/AkT)}; \qquad V \gg \frac{AkT}{q} \tag{21.2}$$

The pure exponential character of (21.2) will produce a straight line on semi-log paper. Thus, if we plot log I vs. V for a semiconductor diode, a straight line region should result, from which I_S and A can be calculated. This procedure is illustrated in Figure 3.1. At low values of voltage, the actual data will depart from the straight line owing to the neglect of the value 1 compared with the exponential in Equation (21.1). At high currents, other effects, such as series resistance cause the actual diode to depart from the characteristic predicted by Equation (21.1).

Exercise 21.2

Using the circuit of Figure 21.2, measure the *V-I* characteristic of a 1N 4003 (or equivalent) semiconductor diode over the current range of 0.1 to 10 mA. Plot the data on semilog coordinates and determine I_S and A.

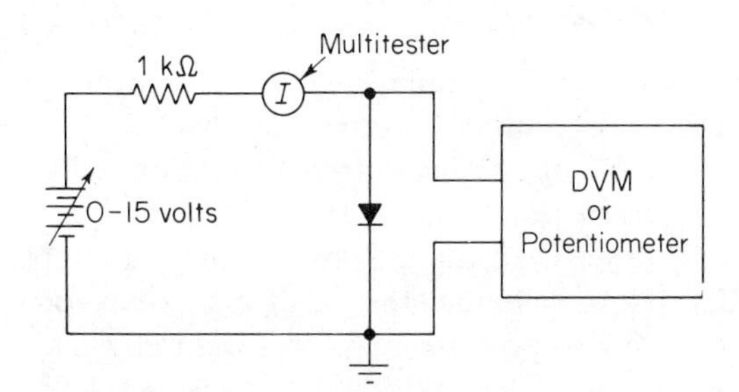

Figure 21.2. Measurement of diode parameters.

The semiconductor diode, unlike the ideal diode, is not lossless. As the forward current in the device increases, the temperature rise due to internal power dissipation limits diode performance. Diodes intended for high-current applications are mounted in large cases and attached to heat sinks to improve their power dissipation capability.

As the reverse voltage of a diode is increased, a point is reached where *reverse breakdown* occurs and substantial current begins to flow in the reverse direction. We have already seen how reverse breakdown can be utilized to produce a stable voltage reference with a zener diode (see Section 13.2.1). However, reverse breakdown in diodes not intended for voltage reference applications usually results in destruction through excessive internal power dissipation.

Exercise 21.3

Using the transistor curve tracer, determine the reverse voltage at which the reverse current in a 1N 4003 or equivalent diode has increased to 1 mA. Set the dissipation limiting resistor to 10 kΩ to protect the diode. How does the "sharpness" of the breakdown compare to that of the zener diode measured in Exercise 13.1.

21.3 transistors

Transistors are divided into two categories: *junction* or *bipolar transistors* and *field-effect* or *unipolar transistors.* The physical mechanisms of operation of the two types are distinctly different, although their output characteristics appear somewhat similar. Although the field-effect transistor was conceived long before the junction transistor, the latter was reduced to practice first and is used in larger numbers and in wider applications. It was technological advances connected with junction transistor fabrication that allowed the construction of practical field-effect devices.

21.3.1 junction transistors

The characteristics and measurement of junction transistors have been discussed in detail in Sections 6.3 and 6.4, and will not be repeated here. However, the aspects of voltage and current limitations will be discussed.

There is no fundamental limit on the current-carrying capacity of a junction transistor, other than the maximum junction temperature due to internal power dissipation. On the other hand, maximum voltage limitations are imposed owing to *reverse breakdown* or *avalanche multiplication* in the collector-to-base junction. These effects both cause a sharp increase in the collector current, nearly independent of the control variable (base current or base-to-emitter voltage), and can destroy the transistor through excessive heating in the same fashion as in a diode. Also, since the output characteristics become less predictable near avalanche, uncertain circuit performance results from operating near this region. Reference 1 discusses the limitations due to avalanche multiplication in some detail.

A typical set of characteristic curves, showing the effects of avalanche multiplication, are shown in Figure 21.3. Notice how the characteristics bend sharply upward, asymptotic to a single value of voltage. That voltage is termed the *sustaining voltage* because the transistor is able to sustain significant collector current with zero base current. The usual data sheet symbol for the sustaining voltage is BV_{CEO}. Notice also that for V_{CE} less than BV_{CEO} the spreading of the characteristics due to avalanche multiplication causes the value of β_F to depend strongly on I_C and V_{CE}.

Exercise 21.4

On the curve tracer examine an *npn* silicon transistor (2N718 or equivalent) at sufficiently large values of V_{CE} that avalanche multiplication occurs. Use as few base-current steps as possible and start with a dissipation limiting resistor of 5 kΩ to protect the transistor. From the characteristics determine the value of BV_{CEO}.

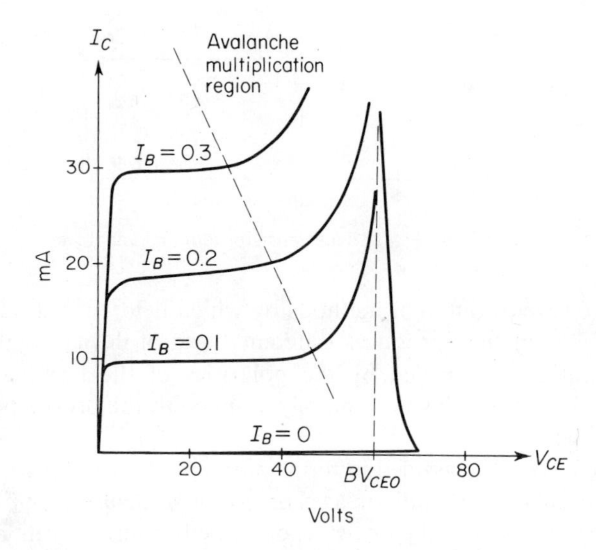

Figure 21.3. Avalanche multiplication in a junction transistor.

The sustaining voltage is normally much less than the reverse breakdown voltage of the collector-to-base diode junction, BV_{CBO}. This can be misleading, for data sheets may list only this breakdown voltage, which makes the transistor appear in the best light for advertising purposes. Thus, it is important to check precisely which value of breakdown voltage is specified before choosing a device.

Exercise 21.5

Using the curve tracer, examine the reverse breakdown characteristic of the collector-to-base junction of the *npn* silicon transistor used in the previous exercise. Leaving the emitter terminal of the transistor open-circuited, connect the base terminal of the transistor to the emitter terminal of the curve tracer. Use a 10-kΩ dissipation limiting resistor. The observed breakdown voltage is BV_{CBO}. Compare it with your previous measurement of BV_{CEO}.

21.3.2 field-effect transistors

Field-effect transistors (FET) are three-terminal devices used for amplifier and logic functions. They are characterized by an extremely high input impedance, ranging from 10^9 to 10^{15} ohms. Because of this high input impedance, the control variable for the output characteristics is the input voltage rather than the input current. In Figure 21.4, the terminal variables for an FET are defined in the common-source connection, the usual method of operation.

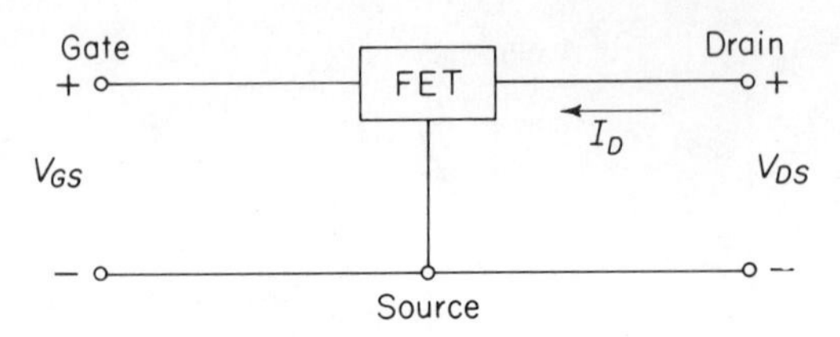

Figure 21.4. Field effect transistor terminal variables.

There are several different methods by which field effect transistors are fabricated. This in turn produces different types of devices and modes of normal operation, controlled by the polarities of the terminal voltages. We will simply present this terminology along with the proper polarities of applied voltages.

First, the device is classified in terms of the method of construction as a *junction* field effect transistor or a *metal oxide semiconductor* (MOS) field effect transistor. Second, the device is described in terms of its basic semiconductor type as *n-channel* or *p-channel.* This determines the normal polarity of V_{DS}, positive for *n*-channel and negative for *p*-channel. Finally, the mode of operation is described as *enhancement mode* or *depletion mode*, which determines the polarity of V_{GS}: the same as V_{DS} in an enhancement-mode device and opposite to V_{DS} in a depletion-mode device. The junction FET operates only in the depletion mode, whereas a MOSFET can operate either in depletion and enhancement mode or in enhancement mode only, depending on the method of construction. These polarities for normal operation are summarized in Table 21.1. The various FET circuit symbols are illustrated in Figure 21.5.

Table 21.1 / Polarities of Terminal Voltages for Normal FET Operation

Type	*n*-Channel		*p*-Channel	
	V_{DS}	V_{GS}	V_{DS}	V_{GS}
Depletion mode (junction and some MOS)	pos	neg	neg	pos
Enhancement mode (MOS only)	pos	pos	neg	neg

A typical set of output characteristics (I_D vs. V_{DS}) for an *n*-channel junction FET is shown in Figure 21.6. It is seen that as the negative voltage at the gate terminal is increased, a point is reached where the drain current is reduced to zero, independent of V_{DS}. The value of V_{GS} that just achieves this cutoff condition is called the *pinchoff voltage* V_p of the FET, and is one of its characterizing parameters. In the case of an enhancement-mode unit, the device will be in cutoff for $V_{GS} = 0$. The value of V_{GS} that sets the device on the verge of conduction is called the *threshold voltage* V_T.

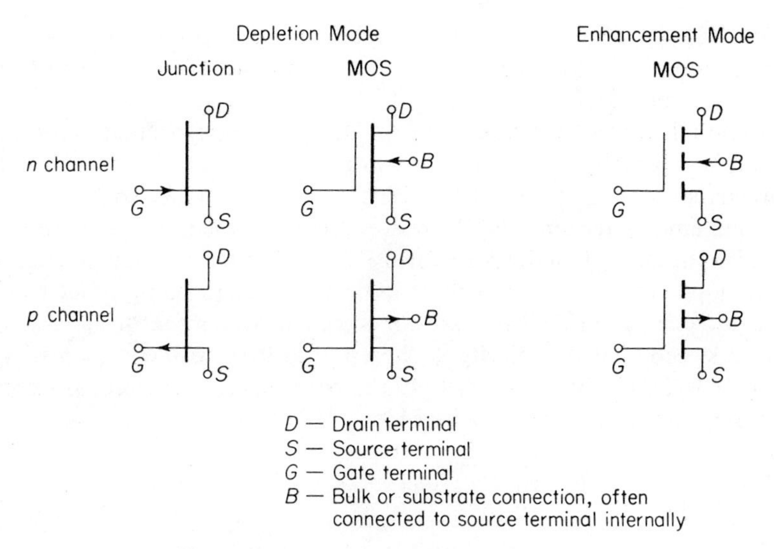

Figure 21.5. Field-effect transistor symbols.

Another characterizing parameter of the FET is the incremental *mutual conductance* or *transconductance* g_m, which governs the performance as a voltage amplifier. A method of determining g_m from the static output characteristics is illustrated in Figure 21.6. Note that g_m is a function of drain current and drain-to-source voltage, and a specification of g_m without the corresponding values of I_D and V_{DS} is incomplete.

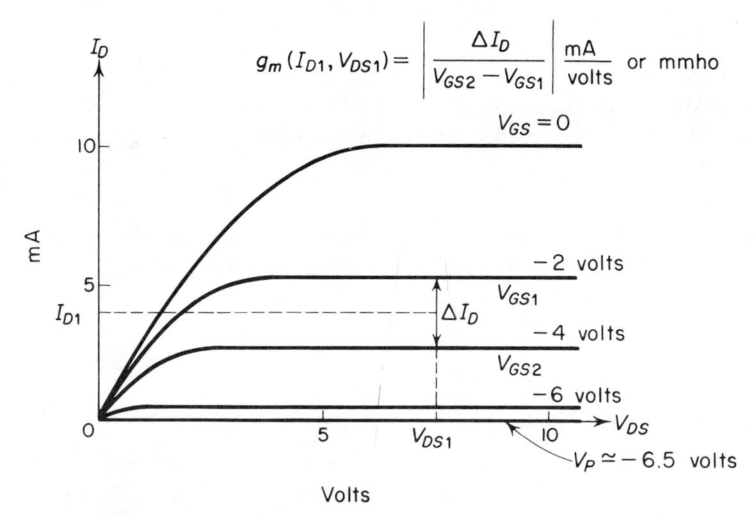

Figure 21.6. Typical output characteristics of an *n*-channel junction field-effect transistor.

In order to display an FET characteristic on a curve tracer designed only for junction transistors, a small circuit modification must be made as shown in Figure 21.7.

Although the curve tracer can provide input (base) voltage steps, their amplitude is usually too small to provide a full display of the FET output characteristics. Therefore, a 1-kΩ resistor must be connected between the base and emitter terminals of the curve tracer and the input step generator placed in the current mode. Since the FET has such a large input impedance, all the input current flows in the 1-kΩ resistor, establishing input voltage steps $V_{GS} = I_B \times 1\text{ k}\Omega$. The collector sweep provides the drain-to-source voltage sweep, and its polarity is chosen according to the V_{DS} polarity in Table 21.1. Similarly, the polarity of the base-current generator is selected according to the required polarity of V_{GS}.

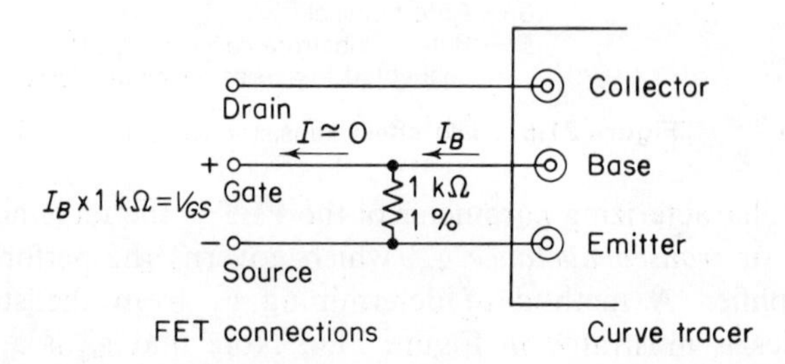

Figure 21.7. Curve tracer modification to display FET characteristics.

Exercise 21.6

Use the curve tracer to display the output characteristics of a junction field-effect transistor (Motorola MPF-105 or equivalent). Determine the pinchoff voltage V_P. From the output curves also find the maximum value of g_m for the device at $V_{DS} = 15$ volts.

As with diodes and junction transistors, the voltage and current limitations for an FET depend on power dissipation and breakdown effects. Since FET's are normally low-power devices, the maximum power dissipation is only a fraction of a watt. Thus, in using a curve tracer to check for breakdown effects at large values of V_{DS} one may inadvertently burn out the device through excessive power dissipation.

With MOS field-effect transistors, there are additional handling precautions. Owing to the extremely high input resistance of such transistors, an accumulation of static charge through handling or testing can produce destructive dielectric breakdown in the unit. To prevent this accidental damage, the leads of the device should be shorted together when the unit is not in actual use. Moreover, the device should be picked up by the case; avoid touching the leads. Finally, do not insert or remove the units from a circuit with the power on, as transient voltages may destroy the device.

21.4 unijunction transistors

The *unijunction transistor* or *double-base diode* is not strictly a transistor in the sense that it is not capable of linear signal amplification. Rather, it is a three-terminal semiconductor device having a stable negative resistance input characteristic which makes it useful in relaxation oscillator circuit and variable frequency pulse generator applications, especially the triggering of silicon controlled rectifiers. Many such circuit applications are discussed in Reference 2.

The circuit symbol and terminology for a unijunction transistor along with a typical input characteristic is shown in Figure 21.8. The limits of the region of negative input resistance are set by the *peak voltage* V_P and *peak current* I_P, and by the *valley voltage* V_V *and valley current* I_V. Although all four of these parameters are a function of the interbase voltage $V_{B_2B_1}$, only the value of V_P is directly proportional:

$$V_P = \eta V_{B_2B_1} \tag{21.3}$$

The constant of proportionality η is called the *intrinsic standoff ratio* and is a characteristic parameter of the device. The value of I_P is normally on the

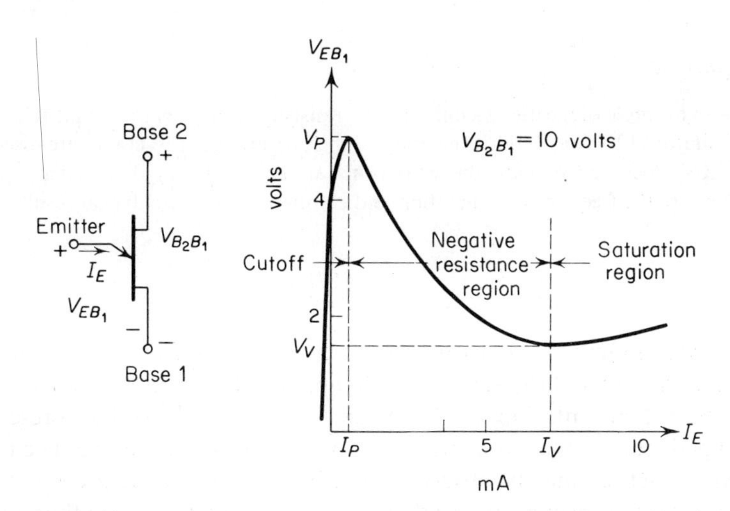

Figure 21.8. Circuit symbol and input characteristics of a unijunction transistor.

order of 10 μA and in most applications may be assumed to be zero. The valley voltage is nominally between 1 and 3 volts, while the value I_V may range from 5 to 20 mA or more, depending on the values of $V_{B_2B_1}$.

Exercise 21.7

Using the circuit of Figure 21.9, display the input characteristic of a unijunction transistor (2N 2160 or equivalent) on an oscilloscope. Because this characteristic has a negative resistance region, it may produce spurious oscillations appearing as a smear over a portion of the *V-I* characteristic. Record the characteristic for values of $V_{B_2B_1}$ of 0, 5, 10, and 20 volts and determine V_P, V_V, and I_V. (*Note:* For some values of $V_{B_2B_1}$ the valley point may not be reached.) From the characteristic curves, determine the value of the intrinsic standoff ratio for the device.

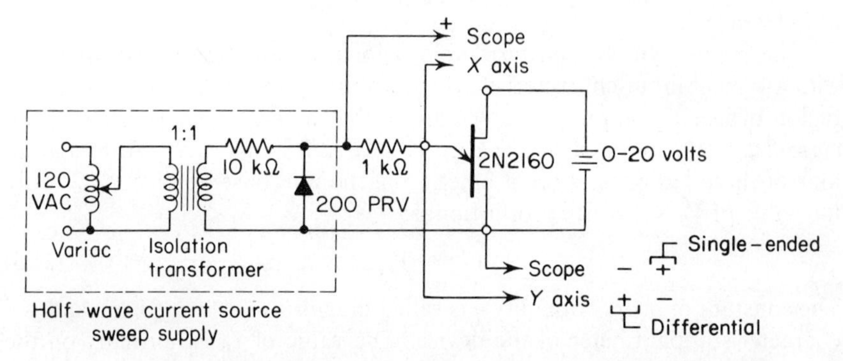

Figure 21.9. Measurement of unijunction transistor input characteristic.

Exercise 21.8

As an example of the use of a unijunction transistor, construct the simple relaxation oscillator of Figure 21.10. The details of the operation of this circuit are discussed in Reference 2. Compare the waveforms at the emitter v_E and at base 1, v_{B_1}. Measure the frequency of operation and compare it to the analytical result

$$f \simeq \frac{1}{RC \ln[1/(1-\eta)]} \tag{21.4}$$

The maximum power dissipation in a unijunction transistor must be calculated in terms of the sum of the dissipation due to emitter current and due to base-2 current, since both may be significant. The emitter presents a low impedance in the saturation region, and hence a maximum emitter current restriction must be observed. The peak value of emitter current also limits the maximum capacitance that can be discharged through the emitter, typically 10 μF or less charged to not more than 30 volts.

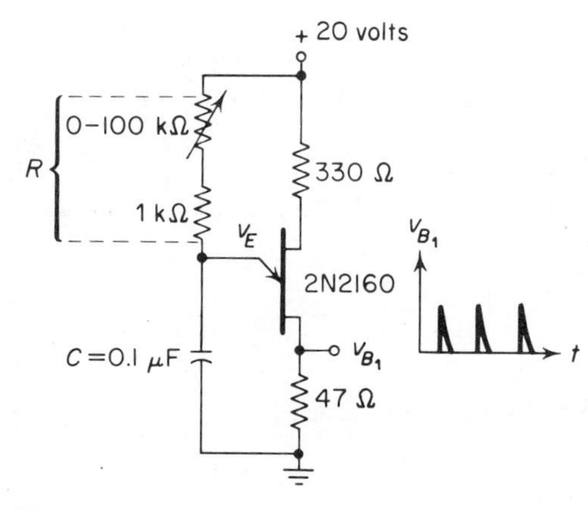

Figure 21.10. Practical relaxation oscillator circuit.

21.5 silicon controlled rectifiers

The silicon controlled rectifier (SCR) differs from the conventional diode in that conduction of current in the forward direction does not take place until a small control signal is applied to a third terminal called the *gate*. Once the gate signal has triggered conduction, the device exhibits a forward characteristic similar to a conventional diode. In the reverse direction, the SCR will block current flow for voltages less than the reverse breakdown voltage.

The significant application of the SCR is its capability to switch tremendous amounts of power efficiently with a very small gate signal. Silicon-controlled rectifiers are available with voltage ratings of 1800 volts and rms current ratings of 1200 amp. Thus control of 2 MW can be accomplished with a few watts of signal at the gate! More common are units that control several kilowatts with a few milliwatts of gate signal.

The terminology, circuit symbol and *V-I* characteristic of the SCR are shown in Figure 21.11. From the characteristic curve, it is seen that in the first quadrant the SCR is bistable and approximates a switch. In the OFF state, the device blocks current flow, while in the ON state, current can flow with only a small voltage across the SCR. Transition from the OFF state to the ON state occurs when, with a positive voltage from anode to cathode, a small trigger current is applied to the gate. Once the SCR is conducting current in the ON state, it will continue to do so after the gate trigger signal is removed and until the forward current is reduced to below a minimum value called the *holding current* I_H. When the current drops to below I_H, the SCR returns to the OFF state. The SCR will also switch to the ON state independent of gate control if the *forward breakover voltage* rating of the device is exceeded.

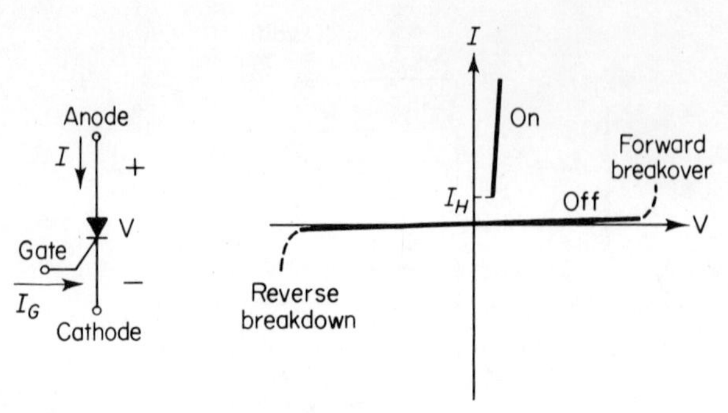

Figure 21.11. Symbol and characteristic of a silicon controlled rectifier.

Exercise 21.9

The circuit of Figure 21.12 may be used to demonstrate the switching properties of an SCR (GE C106B1 or equivalent) and to determine some of the device parameters. With the gate supply at 20 volts, and the anode supply at 40 volts, close the pushbutton and observe the change of state in the SCR. Reset the SCR by reducing the anode voltage to zero. Repeat this process for several values of anode supply voltage and plot the bistable *V-I* characteristic of the device. To determine the holding current, trigger the SCR into the ON state and then reduce the anode supply until the SCR current drops to zero. The minimum value of anode current before the transition from ON to OFF takes place is the value of I_H.

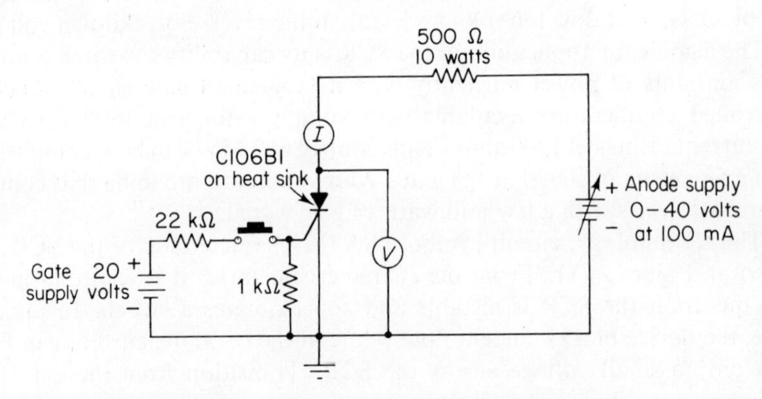

Figure 21.12. Circuit for measurement of holding current.

21.5.1 *SCR power controller*

The ability to switch, and thus control, power leads to many useful applications. In particular the control of AC power is particularly attractive since the SCR will be reset to the OFF state on every cycle. Thus, by triggering the SCR into the ON state at a variable time each cycle, proportional control over the average AC power in a load is achieved. Examples of this control are found in regulated power supplies, motor speed controls, electric oven and furnace temperature controls, battery charging circuits, and electric lamp dimmers.

As an example of proportional control, we will consider the case of an SCR lamp dimmer. The basic circuit is illustrated in Figure 21.13. The power source v_S is a full-wave rectified AC voltage. This permits the SCR to conduct every half-cycle rather than every other half-cycle, since conduction takes place only if v_S is positive. (There are devices similar to the SCR which are bistable for both polarities of v_S and hence do not require full-wave rectified AC for conduction on every half-cycle.) The gate is triggered at a delay time t_d after the start of each cycle. The SCR is then ON from the time of the trigger pulse to the end of the cycle, whereupon the SCR returns to the OFF state through reduction of v_S to zero and hence I_L to a value less than the holding current. By controlling the delay time t_d, the average load current may thus be varied.

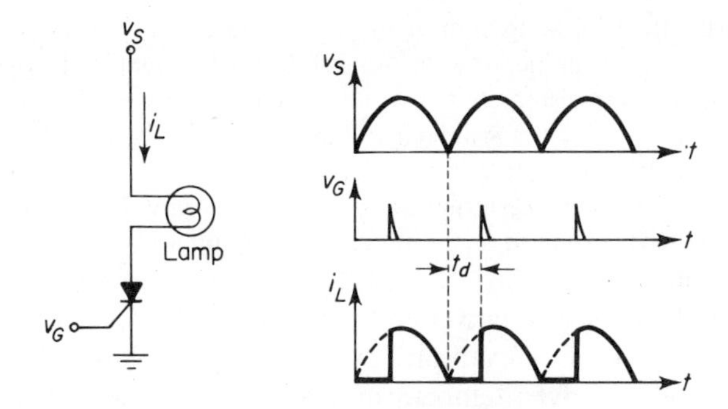

Figure 21.13. Proportional power control with an SCR.

Exercise 21.10

A practical circuit of an SCR lamp dimmer is illustrated in Figure 21.14. *Caution: In this circuit you will be using 120 VAC. Be particularly careful to avoid electric shock and be sure that the circuit is correctly connected before applying the AC power.* Just because this is the last exercise in this text, this is no reason to make it the last experiment you ever perform! (In the case of shock see Section 1.2.1.)

The unijunction transistor relaxation oscillator makes an ideal pulse generator for triggering the SCR. The use of the zener diode and the AC power to operate the trigger pulse generator provides synchronization between the AC power frequency and the unijunction oscillator frequency, such that the first trigger pulse always occurs at time t_d after the start of an AC cycle. This is accomplished by discharging the timing capacitor at the end of each cycle by automatically reducing the relaxation oscillator supply voltage to zero.

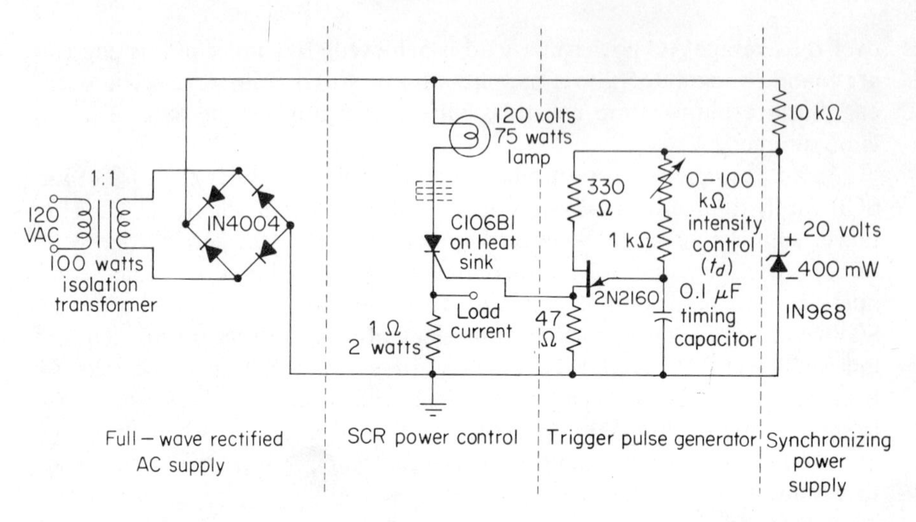

Figure 21.14. SCR lamp dimmer.

Construct the SCR lamp dimmer shown in Figure 21.14 and explore its operation. In particular examine the waveforms of the load current, SCR trigger voltage, and unijunction transistor emitter voltage. A dual-trace or four-trace vertical amplifier in the oscilloscope will be useful in relating the waveforms.

Disconnect the unijunction transistor pulse generator from the zener diode and supply power to it from an external 20-volt DC supply, as in Figure 21.10. Observe the effect of unsynchronized operation on the lamp intensity.

The use of the SCR as a power controller in other applications is similar to the lamp dimmer. For example, by replacing the lamp by an electric oven heating coil, the oven temperature can be controlled. If a thermistor located within the oven is incorporated to control the delay time, then the power to the oven can be automatically controlled to correct for temperature changes. For moderate increases in the AC power handling capability, a larger SCR need only be substituted. The versatility of this circuit is limitless, and Reference 3 is an excellent starting point for those wishing to explore further applications of silicon controlled rectifiers.

21.6 references

1. R. D. Thornton, et al., *Characteristics and Limitations of Transistors*, pp. 44–51, John Wiley & Sons, Inc., New York, 1966.

2. *Transistor Manual*, 7th ed., Chapter 13, General Electric Co., 1964.

3. *SCR Manual*, 4th ed., General Electric Co., 1967.

index

a

b

c